DUTCH LIGHT

Hugh Aldersey-Williams studied natural sciences at Cambridge. He is the author of several books exploring science, design and architecture – including *Tide*, *Periodic Tales*, *Anatomies* and *The Adventures of Sir Thomas Browne in the 21st Century* – and has curated exhibitions at the Victoria and Albert Museum and the Wellcome Collection. He lives in Norfolk with his wife and son.

'Fascinating . . . an impressive piece of scholarship. I learned a lot.'   John Gribbin, author of *Six Impossible Things* and *In Search of Schrödinger's Cat*

'At last – a scintillating biography of Christiaan Huygens, the Dutch mathematician, astronomer and inventor whose splendour has been unjustly eclipsed by the aura of Isaac Newton. After scouring archives, art galleries and museums in both the Netherlands and the UK, Hugh Aldersley-Williams has evocatively illuminated this brilliant polymath who laid the foundations of modern European science.'

Dr Patricia Fara, Emeritus Fellow of Clare College, Cambridge

HUGH ALDERSEY-WILLIAMS

DUTCH LIGHT

*Christiaan Huygens and
the Making of Science in Europe*

PICADOR

First published 2020 by Picador

This paperback edition first published 2021 by Picador
an imprint of Pan Macmillan
The Smithson, 6 Briset Street, London EC1M 5NR
EU Representative: Macmillan Publishers Ireland Ltd, 1st Floor,
The Liffey Trust Centre, 117–126 Sheriff Street Upper,
Dublin 1, DO1 YC43
Associated companies throughout the world
www.panmacmillan.com

ISBN 978-1-5098-9335-5

The right of Hugh Aldersey-Williams to be identified as the
author of this work has been asserted by him in accordance
with the Copyright, Designs and Patents Act 1988.

This book is set in Janson Text, derived from a typeface cut in seventeenth-century
Amsterdam by Nicolas Kis.

1 3 5 7 9 8 6 4 2

A CIP catalogue record for this book is available from the British Library.

Map artwork by Global Blended Learning
Typeset by Palimpsest Book Production Ltd, Falkirk, Stirlingshire
Printed and bound by CPI Group (UK) Ltd, Croydon, CR0 4YY

MIX
Paper from
responsible sources
FSC
www.fsc.org   FSC® C116313

Visit **www.picador.com** to read more about all our books
and to buy them. You will also find features, author interviews and
news of any author events, and you can sign up for e-newsletters
so that you're always first to hear about our new releases.

To Sam

Contents

List of illustrations

List of illustrations

Colour Plates

Author's note

My principal primary source has been Christiaan Huygens's *Oeuvres Complètes*, which contain the letters of Huygens and his correspondents in French, Latin, Dutch and other languages, as well as the texts of his major treatises in their original languages. They also include the editors' summaries of Huygens's scientific achievements and a concise biography in French. I have made my own translations when quoting from this source or from manuscript originals held in the library of the University of Leiden. Many of the secondary sources I consulted are in Dutch (or occasionally another language), and in these cases, too, I have made my own translations, including excerpts of verse, except when a preferable translated text has been available, in which case it is cited in the Notes.

Dates are New Style except where indicated.

THE HUYGENS FAMILY (PARTIAL)

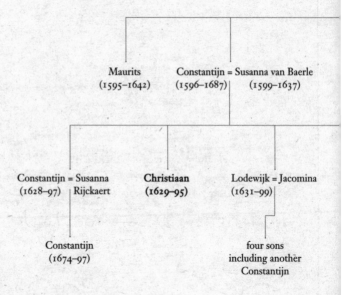

Maurits
(1595–1642)

Constantijn = Susanna van Baerle
(1596–1687)   (1599–1637)

Constantijn = Susanna
(1628–97)   Rijckaert

**Christiaan
(1629–95)**

Lodewijk = Jacomina
(1631–99)

Constantijn
(1674–97)

four sons
including another
Constantijn

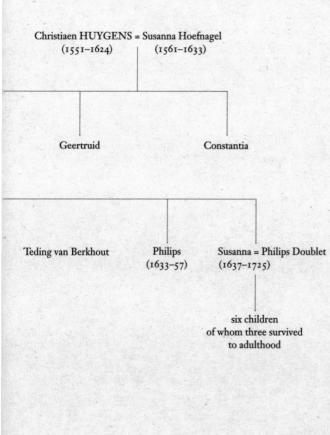

Christiaen HUYGENS = Susanna Hoefnagel
(1551–1624)   (1561–1633)

Geertruid   Constantia

Teding van Berkhout   Philips   Susanna = Philips Doublet
                          (1633–57)   (1637–1725)

six children
of whom three survived
to adulthood

INTRODUCTION

Light lances through the leaded panes and strikes the polished black-and-white tiled floor of the room, imposing one grid at an angle upon another. It crooks as it meets the window surface and crooks again after it has passed through to resume its original direction. It dances and twists, dodges and swerves round imperfections in the old glass. Here and there, tiny blemishes improvise lenses and prisms, which produce magnified and distorted flecks and gobbets of light on the floor, and sometimes a tiny rainbow. The brightest pattern of light is the one cut by the sharp shadows of the lead in the window. This falls off to one side because of the sun's position in the sky. But there is another pool of light, too. Directly below the window, a hazy bluish gleam rises from the tiles, which is the reflection of the light of the sky.

The room is large and bright and was originally used for dinners and musical entertainments. It has windows on three sides, but is even brighter than you'd expect for that. It is almost like being outdoors, which seems odd at first because overhead is not air but a heavy beamed ceiling. Then you realize that this ceiling is almost glowing with light itself, light from yet another source, coming from below the horizon and projected upward through the same window by the water in the moat that surrounds the Huygens house, known as Hofwijck.

This estate – three miles south-east of The Hague – is where

Christiaan Huygens lived after the death of his father, the poet and diplomat Constantijn Huygens, until his own death just eight years later, in 1695. When Constantijn built Hofwijck some fifty years earlier, he wrote that he wanted it to appear 'as if by night, / It grew, like a mushroom revealed in the light'. And so, amid its calm, reflecting waters and with the elevated motorway now roaring too close by, it does, even today. Here, Christiaan finalized treatises on the nature of light and gravity that summed up his prodigious contribution to physics. Here, he set up his telescopes in the spacious grounds and began to speculate about life on other planets.

Christiaan Huygens was Europe's greatest scientist during the latter half of the seventeenth century, until the rise of Isaac Newton, by whom he has been largely eclipsed, especially in English-speaking parts of the world.* This is an unfair judgement of the history of science, for Huygens's achievement exceeds that of Newton in some important respects. A maker as well as an observer and a thinker, he added to both theoretical and practical science in astronomy, optics and mechanics. A supremely talented mathematician, Huygens tackled problems in fields as varied as geometry and probability theory, and became the first to employ mathematical formulae in the solution of problems in physics, the methodology that has become the basis of all modern science. Two centuries before it would become generally accepted, he proposed a wave theory of light. He was first to describe the concept of centrifugal force. Using telescopes of his own design and manufacture, he discovered Saturn's ring system and its largest moon, Titan. He estimated the size of Mars and the distances to many stars. He found a way to incorporate the pendulum to create

---

* I use the word 'scientist' to characterize the vocation of Huygens and many of his peers, although it does not appear in the English language until 1834.

more accurate clocks, realizing the vision of Galileo before him. His innovations in optical instrumentation and timekeeping are still in use today.

Huygens also had range: he was a fine draughtsman, a skill of use not only in designing mechanical and optical devices but also in representing to the world the planetary phenomena he observed through his telescopes, although he was not above making sketch portraits to flatter girlfriends, or little landscapes of the countryside in which he found himself. He was a proficient musician, joining in with performances wherever he went. Occasionally, a few notes of a melody or a lyric for a song are to be found in the margins of his scientific notes. But he also wanted to bring his mathematical science to bear in music, proposing a division of the octave into thirty-one notes, foreshadowing musical innovations of the twentieth century.

More lasting in significance were his efforts to elevate science itself in Europe, both in the Dutch Republic and, more especially and surprisingly, in France, where he was instrumental in the establishment of the French Academy of Sciences. He was also an early fellow of the Royal Society of London, becoming a personification of the potential for science to transcend national borders.

This was not always how he chose to appear to the world, however. When Huygens returned from Paris to The Hague in 1671, he sat for his portrait to be painted by Caspar Netscher, who had already painted several other family members. Netscher's small oil painting demonstrates the artist's mastery in representing fine fabrics. Huygens looks out with wide eyes from amid an ocean of silk and lace. He is at the height of his powers, and yet there is still something about him of the pretty child he once was. If we are in search of a display of learning – perhaps a table nearby with scientific instruments and papers of calculations lying casually on top – then we will have to look elsewhere. This is above all a man of fashion and extravagant taste.

But he was, too, the prototype of a modern scientist. Although he explored many topics, and shuttled between them opportunistically, rather than pursuing what would today be called a programme of research, he conducted his investigations with diligence and rigour, even if, like many others at the time, he did not always rush to publish his findings. His employment of mathematics as well as his awareness of the importance of criteria of reproducibility, verifiability and falsifiability – the understanding that experiments should be repeatable in order to demonstrate their truth, and that experimental results that fail to support a hypothesis must lead to rejection of the hypothesis – reveal the essential seriousness of his project. His topics were well chosen as ones where a breakthrough might realistically be achieved. He did not wander off the track into the realms of superstition, which is more than can be said for some of his contemporaries. So dedicated to learning was he that, on his first visit to London in 1661, he turned down the chance to attend the coronation of King Charles II in order to observe the more interesting transit of Mercury.

This all-round proficiency Huygens undoubtedly perfected with the help of Constantijn, the father who first doted on him and then stood in awe of his son's talents, introducing him as 'my Archimedes' to Descartes and other illustrious visitors to his household. The long-lived Constantijn was a strong moral and intellectual influence for almost the whole of Christiaan's life – so much so that when he eventually died at the age of ninety, Christiaan, by then fifty-eight years old, morosely had himself painted in the garb of an orphan.

t, composer, diplomat, architect and artist, Constantijn was
less remarkable a figure than his son, which is why I devote
ostantial parts of this book to him. Born in 1596, he served as
e secretary to a succession of stadholders, as governors of prov-
ces of the Dutch Republic were known. Constantijn was what
the Dutch called a *kenner* – not an amateur, a mere lover of his
studies and pastimes, but one who has sweated to know them
completely so that he has become a master himself, even though
he will never need to rely on these skills to earn his living.

So it was that Constantijn Huygens learned to draw and paint
from one of the best draughtsmen of the day, and was able to put
his first-hand expertise to good use when he identified the raw
talent of the young Rembrandt, whose work he believed might
fittingly adorn the stadholder's court in The Hague. Constantijn
effectively launched the artist on his career, although their friendship petered out in a trail of truculent correspondence as an
increasingly bumptious Rembrandt chased his fees.

The elder Huygens was also enough of an architect to take a
leading role in designing his own houses. He slotted easily into the
circle of leading Dutch poets, and documented his long life in verse.
When, for instance, he proposed that a grand avenue be built from
The Hague to the sea at Scheveningen, he wrote a poem about
that, too. He composed hundreds of musical works and played
many instruments. Unsurprisingly, his diplomatic missions attained
a cultural dimension that did him no harm, and long before he was
forty he had been knighted both in England, where he had pleased
James I with his lute-playing, and in France, where it was his literary
accomplishments that impressed Louis XIII.

With these diverse interests, it is no surprise that learn that
Constantijn was absorbed by scientific questions, too. The *quality*
of his engagement with science presents an instructive contrast
with that of his son, which is another reason to spend some time

with him here. If Christiaan was the prototype of a mod
scientist, then Constantijn was an exemplar of a foregoing t
known as the 'curioso' – the man who wishes to know more ab
natural phenomena but who does not always ask the right que
tions or have the necessary equipment to answer them. Th
paternal fascination was surely formative for his children.

Constantijn and his wife Susanna would have four sons and
a daughter. The *kenner* instinct was inculcated in all of them, but
especially in the two older sons: the eldest, another Constantijn,
became a skilled grinder of lenses while Christiaan learned how
to construct the intricate mechanisms for his own horological
inventions and optical devices. Perhaps, too, their father's patho-
logically impaired vision was a factor in the path that the boys
chose. The younger Constantijn eventually succeeded his father
in a diplomatic career by becoming secretary to the stadholder
William III, and accompanied him on his triumphal journey
through England in 1688, to join his wife Mary II on the English
throne. Like Samuel Pepys, Constantijn also kept a diary that
recorded every aspect of his personal and political life, from sexual
gossip to the progress of the 'Glorious Revolution'.

The Huygens family milieu in The Hague, then, is essential to
an understanding of who Christiaan Huygens was able to become.
This was a household not only where Descartes and Rembrandt
were among the guests, but also one close to the power and influence of the government of the Dutch Republic. All this provided
introductions, but Christiaan's brilliance did the rest, as he found
his place in a European intellectual firmament that included Blaise
Pascal, Pierre de Fermat, Marin Mersenne, Jacob Bernoulli, Robert
Hooke, Robert Boyle, John Wallis, Giovanni Cassini, Thomas
Hobbes, John Locke, Gottfried Leibniz and Isaac Newton.

Was there something about the light? That's what everybody wondered about the painting of the Dutch 'Golden Age'. The sun was softer, the colours less harsh, the contrasts less violent than in Tuscany or Madrid. The windows were larger, the interiors less dim. Dutch light led artists to revel for the first time in the domestic scale of daily life, with its tranquil landscapes, unpretentious rooms and polite encounters.

By the early seventeenth century, Dutch artists were already benefiting from the scientific understanding of perspective, and knew about the camera obscura, a device whereby an image of a scene could be projected through a small hole in a wall or other barrier onto some kind of screen. This novelty became creatively interesting when it was discovered that the exterior scene could be optically altered by placing a lens in the hole, allowing artists to compress impossibly wide panoramas onto canvas. It almost seemed as if they were able to trap light itself, releasing it again between gilded frames, as visions of the world made new in glowing quadrature. But the scenes they captured and the light they worked with – those were *natuurlijk*, natural, that is to say both devoid of artifice and composed of nature. They sprang from local earth and air and water. Theirs was art of its place.

Did the sciences also benefit by native circumstance? We know the unsurpassed art of the Dutch seventeenth century: the landscapes of Ruisdael, the portraits of Rembrandt, the interiors of Vermeer. But the science was its equal, and we should not be embarrassed to think of the two together. As the Dutch painter and art scholar Samuel van Hoogstraeten decreed in 1687: 'The Art of Painting is a science for representing all the conceptions or impressions which the whole of visible nature can offer and for beguiling the eye with outline and colour.' And what tool was more essential to the beguiler than an ability to handle light, the light that illuminates visible nature, the light that permits us to see it? Light is surely the common factor that unites the interests of art and science.

Much of the science of this period is directed towards understanding that light. Willebrord Snel of Leiden measured the size of the Earth and restated the law of refraction that still bears his name as Snell's law. Then in nearby Delft, and in Alkmaar and Middelburg, Anthoni Leeuwenhoek, Johannes Swammerdam and Cornelis Drebbel made some of the first investigations using microscopes of their own design. The first telescope was demonstrated from a tower in The Hague. There followed, in addition to Huygens and his family, many other Dutch grinders of lenses and assemblers of experimental optical devices, not least the philosopher Baruch Spinoza, who earned his modest keep as a *lenzenslijper* after he was expelled from his religious community in Amsterdam.

Spinoza, the child of immigrant Jews from Portugal, was fortunate to be born in the relatively liberal Dutch Republic. But others came purposely for its freedoms. The most notable of all was René Descartes, who, in 1628, fled religious turmoil in France for the intellectual liberty of the new Dutch universities, and then to find a quiet place to write his philosophical masterwork, *Discourse on Method*. Many regard this treatise, in which Descartes introduces his famous *cogito* argument, 'I think, therefore I am', as the foundational work of modern philosophy. But Descartes saw it rather differently, as the theoretical preamble to a series of works concerning the nature of the world in all its aspects, including *La dioptrique*, his exploration of the physics of optics, the nature of light and the anatomy of the eye. Both of these great philosophers, too, were guided by the light.

The light made the United Provinces a place for looking. The liberal, inquisitive times permitted it. And the place insisted upon it: no shadows on level ground.

What things might you wish to see? Look out first of all. There might be danger in these times of recurrent war and tenuous peace. Across the flat landscape, you might notice approaching strangers or enemy soldiers. At sea, distant ships, unfamiliar flags and wrecking sand bars just pricking the surface of the brine. Danger, perhaps, and opportunity, too. Remote shores to claim for trade and empire. Any place on a map or a chart or, better still, not marked on one yet. A place as yet unseen. Visualize it. Is a hoped-for gain not called a prospect?

Now look around. What is this Dutch country? Of what is it made? Water. So much water, flat *zee*s and *meer*s, sometimes gleaming, sometimes dark when they are scuffed by the wind. Light always changing. Flat fields with ditches and dykes all the way out to the low horizon, with here and there perhaps a few stands of trees roughly shaken by the breeze. Scoured beaches, open heaths, scrubbed towns and villages identifiable from afar by their church spires.

Or look up. You might nurse the ambition to grasp the unreachably distant, the heavens, the stars and planets and the voids in between them. Can you make darkness visible? What is out there? Look down. Look *in*. Look closely – it is the first time any human has been able to do so – at nature's tiny miracles, seeds, insects, moulds, swimming animalcules. Do you dare to look at the human body? Your own in a mirror larger and more perfect than any you have known before, or somebody else's (poor wretch) at a surgical praelector's anatomy lesson. Both of these spectacles, the one domestic and private, the other theatrically public, were becoming technically feasible and socially acceptable as shame and superstition about the human body were swept away. You might be curious about the origins of life, and wish to spy on your potential descendants *in semine* or *in utero*. Keep looking if you can bear to. At our dirt and dust. At human waste. What is lost of us when we spit and defecate? In the seventeenth century, it's suddenly all there to be seen. The very farthest thing. The very smallest thing. Things of beauty and wonder. Cosmos and microcosmos.

Then you might wish to see the quality: the fineness of the weave, the clarity of the diamond, the skill of the artist with the brush. The quotidian, the ordinary, the small: the eye of the needle, the end of the thread, the missing *stuiver* or *duit*, the bottom of a pocket, the beginning of a hole. You might need help to read the written word, as Papa Constantijn did. You might long to see again the things you once could see unaided.

Look about you. Regard your fellow citizens. How well they are doing! You might even wish to peer into your neighbours' rooms, if only to be sure they have nothing to hide. And they will tempt you with their large polished windows. 'The country is flat,' explains the novelist Cees Nooteboom,

> which leads to the extreme visibility of people, and that in turn has become visible in their conduct. The Dutch don't go towards each other, they come up against each other. They drill their radiant eyes into one another, and weigh their soul. There are no hiding places. Not even their houses. They open their curtains, and consider it a virtue.

Go on, take a peek, have a snoop. Those words are Dutch, too.

All of these things are better discerned by using the right optics. Thanks to the new dissectors prepared to press sharp steel into slimy viscera, it became understood that the lens of the human eye is not in fact the organ of sight, only its enabler, a kind of secretary, sorting incoming information into order for executive processing by the brain. If your own cornea, aqueous humour and lens are not up to the job, then you will need the assistance of additional lenses in the form of spectacles or a magnifying glass. To see new worlds, you will need new instruments: a spyglass or a perspective tube for the great and distant; a thread-counter or a 'flea glass' for the small and close. There was nowhere better equipped to manufacture these devices than the tech hubs that were the thriving cities of the Dutch provinces.

Then who, but an artist, do you summon to record the images you see in the eyepiece?

Unless, of course, you are enough of an artist to do the job yourself, which Christiaan Huygens certainly was.

Christiaan Huygens offers so much: he made major discoveries in astronomy and physics; he showed how mathematics could be used to describe the operation of nature, without which modern science could not function; he was an inventor and fabricator of ingenious devices. He clearly has a story that needs telling. To say that it is necessary to reclaim him from the margins of history ought to be putting it too strongly, since his work is so thoroughly enmeshed with that of better-known major scientific figures. Yet even many historians of science have contrived to ignore him. And who else has heard of Huygens now? Would it not be good to be reminded why he would have been 'the greatest scientist of his generation', in John Gribbin's words, 'if he had not had the misfortune to be active in science at almost exactly the same time as Isaac Newton'?

This kind of amnesia is not uncommon in the anglophone world. Yet there is something not merely neglectful, but actively unjust about the way in which Huygens's reputation has been effectively annihilated by the towering presence of Newton. Writing about Huygens therefore involves taking a stance. The historiography of science requires that those whose theories are overtaken by better theories shall be forgotten. But this is not quite what happened in Huygens's case. His discoveries for the most part still stand – your analogue wristwatch runs using his mechanism; you understand that light travels as a wave. It is rather that Newton radiates such a dazzling light that Huygens is simply outshone.

So how to tell Huygens's story? Should I pretend that Newton did not exist? That might be possible. Huygens worked in the Dutch Republic and in France. Tempting, perhaps, to bring down a fog in the English Channel so that the islands of Britain cannot be seen and to screen out English science for the duration. But Huygens and Newton were contemporaries: Newton was thirteen years Huygens's junior, and outlived him by thirty-two years. More to the point, they were aware of one another's work. They corresponded and even met. In fact, Huygens was one of the few people whose scientific opinion Newton valued. That certainly would not do.

Perhaps, therefore, I should position Huygens in opposition to Newton. But then it would still be Newton's story. Huygens's story would be the antithesis, an un-story. He would be the one who didn't explain gravity, who didn't use calculus, who didn't split light with a prism. That telling would insist on a kind of parity between these discoveries and Huygens's own – centrifugal force, mathematical formulae, light as a wave – which would be tedious as well as tendentious to read.

Or, the scientifically minded might interject, I could just write objectively the story of Huygens's life and bring Newton in where he is relevant and leave him out where he is not. Wouldn't that be best? Why complicate matters? Certainly, Christiaan's life seems to have a convenient arc. By the age of seventeen, in 1646, he had demonstrated a mathematical prowess that brought him to the notice of the greatest practitioners of the day. In 1655, Saturn's moon, Titan, and that planet's mysterious ring became his calling card. In 1658, he presented his design for a more accurate clock to the Dutch state. In 1659, he diagrammed the antecedent of the slide projector, which he called a 'magic lantern'. The late 1660s he spent largely in Paris, consolidating his professional relationships. After further work on light, establishing that it must have a wave-like nature, and new experiments with telescopes and microscopy, as well as improvements to clock mechanisms, he published major treatises on time in 1673 and on light in 1690. From his prodigious youth, then, he advanced with luck and skill (fortune favours the prepared mind, as a later scientist said), seizing his opportunities to make some surprising discoveries, settling down in dogged pursuit of other goals, before achieving a remarkable synthesis in maturity with his major works, and finally withdrawing into a period of eccentric decline.

That would make a satisfying narrative, would it not? But closer inspection reveals this to be a gross simplification. Huygens's life was nothing like as well shaped as this storyline suggests. The world gets in the way: he pursues many projects at once; they stop and start; he is here (The Hague), he is there (Paris); he is well, he is ill; there is peace, there is war. All this makes it impossible to tell Huygens's story as a simple sequence of events.

Besides, Huygens has other stories. He is not only the overshadowed genius. He is the polymath – curious enough and competent enough to make progress in several different fields at once. This story has allure, especially in these days when it is hard enough to be an expert in one tiny area of specialization and polymathy seems like a dream from a world before some kind of Fall. He is the diligent correspondent, the networker, the diplomat, too.

Another of Huygens's stories clearly revolves around his

extraordinarily attractive family: his prolific, ubiquitous and apparently immortal father – the poet, composer and right-hand man of the stadholders – and his older brother, also called Constantijn, who succeeded their father to accompany William III of Orange on his peaceful invasion of England, and recorded the adventure in a remarkable diary. If John Donne, the doyen of English metaphysical poets, had fathered both Isaac Newton and Samuel Pepys, one could not have devised a more propitious family constellation.

And there is more. Christiaan is the internationalist, heroically ignoring national differences and distinctions because they have no meaning in his world. He is a key figure – perhaps the key figure – in the construction of scientific research as an *international* project during the seventeenth century. This is a rather different story. Early modern science was mostly an enterprise of brilliant individuals who enjoyed the patronage of enlightened rulers here and there. Before the emergence of many nation states, it was certainly not organized at a national level. This is what makes Huygens's years in Paris truly remarkable. Huygens's brilliance as a mathematician and astronomer was readily acknowledged by most of the leading French natural philosophers, and he responded with alacrity to the initiative of Louis XIV's influential minister Jean-Baptiste Colbert in 1666 to establish a scholarly academy to advise on the scientific improvement of the French state. Without Huygens – and one or two other cosmopolitan intermediaries such as Henry Oldenburg at the Royal Society in London – science might have remained for rather longer a rudderless amusement confined within isolated courts.

Holland, as a land with neither a monarch nor even, for much of Huygens's career, a stadholder, nevertheless prospered during the republican interregnum, owing to colonial conquests and the expansion of maritime trade. This economic growth brought with it a kind of freedom unknown elsewhere. Huygens could only have envisaged a more open way of doing things. Yet the Dutch Republic at this time could never have furnished its own scientific centre, because of the intense rivalry among its cities and provinces. Huygens was fortunate to find Paris so open to him, and France was fortunate to find him.

This achievement is all the more remarkable when one remembers that the second half of the seventeenth century, when Huygens flourished, was a period of great religious and political turmoil and almost constant warring between European nations. The ferment accelerated the spread of new ideas, as well as raising practical obstacles to intellectual progress. At different times, the English philosophers Thomas Hobbes and John Locke, both well known to the Huygens family, found themselves respectively in exile in Paris and Amsterdam, for instance, while Dutch-trained painters such as Peter Lely and Godfrey Kneller sought work in England where patronage became more abundant. Huygens sought English patents for his maritime clock even as the Dutch navy sailed up the River Medway to torch the English fleet at Chatham in 1667. And he continued to work with French colleagues when Louis XIV invaded the Low Countries five years later, a campaign during which the Dutch were driven to flood their own precious farmland as a defence against the French advance.

Two factors, then, explain why Huygens has a strong claim to be regarded as the leading actor in 'the making of science in Europe': first, his bringing of mathematical rigour to the description of physical phenomena; and second, his resourcefulness in developing institutional frameworks for scientific research on the continent. It is impossible to imagine progress in science today without either of these foundations being in place. As Huygens is purported to have said: 'The world is my country, science my religion.'

While wandering among the dunes that guard Dutch cities against the incursions of the North Sea, I began to realize that Huygens's geographical location might have a more material relevance to his work. It was these dunes that provided the sand for the glass from which lenses could be ground. Did this explain the extraordinary flowering of optical science that happened here, dominated by Huygens but preceded by Snel, and by the numerous claimants to have developed the telescope, and then followed by the microscopists Leeuwenhoek and Swammerdam, and by the philosophical lens-makers Descartes and Spinoza? Perhaps it was true, as Constantijn Huygens had written, that 'The Lord's benevolence shines from every dune'.

This suggested another kind of story. Not the old one of the lone genius with his flashes of brilliance striking out of nowhere, but instead one of an investigator deeply engaged with his place, for whom light and sand were raw materials.

This is the world I hope to describe in *Dutch Light*. I have followed in Huygens's footsteps through the Netherlands and to Paris and London. I visited his houses, or more often the sites where they once stood. I sat in the grounds at Hofwijck where he sketched, and watched indoors as the sunlight tracked across the chequered floor. I peered at – and occasionally through – the few surviving instruments that he built and used. I inspected his portraits for clues to his personality.

My principal source has inevitably been the twenty-two volumes of Christiaan Huygens's *Oeuvres Complètes*, compiled with extraordinary labour by an international team of scientists and historians through two world wars from 1888 to 1950. Here, I could hear the authentic voice of the scientist. And in their letters, diaries and poems, I could hear the voices of his father and siblings. For a closer personal connection, I turned to Christiaan's original notes in the library at the University of Leiden, which revealed additional facets of Huygens's character in their dense tapestry of working sketches and calculations, showing how he pursued many thoughts – mathematical, mechanical, astronomical and musical – at the same time. There, I held in my hand Christiaan's little sketch of Saturn and its ring (trying to ignore the wobbly pencilled line that the editors of the *Oeuvres Complètes* had crudely looped around it in order to select it for illustration). The perfect circle of the planet's outline was precisely inked, as were the edges of the ellipse that girdled it. The space around the planet, and crucially also the spaces between the planet's surface and the inner edge of the ring, were delicately shaded in a dove-grey wash of ink, and the whole was hemmed in by busy handwriting. Huygens had made this beautiful little drawing in 1659.

My voyage of discovery has not been obviously epic. Rather, it has been an interior journey, into a world of luxury and leisure, but also into a world of curiosity, seriousness and purpose. A modest world, in its way, but not a smaller one. Like a 'Dutch interior' painting, it turns out to contain everything.

I

SAND, LIGHT, GLASS

Had anything faster ever been made by man? It seemed more to fly than to trundle along the sands at Scheveningen on the coast close by The Hague, reaching speeds of twenty-five miles per hour. A chassis with four large cartwheels supported the hull, which could carry a score or more of madcap passengers. Two masts bore large square-rigged sails perfect for propelling the vessel downwind and, when the Stadholder Prince Maurits, second son of the assassinated William the Silent, took the helm, the standard of the House of Orange flew from the masthead. In a stiff breeze, the *zeilwagen* shot along the beach, scattering the gulls, going as fast as the wind itself so that those aboard no longer felt its breath on their faces, according to a French observer in 1606.

The astonishing craft was the design of Simon Stevin, who was probably inspired by illustrations of much older Chinese sand-yachts that had appeared, based on recent explorers' accounts, in the magnificent atlases of the Flemish cartographer Gerard Mercator. In Laurence Sterne's *Tristram Shandy*, Tristram's uncle Toby reminisces over the 'celebrated sailing chariot, which belonged to Prince Maurice, and was of such wonderful contrivance and velocity, as to carry half a dozen people thirty German miles, in I don't know how few minutes', although he can hardly have seen it if he truly also fought at the Siege of Namur (1695), as he so persistently claims.

1. Simon Stevin's sand-yacht racing along the beach at Scheveningen with Prince Maurits aboard and his standard flying from the masthead.

Stevin was the prototype of the Dutch scientist. He had an excellent command of mathematics, but was always concerned with the practical utility of his work. A refugee from Bruges, he had settled in Leiden in 1581 at the age of thirty-three, and began to publish copiously – books of arithmetic, geometry, measurement, bookkeeping, mechanics and hydrostatics followed in quick succession, as well as applied texts on topics such as the ideal design of fortifications and sluices. He held no university position, earning his living through various windmill and hydraulic engineering projects, but this was no hindrance to his achievement.

Indeed, his situation was somewhat representative of the national condition in which 'the lack of entrenched scientific and philosophical beliefs . . . offered ample opportunity for scientific innovation'. For it was the Dutch Republic's dense commercial and cultural network of proud cities rather than its academic institutions that did most to promote the emergence and spread of new ideas.

And what ideas they were. Surely the sand-yacht of accountancy was Stevin's recommendation, building on ancient Chinese and Arab ideas, to use decimals in arithmetical calculations in place of vulgar fractions which could have any denominator. Previously, a trader totting up his takings for the day might have to juggle prices denominated in thirds, eighths, fiftieths and sixtieths in a range of currencies. In *De Thiende*, or 'The Tenth' (1585), Stevin deployed a zero within a circle next to a whole number to denote the integer part of a fractional number (showing in effect that the integer is to be divided by ten to the power of zero, or one – in other words left unchanged). A one within a circle indicated tenths (divide by ten to the power one), a two in a circle hundredths (divide by ten to the power two), and so on. Thirty years later, the Scottish mathematician John Napier streamlined Stevin's notation by introducing the familiar comma or point to separate off the fractional part of any number, making these circled digits unnecessary, but the principle of displaying this part in sequential powers of tenths was established. Thus, the irrational number π that is represented as 3.1416 . . . in Napier's system appears as 3⓪1①4②1③6④ in Stevin's notation.

In 1590, Stevin entered the employ of the Stadholder Prince Maurits as an army quartermaster. He took on military survey work, adapting the design of Italian Renaissance forts to local conditions, introducing sluices and water-filled moats in place of the dry moats of the originals, to create Europe's most advanced fortifications along the Dutch Republic's ever-shifting southern border with the Spanish Netherlands. In 1600, at Maurits's request, he set up a school of engineering at Leiden, where he decided that the teaching should be in Dutch rather than Latin. The field of engineering thus taught came to be known as *duytsche mathematycke*, or Dutch mathematics. The school's most illustrious teacher was Frans van Schooten, whose son, also Frans, would later be the tutor in mathematics to Christiaan Huygens.

Stevin's first thought was always for how his work might be applied in practice, and here he turned pure mathematical disciplines such as geometry to the good of the nation. He hoped that his example would be widely followed through the textbooks that he continued to produce. One such volume, *Van de Deursichtighe* ('Perspective'), written for Prince Maurits himself, is notable for its cool analysis of artists' secrets, including a design for a perspective-capturing device whereby one could record on a glass viewing plate an accurate impression of a scene behind; the idea so impressed Maurits that he had one made.

Stevin's greatest legacy is not the sand-yacht or decimal notation or his defensive structures, however, but the very language of the sciences in the Netherlands. He published his works in Dutch in order that they would be of the greatest use to local builders, craftsman and traders. *De Thiende*, for example, he dedicated 'to stargazers, surveyors, carpet measurers, wine gaugers, measurers of bodies in general, mint masters and all merchants'. Writing in Dutch only a few years after the secession of the northern provinces from the Spanish crown was, too, a proud statement of a still fragile independence.

But Stevin went further. In *Beghinselen der Weeghconst* ('Principles of Weighing', 1586) he set forth his conviction that Dutch was the best language for expressing scientific ideas, a conclusion that he reached based on its unusually high proportion of monosyllabic words, which could be compounded into longer words as necessary. He even believed that men had conversed in Dutch during an imagined *wijsentijt*, the Age of Wisdom long before ancient Greece, when all was once known.

The words he gave to the sciences themselves illustrate Stevin's point. Physics is *natuurkunde*, or knowledge of nature. Dutch is practically unique among languages in having its own word for physics rather than one derived from the Greek *physika*. Other words reveal a more ambitious agenda, not only taking ownership of a science by discarding its Greek- or Latin-derived name, but also drawing a subtle connection between what is demonstrable and what is scientific. Thus, mathematics becomes *wiskunde*, which can be translated as certain knowledge. Chemistry becomes *scheikunde*, the art of separation, an acknowledgement of the beginnings of a shift towards an analytical science, and a useful alternative to *chemie* that severs the etymological connection with disreputable alchemy. For comparison, the *Oxford English Dictionary* gives 1605 for the first English usage of 'chemistry', contemporary with Stevin's scientific vocabulary. The word's definition at first included alchemy as well as the emerging science of chemistry. English 'chemistry' must still live with its unhelpful secondary meaning of 'mysterious agency or change', and indeed its later concomitant sense of 'instinctual attraction' between two people. 'Astronomy' is similarly tainted. Originally, as one subject of the classical *quadrivium*, it meant the science of the motion of the stars and planets *and* of their supposed effect on nature and man – that is to say, also including the pseudoscience we now call astrology. English usage has constantly had to fight off this older double meaning; Stevin's Dutch word *sterrenkunde*, knowledge of the stars, simply removes the difficulty.

Stevin also invented or promoted the use of many straightforwardly descriptive terms in the disciplines he named, such as *driehouck* (three-corner) for triangle and *lanckrondt* (long circle) for ellipse. A trapezium becomes a *bijl*, the Dutch word for an axe, and now everybody knows its shape. Perhaps nothing displayed his devotion to science as above all a practical activity better than his mischievous choice of word for 'theory', *spiegeling*. Literally 'mirroring', *spiegeling* might be translated tactfully as 'reflection'; more provocatively, it might just be 'mirror-gazing'.

If Stevin was the prototype, then Christiaan Huygens would be the finished article. Stevin showed that it was possible to flourish intellectually outside the university, if you could find alternative patronage. Like Stevin, Huygens was concerned with the practical benefit of his work. Like Stevin, his work was grounded in sound mathematical methods at the boundaries between mechanics, geometry and optics. So one Dutch polymath lays the path for another, naming the very sciences in which his successor will become the paragon. A half-century later, Huygens will be acclaimed as the greatest mathematician, astronomer and physicist of the age.

Middelburg finally lived up to its name for a few heady decades as the sixteenth century rolled tumultuously over into the seventeenth. It found itself not only, as its name indicates, the central fortified town on dune-walled Walcheren, one of the six islands that make up the majority of the province of Zeeland, but also briefly the centre of commercial and creative activity in the Low Countries. Art and science flourished together.

The population of Middelburg doubled to 20,000 in the thirty years up to 1600, and added another 5,000 by 1622, overtaking Delft and Dordrecht to become the fourth largest city in the northern Netherlands after Amsterdam, Leiden and Haarlem. The city prospered because of its geographical location – in the middle again – during the Eighty Years War that led to the emergence of the independent Dutch Republic.

The war had begun in April 1568 when William I of Orange returned from self-imposed exile in Germany with an army of mercenaries in an attempt to force out the Duke of Alva, King Philip II of Spain's governor-general of the Netherlands. Assaults led by insurgents known as the Sea-Beggars soon captured some weakly defended ports from the Spanish forces, and increasing numbers of towns came under Protestant control. After a long siege, Middelburg chose for Protestantism and Prince William in 1574. The great abbey at the heart of the city was seized, and turned over for use as the state offices of Zeeland. In the Zeeuwsmuseum that now occupies part of the monastic complex, an anonymous allegorical painting shows the Dutch provinces as blonde women kneeling in chains before the Duke of Alva, who sits under a canopy decked with his instruments of torture while a red-dressed cardinal puffs up his ego and his ire with a bellows.

Alva's repression continued, with attacks on cities that showed signs of tolerating Protestantism. It was especially brutal in Antwerp, the most powerful commercial centre in Europe, where on 4 November 1576 underpaid Spanish troops ran amok in the 'Spanish Fury', plundering and burning the city and killing many hundreds of its inhabitants. After that, Antwerp joined in with the north and the fragile 'religious peace' promised by William, who in 1581 had been proclaimed stadholder over all of the seven Dutch provinces.[*]

---

[*] The Dutch *stadhouder* translates literally as place-holder, congruent with lieutenant in French. The title is conventionally anglicized as 'stadholder' and is similar in rank to a high steward or a governor. A stadholder might be the governor of one or more provinces. The seven provinces at this time were, working broadly northward, Zeeland, Holland, Utrecht, Gelderland,

The Spanish retook Antwerp in 1585, when the city was brought close to starvation by a blockade of its trade artery, the Scheldt river. Its Protestant inhabitants were forced to sell up, and nearly half the population of 80,000 printers and financiers, goldsmiths and painters, spice merchants and cloth traders – the flower of more than a hundred guilds – fled north over the next four years. The first place that they came to where they might hope to carry on their trades while they waited for normality to return was Middelburg, barely fifty miles away.

Long a centre for wine importation, Middelburg now diversified into new 'rich trades', taking advantage of this influx of artisans and merchants. An 'illustrious school', a lesser form of university, was established in 1592. Offices opened of the new East and West Indies trading companies, whose ships brought in spices, porcelain, curios and unfamiliar animals and plants. The well-to-do of Middelburg shared the excitement of these discoveries, leading to an eruption of interest in topics from alchemy to botany, and from mathematics to the culinary arts. One citizen, Jehan Somer, did not allow the fact that he walked with a crutch deter him from undertaking sea voyages around the eastern Mediterranean. He returned from his explorations with various exotic flowers which he planted in his Middelburg garden; one of them was the tulip, whose bulbs enthusiasts would later famously trade for astronomical sums of money.

Overijssel, Friesland and Groningen. Their political assembly was termed the States General. Various other lands that did not fall within the provinces were directly governed, in the same way as overseas colonies, by the States General. Most of these lands joined the republic following the Treaty of Westphalia at the end of the Eighty Years War, becoming the province of North Brabant.

Fine glass had been made in Antwerp since the 1540s by artisans from the Venetian island of Murano where the secrets of the technique had been kept for centuries.* By 1582, Middelburg became the first city in the northern Netherlands with a glass-works, when the Antwerper Govert van der Haghen migrated north with his Italian employees, and set up shop on the Kousteensedijk at the edge of town (trades based around kilns or furnaces were often sited away from city centres because of the risk of fire).

Prior to this, Low Countries glass had been 'rude and coarse'. Glass produced in the Venetian manner, by contrast, was called *cristallo* because of its resemblance to rock crystal (quartz) in its white transparency. The principal ingredient of all glass is sand, which is heated – probably by peat fires in Low Countries glass-works – until it is fused together. The melting point is lowered by mixing the sand with an alkaline flux such as a plant ash, which contains potash or soda. The ash is also rich in calcium, which makes the resulting glass less brittle. More calcium may be provided by adding seashells to the mix. Finally, a mineral containing manganese is sometimes included, as its purple tint helps to cancel out the greenness that often arises from the presence of iron compounds in the sand.

The Murano glass-makers favoured specific sources of these ingredients – sand from the River Ticino, a tributary of the Po that flows from the Alps to the west of Milan; ash from a variety of saltwort called *barilla*, which was imported from Syria; and manganese from ores obtained in the Piedmont mountains. These were hardly accessible to northern European glass-makers. Fortunately, though, glass is a conceding material that is not

---

* Glass-making spread very slowly through Europe in part because the Venetian Republic imposed harsh penalties on workers who emigrated, in some cases even sending assassins after them.

reliant on critical ingredients or precise admixtures. Local seaweeds and marsh plants were able to provide the ash. Oxides of manganese were present in the painters' pigments that were in such prolific use in the Low Countries – painters and glass-makers were members of the same guild. As for sand, you only had to look around. There was no shortage of that. Great dunes rose high as trees in all directions only a few miles away from the city.

It is tempting to believe there must be an autochthonous explanation for Middelburg's sudden and unexpected contribution to optical science, and that it was its particular variety of sand, that material 'so characteristic of the Dutch landscape', as the chemist and philosopher André Klukhuhn has put it, that made all the difference. But to build a case for the specialness of Middelburg, it is necessary to face the fact that *most* of the towns and cities in the coastal regions of the Low Countries are built on sand and protected by dunes. Why were these discoveries not made in Bruges or The Hague or Haarlem? The main requirement for the manufacture of fine glass for the table or for optical purposes is a source of sand that is low in iron, otherwise only dark green or brown 'forest glass' results, which is good only for bottles. Any of these places might have furnished this. But they did not. And even though sand might conceivably be transported from far away – it was often carried as ballast by ships – it is most likely that Govert van der Haghen used local sand in his kiln on the Kousteensedijk, and that his Murano-trained workers knew how to do the rest. His works was soon turning out highly trans-parent glass to be made into mirrors, beads and spectacles for the delighted citizens of Middelburg.

More important than any indigenous raw material, however, in the encouragement of innovation in Middelburg was the economic and cultural climate that had arrived along with its cosmopolitan influx from Antwerp, a diaspora that would continue to diffuse throughout the Netherlands, sowing the seeds for the emergence of the Dutch 'Golden Age'. At the end of the sixteenth century, the worthies of Middelburg were smart enough to recognize that their moment had come, and to do what they could to prolong it. In 1591 the States of Zeeland awarded the local patent for glass-making to Govert van der Haghen and gave him an interest-free loan to ensure that he did not move on to Amsterdam. Their timely action would assure their city's place in scientific history.

The Middelburg glassworks was well established by 1594, when a German immigrant, Hans Lipperhey, living hard by the abbey's St Nicholas church, set up shop as an optician and began to buy glass locally for making spectacles. He was ideally positioned to find customers. The merchants' exchange, the city mint and the illustrious college all fell within the shadow of the abbey towers. Most of those whose work required them to read, and who might therefore need spectacles, were on his doorstep, and his growing roster of clients soon placed him at the centre of a web of connection with the powerful and scholarly that stretched through the Dutch provinces and beyond.

Spectacle lenses could be made in a number of ways. One starting point was to break a blown globe of glass into curved fragments, from which dish-like roundels were then cut (thin glass can be roughly cut with scissors, which is often done under water to prevent shards flying). These roundels could then be ground on one side or the other to produce either a convex or a concave lens. However, the surface of the original glass globe is not necessarily uniformly curved, which can lead to optical distortions. An alternative was to start with mirror glass, where one side was guaranteed to be perfectly flat and thus free from such distortions. The other side could then be ground to the required shape with a greater chance of producing a good lens.

Lipperhey must have experimented at length with these possibilities. He found he could produce lenses that were of good quality towards the centre of each glass disc, but that the edges were less perfect. However, he noticed that if he simply blocked the light passing through the outer part of the lens by means of an opaque ring, the overall clarity of the observed scene improved. It had been known since the mid sixteenth century that enlarged images could be produced by placing two suitable lenses some distance apart from each other and placing one's eye up close to one of them. But such images were never fully in focus. Lipperhey constructed a tube with two lenses – a convex objective lens and a concave eyepiece – and introduced a partial diaphragm to block the light passing through the outer part of the objective lens. Again, he found that, although the brightness was reduced because less light was transmitted through the lens, the clarity of the magnified image was never-theless greatly improved. The simple – if counter-intuitive – act of inserting a diaphragm into the path of the light had led Lipperhey to the creation of a practical telescope.

On 25 September 1608, Lipperhey received a letter of recom-mendation from the Zeeland authorities giving him leave to present his 'sights involving glasses' to the national government in The Hague, where he hoped to obtain a patent. A few days later, he found himself climbing the great oak staircase of the new tower at the Binnenhof, the stadholder's palace in The Hague, along with the stadholder himself, Maurits of Nassau, the future Prince of Orange, and his half-brother Frederik Hendrik who would succeed him in this position. Also in attendance were Prince

Maurits's key advisor, Johan van Oldenbarnevelt, who had negotiated the fledgling nation's triple alliance with France and England
against Spain, and Ambrogio Spinola, the Genoese commander
of the Spanish forces in the Low Countries, both present for talks
that had already run on for the best part of a year to arrange a
truce in the long war. The chance to be a part of a scientific
demonstration must have come as a welcome diversion to the
assembled ambassadors and generals.

The demonstration proved a triumph. It was reported that
the men had been able to see the clock tower in Delft, more than
five miles away, and – catching the evening light perhaps – church
windows in Leiden, which was twice as far. It was possible to pick
out smaller objects a mile or more away, and stars in the night
sky previously invisible to the naked eye. Had Maurits known in
advance how powerful the instrument was, he surely would have
arranged the demonstration without his enemy present, for all
those standing on the Binnenhof tower immediately saw the
military potential. Spinola remarked: 'From now on I will never
be safe, because you can see me from afar.' Maurits promised that
while his soldiers might indeed see him, they would be given
orders not to shoot.

A week later, after further inspections, the States General gave
Lipperhey 300 guilders for the device, requesting that he produce
a binocular version and utilize glass of higher transparency. Within
a few months, he had delivered the binocular telescope, and was
rewarded with a further 600 guilders. The French ambassador
lost no time in asking Lipperhey to make an instrument for him,
too. However the terms of the Middelburger's deal with the States
General sensibly forbade him from working for other parties.
Lipperhey had attempted to negotiate a price of 1,000 guilders
for each device, but 300 guilders was in any case a substantial
sum. The total payment was enough for him to immediately buy

his neighbour's house in Middelburg, which he renamed De Drie Vare Gesichten (The Three Telescopes).

Despite the Dutch precautions, news of the innovation spread rapidly, sped on its way around Europe by diplomats returning home from the peace negotiations. A notice quickly circulated in The Hague 'concerning certain "lunettes" which had been presented to Count Maurice, by which one could distinctly perceive objects three or four leagues distant as if they were 100 paces away'. Within a few weeks, the notice found its way to Paris. Very soon, emperors, princes and scholars across the continent were in possession of their own instruments. Having failed to engage Lipperhey, the French ambassador took the simple expedient of seeking out a French soldier on active service against the Dutch who had used the 'far-sight' in the field and obtained the secret of the design that way.

By the following spring, spyglasses were on sale in Paris, from where one was taken to Galileo in Padua. He tweaked the design, claiming to have devised the principle himself already, and made dozens of prototypes, passing on his rejects so that very soon even more people were made aware of this instrument capable of bringing the distant close. In January 1610 he used his *perspicillum* or *occhiale* to observe four moons in orbit around the planet Jupiter, the first new objects to be discovered in the solar system since antiquity. The following year, Johannes Kepler in Prague published the first theoretical analysis of this highly practical invention, which now acquired the name 'telescope'.

Middelburg's role in this story is not yet concluded, however. The States General did not award Lipperhey the patent protection he sought for his invention, with good reason, as they were already aware of petitions from other claimants. One of these was Lipperhey's near neighbour in Middelburg, Zacharias Jansen, who had a reputation as a drunkard and a counterfeiter. His claim to have made the invention substantially earlier, around 1590, was once given credence, thanks in large part to the myth-making efforts of his son, Johannes Sachariassen, who pursued his own lens-grinding and spectacle-making trade decades later. However, Jansen was just a child in 1590. It seems more likely that this disreputable individual quickly copied what he saw of Lipperhey's work in 1608 and saw an opportunity to profit by it.

It seemed the whole city was a hotbed of optical innovation. Supported by the market in spectacles, makers and users alike had the freedom to explore the deeper potential of the technology to provide them with amusement and instruction. The learned of all professions found their curiosity piqued, from progressive

CVPIDO BRILLE-MAN·

2. This engraving of Cupid and Venus scanning the horizon for love is thought to be the first depiction of a telescope in the Low Countries.

Copernican vicars such as Philippus Lansbergen to the poet Jacob Cats, who became the city magistrate in 1621. In 'Cupido brilleman', Cats left a humorous verse impression of a Middelburg telescopist, which depicts Cupid taking up the instrument in order to further his matchmaking efforts: 'He now is just a spot, who first the spotter was.' The pun is optical, referring to the apparent change in size of subject and observer as they view one another, but also satirical, since *spot* is Dutch for a joke or for the person who is the object of one.

Just two weeks after Lipperhey's audience in The Hague, a more plausible claim came forward from Jacob Metius of Alkmaar, who was the son of William I's military engineer Adriaen Metius and brother of the professor of mathematics at the University of Franeker. It seems likely that his instrument was less powerful than Lipperhey's, because he was sent away with just 100 guilders and orders to improve his design. It is possible that Metius, too, had paid a visit to Middelburg, although it lies a long way from Alkmaar, as he had complained about the quality of glass available generally, and this was the city known for having the best. Perhaps, on such an occasion, he casually showed too much of his idea to the best spectacles-maker in town, Hans Lipperhey.

The suggestion that Metius was the true inventor of the telescope was lent credibility later when René Descartes named him in his 1637 treatise on optics, probably because he heard it from Metius's brother, with whom he had studied optics at Franeker.* Descartes was nevertheless troubled by the idea that a

---

* In a couplet of 1632, Constantijn Huygens also acclaimed Metius, although he was surely aware, too, of Lipperhey's demonstration for the stadholder, which happened when his father was secretary to the council of state. Christiaan Huygens later went against both Descartes and his father when he credited Lipperhey with the invention of the telescope in his own later treatise on optics (Huygens ed. Worp, *Gedichten* vol. 2 236; OC13 438).

mere artisan might have stumbled across such an important break-through without the benefit of any scientific theory to guide him.

A firmer connection between Metius's Alkmaar and Lipperhey's Middelburg was certainly made by another man, Cornelis Drebbel, one of the most extraordinarily fecund inventors of the seven-teenth century, who would go on to make his name by showing 'perpetual motion' machines and other wild ideas in the courts of London and Prague. Drebbel was born in Alkmaar, where he was apprenticed to a local engraver. He was an exact contempor-ary of the Metius brothers, and may have first learned about working with glass from them. Around 1600 he moved to Middelburg, where he lived for a few years. He designed a city fountain and continued to develop his skills as a glass-worker, becoming highly proficient in grinding lenses, using his inventor's mind to devise machines for doing the job that would yield more

3. Cornelis Drebbel, a prolific Dutch inventor and engineer who demonstrated a submarine in the River Thames and tutored Constantijn Huygens in lens-grinding.

consistent results than manual labour. Although Drebbel had moved on to London by 1608, when the rival patent applicants clashed in The Hague, he too later established himself as a maker of telescopes, believing Metius to have been the inventor.

Drebbel played a significant role in the development of optical technology through his instruction of younger men in the art of lens-grinding. One of his pupils was Isaac Beeckman, whose family, like so many others, had fled to Middelburg from the south during the first phase of the Eighty Years War. Another polymath, Beeckman made investigations in hydraulic engineering, medicine and meteorology, among other fields, and became an important friend of Descartes, encouraging him in his mathematical study of the natural world. Exhorted by the clergyman-astronomer Lansbergen to develop a better telescope based on the design of Galileo, Beeckman took the art and science of lens-grinding to new heights during the 1620s and 1630s, eventually surpassing the skill of the established lens-grinders such as Sachariassen.

Another who learned from Drebbel was Constantijn Huygens, who met the Dutchman in London in 1621. Though Drebbel 'looked like a Dutch farmer', as Huygens wrote, it was clear that he 'aimed to surpass the ancient Greeks and Romans in knowledge'. He impressed the young diplomat with his arcane knowledge and his marvellous mechanical experiments, and taught him how to grind lenses – a skill that Huygens would later transmit to his sons.

This is a story of water and ice and the flat, flat land.

In the winter of 1622 a hard freeze came to all of Europe. Even the Golden Horn of Istanbul froze over. Onto the thick ice of the meres and rivers and canals and ditches around Leiden stepped the students of Willebrord Snellius with their measuring devices. The level surface of the frozen water was an invitation to do science.

For many years, Snellius had been engaged in a grand project to survey the young country. Using a method of triangulation devised by his countryman Gemma Frisius, he had placed the city of Alkmaar under a mathematical net of triangles by means of which he could begin to calculate long distances with great accuracy. Distances between the points of each triangle were measured with an iron chain, and angles were calculated using a large brass quadrant with fine graduations of two minutes of arc, one thirtieth of a degree. Eventually, by sighting from church tower to church tower, his survey lines reached all the way to Bergen-op-Zoom, nearly 150 kilometres to the south along approximately the same meridian.

Using astronomical methods to determine the latitude at each end point of a line of known length along this meridian, he was able to calculate the size of the Earth. He calculated that one degree of latitude came to 28,500 Rhineland rods, which put the Earth's circumference at 38,600 kilometres – about 4 per cent less than its known value today, a remarkably close estimate. He published this result in a work called *Eratosthenes Batavus*, named after the Greek mathematician who estimated the size of the Earth by comparing shadows in different places, and the ancient tribe of Holland during the time of the Roman Empire, the Batavi. This was one of the first practical surveys of a significant area of land, and it soon inspired similar exercises in England, Italy and France. However, Snellius was unhappy about inaccuracies in his earlier measurements, and it was the opportunity to obtain more accurate readings on the horizontal ice that now drew him and his assistants out into the crisp air.

Snellius, or Snel van Royen, was born in Leiden in 1580. His father Rudolph was the first professor of mathematics at the university founded there by William I in 1575, and Willebrord would eventually succeed him. After some years abroad learning astronomy from Tycho Brahe and Johannes Kepler in Prague as well as law in Paris, Willebrord returned to Leiden in 1604, where he studied mathematics with his father and set about translating the work of Simon Stevin into Latin so that it might be read more widely. He wrote his own books on the conic sections and on navigation, and discussed methods of calculating π to greater accuracy, before embarking upon his great land survey.

Snel's fame rests on none of this work, however, but on his eponymous law of refraction. In modern terms, Snell's law (the spelling of his name is conventionally anglicized in this context) states that the ratio of the sines of the angles of incidence and refraction when light passes from one medium to another is inversely proportional to the ratio of the refractive indices of the two media. His ideas probably began to take shape from 1618, when he began working with telescopes, for example deducing, by means of parallax from observations made at two different points, that the path of a certain comet must lie far beyond the moon. The use of lenses in such equipment demands at least an intuitive appreciation of the phenomenon of refraction. But Snel formulated his law by considering the simpler circumstance of a point of light under a layer of water, such as might be made by a coin catching the sun at the bottom of a canal. To the human eye in the air above, the coin appears to be lying in shallower water than it actually is because the path of the light is bent at the water surface. By 1621, after many experiments, Snel established that there was a definite mathematical relationship between the lengths of the real and apparent paths of the light under the water, and that this was in proportion to the ratio of sines of the angles made as the light passed through the water surface.

As often happens in the history of science, however, the full story is more involved than this, for in fact 'Snell's law' was discovered repeatedly, both before and after Snel did his experiments.* Because Snel himself did not announce his discovery at the time, and his original manuscript was later lost, his rightful place in the sequence of events has become hard to discern. The English mathematician Thomas Harriot discovered the essence of the sine relationship in 1601, using an astrolabe suspended in water, but he did not publish either, and his work, too, was revealed only later. It was not until René Descartes's treatise on refraction, *La dioptrique*, appeared in 1637 that the law of refraction entered print.

It seems that Descartes discovered the law independently by a geometrical analysis similar to that of Snel, though without performing any direct experiments or measurements himself, while he was living in Paris in 1626. He told his friend Isaac Beeckman about it when he came to Holland two years later, although later Dutch figures such as the scholar Isaac Vossius, who had read Snel's work and made use of it for his own treatise on the nature of light, believed that Descartes must have had sight of it too. Christiaan Huygens echoed this view in his own treatise on optics, and in 1693 he confirmed to the philosopher Pierre Bayle 'that the laws of refraction are not the invention of Mr Descartes despite all appearances, for it is certain that he read the manuscript book of Snellius, which I have seen too'.

Ironically, by the time that Descartes published the law of

---

* As is also increasingly the case in the modern history of science, it is now recognized that the discovery was effectively made by somebody else entirely, working much earlier, much further east. The mathematician Abū Saʻd al-ʻAlāʼ ibn Sahl employed the same geometric formula for making more efficient burning glasses in tenth-century Baghdad (R. Rashed, *Isis*, vol. 81 (1990) 464–91).

refraction, its validity appeared to be threatened by new discoveries about the behaviour of light, such as its splitting into colours and its strange behaviour in relation to certain transparent materials such as the mineral known as Iceland spar. For a generation, it seemed that the law of refraction might not have the force of a law at all, until further work by Newton and Huygens was able to show that it did in fact remain valid in relation to these effects.

It is Rembrandt's earliest known work, and how appropriate that he should begin with sight.

De Brillenverkoper is one of a set of five allegories on the human senses, painted in 1624 when the artist was just eighteen years old. The tiny canvas shows a gypsy spectacles seller attempting to persuade an old couple to buy. He is dressed in an exotic tunic and wearing a rich purple turban. A tray of wares is strapped around his shoulders with many pairs of glasses in it. His assortment may have been laid out according to optical strength, in order to help him pick out the glasses most likely to be suitable for customers of a certain age. With a smirk, he proffers one set to the old man, who points to his wine-shot aubergine of a nose as if his greatest worry is finding a pair that can bridge it rather than one to correct his vision. The man's wife stands by his side, her eyes half closed, her hand reaching tentatively forward, suggesting that she is virtually blind, glasses or no.

Not ideal customers, maybe, but where is the allegory? Well, the Dutch expression 'to sell glasses' – *brillen verkopen* – once meant to deceive, because the buyer is in no position to see what they are buying. The seller is a swindler.

The painting is nevertheless revealing about the nature of this trade in the early Dutch Republic. Spectacles were slow to catch on for a number of reasons. Poor eyesight was regarded primarily as a medical complaint rather than a problem that might be remedied by optics. Compared to the thick lenses of 'reading stones' or cloth traders' thread-counters, the weak lenses that most people required for spectacles were challenging to make because their lesser curvature had to be very carefully controlled. Although wealthy people had been able to buy spectacles since the thirteenth century, it was the advent of printing that did most to increase popular demand. However, there were no eye tests or personal prescriptions. Finding spectacles to improve your vision was a matter of trial and error. This is why the *brillenverkoper*'s tray is so heavily laden; he is offering at least a choice if not a perfect solution for his customers' dysopia. The glasses themselves, with unframed lenses, rest almost invisible against the velvet lining of his box, their presence indicated by the barest stroke of lead-white paint where the light catches the circumference of each glass roundel. The transparency of the lenses suggests that these examples are made of superior *cristallo* rather than green-tinted forest glass. (The Dutch word *bril*, meaning a pair of spectacles, is derived from the Latin word for beryl, an almost transparent crystalline mineral rather than a glass.) The business grew rapidly into a thriving trade, with glasses sold door to door as well as in shops and markets, the volume of sales boosted by the never-ending search for the right prescription. In England around this time, for example, one Lord William Howard bought at least twenty-seven pairs of spectacles in thirteen years, presumably because of the difficulty in matching his needs. A popular joke in Holland, meanwhile, reflecting both the vibrancy of this market and the generally high level of literacy, centred on a man trying on many spectacles, none of which will do; it turns out that he simply can't read.

In the space of little more than a generation, it seems, glass lenses shaped to ameliorate human eyesight attained a degree of familiarity sufficient to make them a topic of satire, while more specialized optical instruments began to spread among the cognoscenti, and the first efforts were made towards understanding the fundamental principles of how these objects worked. Into this bright world of vision will step first dim-sighted Constantijn and then, by turns, his four sons.

2

THE *KENNER* OF ALL THINGS

In 1578, the twenty-seven-year-old Christiaen Huygens was appointed as one of four secretaries to Prince William of Orange.[*] It was a position for which he was well qualified, coming from one of the most important families in Brabant and having trained dutifully in the law, and an honour to enter the service of the man who only months before had triumphantly brought several new provinces into alliance with Holland and Zeeland, and who would soon be acclaimed as the 'father of the fatherland'.

Christiaen was born 'between ten and eleven o'clock in the evening' on 22 April 1551, according to his son's family memoir, in Terheijden, close to the city of Breda, where the House of Orange-Nassau held court. The infant's father was already dead

---

[*] Like the Buendía family in Gabriel García Márquez's *One Hundred Years of Solitude*, the Huygenses confusingly make use of very few given names down the generations. The Christiaen Huygens here (1551–1624) is the father of the poet and diplomat Constantijn Huygens (1596–1687). Constantijn's first son was also Constantijn (1628–97), the second was Christiaan (1629–95), named for his grandfather, who is my principal subject. I have used the earlier spelling Christiaen to refer to the grandfather, and Christiaan for the grandson. Where they are liable to be confused, I have occasionally added 'the elder' or 'the younger' to indicate which Constantijn is meant, although the family themselves used a variety of nicknames to keep things clear. To make matters worse, the Huygens wives of these three generations were all called Susanna or Suzanna, as was the elder Constantijn's daughter, but here context is usually sufficient to avoid ambiguity.

by then, and his mother was to die when he was just five years old.

By this time, William had inherited estates from his birth-place at Nassau in the Landgraviate of Hesse as well as the city of Orange in France and various lands in Holland, Flanders and Brabant, where the Lutheran boy had received a Roman Catholic education. Most of the territory that is now the Netherlands had been ruled from Spain under the Habsburg dynasty of Charles V since the beginning of the sixteenth century, with provincial governors known as stadholders appointed by Spain. In 1559, at the age of twenty-six, William was appointed by King Philip II of Spain as the stadholder of Holland, Zeeland and Utrecht.

Here, he sought to balance the competing religious interests, maintaining the loyalty to the Catholic Church demanded by the provinces' Habsburg Spanish rulers while also tolerating Protestant worship. Despite William's considerable political skill, which earned him the byname 'the Silent', it proved an impossible task. Ever bolder displays of Protestant devotion led in 1566 to the iconoclastic fury known as the *beeldenstorm* and, in the following year, to the violent reassertion of Spanish control under Philip's hated governor-general the Duke of Alva.

With Protestants fled abroad or driven underground by Alva's fanatical terror, the Low Countries were now primed for revolt. William, who had withdrawn to Germany, at first held back from armed conflict, but he soon found himself the focal figure of the resistance, gradually knitting a coalition between towns, cities and provinces that paid due heed to their religious and commercial particularities. However, the Dutch proved unable to match Alva in battle during the opening skirmishes of the Eighty Years War. After several years of advance and retreat, sieges and strategic floodings, the Spanish controlled the Catholic southern Netherlands while the north was consolidated in its Protestantism. William believed the best hope for regaining control of Brabant and other southern areas was to promote a regime that would permit both Catholic and Protestant worship, and in September 1578 in Brussels he proclaimed a 'Religious Peace'.

It was into this divided world that Christiaen Huygens cautiously entered. He was able to show that he possessed the necessary qualities for a diplomatic secretary on a particularly delicate mission in 1581. According to a story related by his son Constantijn, the Spanish ambassador in London had engaged a Dutch spy, one Willem Janszoon van Hooren, on the promise that he would secure for the Spanish the port of Vlissingen, then an English garrison. The man's ten-year-old son was held as a bond in London while he sailed for Vlissingen. When van Hooren double-crossed the ambassador and informed William of the plot, Huygens was quickly briefed to go to London and obtain the boy's freedom. As he moved to leave the room, Huygens paused at the door and became tangled in the rope-and-pulley arrangement used to hold it open. William laughed that it was a warning: Christiaen must mind not to end up on the rope doing his job. Fortunately, the daring mission was a success. Huygens managed to detain the ambassador in conversation while an accompanying soldier seized the boy; the attack on Vlissingen was averted.

On 10 July 1584, William I of Orange was shot at the foot of the stairs in his Delft residence, the Prinsenhof, by a Burgundian Catholic, the first assassination of a head of state with a handgun. By this time, however, Huygens had demonstrated loyalty and ability, and he continued in service as secretary to William's successors in the House of Orange, and established a family career path that would be followed first by his son and then by his grandson for more than a hundred years, so that eventually it must have seemed there had never been a time when the Huygenses were not tied to the Dutch royal house.

If Constantijn Huygens inherited his diplomatic vocation – and his fine-boned looks – from his father, then his affinity for art came undoubtedly from his mother. On 26 August 1592 Christiaen married the thirty-year-old Susanna Hoefnagel. The Hoefnagels were a leading Antwerp family of painters. Susanna's brother Joris became famous for detailed cityscapes and miniatures. His highly accurate depictions of natural history, especially insects, give him some claim to be called the father of scientific illustration. His son Jacob followed in his father's steps and became the court painter to the Holy Roman Emperor Rudolf II in Prague.

These parents must have set a remarkable example to their children by the circles in which they moved, but they also took unusual steps to guide the education of their four daughters and especially the two sons, Maurits, born in 1595, and Constantijn, born a year later on 4 September 1596. They imitated the 'attractions of children's games' to make instruction in music, singing and dancing, and later in writing and languages, more appealing. French was memorized by reciting prayers in the language, while natural conversation was encouraged by arranging classes with French Huguenots exiled in The Hague. It was a model that worked well for Constantijn Huygens, and he later adopted the same methods in the education of his own children.

Painted by Pieter Isaacsz around 1605, a splendid harpsichord lid in the collection of the Rijksmuseum shows the mythical Maid of Amsterdam, her hand poised proprietorially on a globe, surveying a scene of riches from the continents of the world. A bounty of exotic people and produce and elaborate cultural treasure has washed up seemingly on the Dutch shore, for the lid is also the shape of the country. The work is called *Amsterdam as the Centre of World Trade*.

By the beginning of the seventeenth century, Dutch arts had begun to prosper in the wake of the Revolt, stimulated by an influx of refugee artists and the rapid expansion of trade, leading to a flowering of art on a scale not seen anywhere before or since. Works of art and decorative objects were no longer the exclusive preserve of aristocrats but now the prerogative of merchants and artisans.

But *caveat emptor*, and *caveat venditor*, too: not all artists were any good, and not all buyers were motivated by fine feeling. To navigate this cornucopious new world would require artistic judgement and discernment, and Constantijn Huygens's education gave him these skills across a range of arts. Along with riding, swordsmanship and correct dress, the gentleman of the day cultivated not simply an interest but a comprehensive knowledge of the arts – he aimed to become a *kenner*, a connoisseur, one who truly knows. To appraise a painting properly one had to be able to paint; to appreciate a piece of music one had to be able to play, and preferably to compose. And Constantijn could. 'I would have achieved a certain repute among artists of the middling sort,' was his own verdict on his ability as a painter. In time, he would pass on a sense of the value of such hard-won authentic knowledge to his children.

His earliest education at home laid the groundwork: he learned languages, singing and dancing from the age of four. At six, he began on the viol, and other instruments soon followed. Huygens's facility with a quill pen also came early on, important for a man who is thought to have written up to 100,000 letters during his long diplomatic career. His small, neat handwriting, unpretentious and businesslike, displayed his calligraphic skill, but was also essential, as Huygens was often required to write on tiny scraps of paper that could be tightly folded and sent secretly. Soon he began writing verse too, at first in Latin, inspired by his reading of the Roman poets, and later in French and Dutch.

Drawing and painting he learned from Hendrik Hondius, a noted cartographer and engraver, whose cold architectural style nevertheless did not much appeal to Huygens, who would have preferred to be instructed by the freer hand of the painter Jacques de Gheyn II. He explored various applied arts too, even trying ivory-carving, for which he had his own lathe. Through masters such as these, the young Huygens acquired a thorough knowledge of artistic theory and practice that would later enable him to put forward promising artists for patronage and to acquire and commission works for his employers at the court. He became a *kenner*.

It is hard to translate this word adequately. Superficially, *kenner* corresponds exactly with our (and the French) *connoisseur*, but that word has become debased in contemporary usage and no longer carries the implication of practical expertise in addition to schol-arly interest. The word is most useful as a term of distinction from the ill-disciplined and often uninformed enthusiasm of the amateur or *liefhebber*, of whom there were great numbers in the newly rich land. These were the kind of people whose pursuit of aesthetic novelty would leave them with burnt fingers in the notorious tulip mania of 1637, and who were disparaged by real artists and critics such as Samuel van Hoogstraeten. But even the *liefhebber* was better than the *naamkoper*, who bought art solely on the strength of the artist's name.

The painting *liefhebber* might occasionally dabble, but the *kenner* knew how to grind pigments, mix paints and use a brush, and was able to apply this detailed knowledge not only to paint, but for example to establish the authenticity of works and to

resolve questions of attribution. His procedure was based, therefore, not purely on visual analysis, but involved a more complex
sensorium in which his motor memory of using a brush was
activated by the sight of brushstrokes on the canvas.

Constantijn Huygens was called upon to employ this skill on
many occasions. Most notably, he was able to identify the supreme
talents of the young Jan Lievens and Rembrandt and help them
towards early commissions. But he was also able to detect professional malpractice. When the Flemish painter Gonzales Coques,
commissioned by Frederik Hendrik of Orange for a series of
paintings on the theme of Psyche, presented works that he had
subcontracted to a lesser artist, Huygens was able to show, to the
hilarity of the court and to Coques's acute embarrassment, that
the works did not bear the right *'kennelyke'* hallmarks of the artist,
and were in fact copied from some prints after Raphael.

The range of Huygens's artistry must have done much to
augment his work as a diplomat, whose professional life is so
much concerned with orchestrating the exchange of refined gifts.
On occasion, he was even able to employ his own creative powers
directly. He always believed that his ability as a musician was a
significant factor in gaining him access to influential people. But
he also made material offerings. A short verse addressed to the
specialist in floral still lifes, Daniel Seghers, indicates that Huygens
felt himself proficient enough to dare to enclose a small painting
or drawing of his own with a letter to the artist dictated by
Frederik Hendrik thanking him for a painting of a vase with
flowers.

> Cast a compassionate eye on my wilted blooms,
> Flower maker under God. They could not presume
> To be anything Seghers-like . . .

Huygens clearly regarded his artistic facility as something to be nurtured lifelong. He went to his cousin Jacob Hoefnagel to learn the art of painting watercolours, for example, and as an old man he took up the guitar in addition to his other instruments. Such a high level of accomplishment naturally brought the reward of creative satisfaction for its own sake, too. Simply turning to a new medium may have helped Huygens overcome a creative block in another. Spoilt for choice in his creative faculties, Huygens may also have fretted that too much prowess displayed in one art might cause people to underprize him in others.

In addition, Huygens was able to exert an enviable degree of control over his aesthetic environment. He commissioned portraits of himself and members of his family from fashionable artists such as Antoon van Dyck, Jan Lievens, Caspar Netscher and Michiel van Mierevelt, and owned other works by painters as varied in style as Rembrandt and Pieter Saenredam. These were hung in the house on the Plein in The Hague, designed to Huygens's expertly informed brief by two of the best Dutch architects of the age.

The 'embarrassment of riches', to reuse the phrase taken by Simon Schama for his thrilling study of Dutch culture in the seventeenth century, was a discomfort keenly felt in the increasingly Calvinist republic. An informed knowledge – a *kennis* – of the world's cultural treasure offered a kind of possession that did not offend against public frugality. But correct Reformed conduct was not to be equated with simple abstemiousness; too many Dutch were too wealthy for that. This is where the discernment of the *kenner* counted for something. Luxury objects would be deemed more acceptable if they were 'authentic' rather than merely pretentious. The house that Huygens built on the Plein was large and grand, but it was not as ostentatious as it might have been. Indeed, its restrained classical style appeared almost severe.

The pursuit of science provided a more discreet way to squander one's riches. Telescopes and astrolabes, instruments for mensuration and drawing, collections of minerals and stuffed animals, all made a display of serious living and learning. Huygens had not neglected the sciences in his education, studying aspects of physics, medicine and alchemy. He knew enough about them to be able to help Descartes in making telescope lenses, and he was later closely involved in his sons' astronomical activities. He had his own scientific interests, too, for example in herbal medicines. He also developed a fancy for perfumery, copying down long lists of the spices and other ingredients needed to mix them and distilling his preparations, presumably in some kind of home laboratory.

Constantijn Huygens was one of the last of the generalists who could legitimately claim to be proficient in many fields. The enormous expansion of the arts and sciences during his lifetime would see to it that this was no longer a realistic aspiration for his children. Christiaan was to be one of the first specialists, becoming as close to being a professional scientist as was then possible.

At Leiden University, Constantijn Huygens broadened his learning still further. He acquired a solid grounding in medicine and the sciences as well as in aspects of philosophy and theology. He heard leading exponents of Latin verse such as Daniel Heinsius and Caspar Barlaeus, with whom he would later find himself in sportive rivalry. But he found many of his fellow students to be arrogant and boastful young men who scoffed at their more virtuous peers; perhaps Huygens himself came across as a little superior and priggish. He did form some friendships, however, including one with Cesar

Calandrini, with whom he shared lodgings and who later moved to London and became a stern Calvinist priest.

After graduation, Huygens hoped to go to England too, making use of family contacts in The Hague at a time when various English representatives sat on the Dutch Council of State following the two countries' alliance against Spain. He got his chance in the summer of 1618, travelling with Sir Dudley Carleton, the English ambassador in The Hague. With boils and carbuncles on his feet and fearing the plague, he hobbled round picture collections, visited Windsor and Greenwich, Oxford, Woodstock and Cambridge, embraced his friend Calandrini, and wrote home with verses and complaints about the costs of travel and medical care. Coming from a new-minted republic, Huygens was keen to see what a real king was like, and on 10 July his curiosity was satisfied when James I paid his customary annual visit to Huygens's host in England, the long-serving Dutch ambassador Noël de Caron, in order to sample the ripe cherries in the garden of his residence in Lambeth, a ritual that involved the setting up of carpeted steps to the tree so that king could reach the best fruit.

A few weeks later, Huygens journeyed with Caron to the king's palace at Bagshot in Surrey, where they stayed for a week. He wrote to his parents enclosing some venison pâté from a deer that the king had hunted during the course of one long day until it blundered into a pond and was captured. He gave an unsparing description of the beast's demise, 'from which you might deduce that it should be eaten with reverence and attention'. He also described how on this occasion he found himself playing his lute for James,

in which – I know not what right-hand demon encouraged me to make something worthy of the ear of a great king – I acquitted myself well enough for half an hour that this prince, little given to music, was forced to interrupt his card-game to listen to me, as he did with good grace, without sparing his usual remarks to assure me of the contentment that he took, even doing me the honour of speaking to me with a kind and smiling face. Upon my having kissed his hand, I took my leave in all good style, well satisfied with the good success of my little business. Here was an Iliad in the space of half an hour.

In the autumn, Carleton was sent back to Holland in response to a new religious and political crisis there. Making his own way back to The Hague, Huygens was able to sit in on some of the meetings at the Synod of Dort (Dordrecht), the international Protestant convention that consolidated points of doctrine in the Dutch Reformed Church and made it a canonical aspect of the faith that humans are fatally and irremediably sinful creatures. The event prompted a poetic confession from Huygens himself, who interleaved his verses with the lines of the Apostles' Creed. Under the line 'He will come again to judge the living and the dead', he wrote in celestial images:

> When water, air and earth shall tremble and shake,
> When sun and moon both their brightness shall forsake,
> When heaven is without light, and stars no longer shine
> Of the living God shall these things be the sign.

Huygens must have proved his worth in England, because he was selected to accompany another diplomatic mission the following year when the States General signed a treaty with that other trading European republic, Venice, in December 1619. The tour was a formative experience for Huygens, opening his eyes and ears to the art, architecture and music of Italy in ways that would transform his own creativity. In Vicenza, he visited the Teatro Olimpico by the architect Andrea Palladio, while at St Mark's in Venice he heard Claudio Monteverdi conducting his own compositions.

By January 1621, Huygens was in London once more, having been officially taken on as a secretary in the Dutch embassy to England. The Dutch aim was to enlist James's help in bolstering Protestant Bohemia, where his son-in-law had been deposed as king, against the expansionist ambitions of the Habsburg empire. However, since England did not want to provoke Habsburg Spain, it was a hopeless cause. Huygens's task was to take notes in the international conferences and bilateral meetings between diplomats, but his duties also took in a ceaseless round of entertainments, from jousting tournaments to ballets, that reflected the historical moment, poised between the medieval and the modern worlds.

As tensions rose with the spread of the Thirty Years War in central Europe and the renewal of Dutch attempts to shake off their Spanish oppressors, the diplomatic stakes increased, too. Following a brief return to The Hague, Huygens went back to England, trying to put the Dutch side of the argument in disputes that had arisen between the rival English and Dutch Indies trading companies, and hoping to turn James away from his growing rapprochement with Spain, which now included the prospect of his son Prince Charles's marriage to a Spanish princess. For him, it was a routine marked by continual frustration, with only occasional indications that it might be worthwhile to stick at it. One such moment came when he was sent out to apply for an audience with the king at his hunting lodge in Royston, Hertfordshire. He was surprised to emerge from the meeting with a knighthood bestowed upon him. On the ride back to London, however, he and his horse were forced off the road by a passing convoy of carts and became trapped until they were rescued. Later in life, it was his deliverance from this mishap rather than the royal honour for which he gave thanks.

Undoubtedly, Huygens owed his rapid preferment in England not only to his own considerable charms and abilities but also to the influence of the aged Caron, who had cemented the Anglo-Dutch alliance with Queen Elizabeth I as long ago as 1585 and had retained his post under her successor. Huygens was quick to acknowledge such help when it was given, which facilitated his further rise up the social ladder.

He nevertheless viewed his profession with a certain ironic detachment. In a verse reflecting on his embassy trips to England, he described the job of an ambassador:

> He is an honest Spy; abroad a canny dealer;
> And here an all-round dangerous indweller;
> A Prince's longest Arm; an uninvited guest,
> Who must locate their place however among the best;

Compare this with the famous remark of the Huygenses' sometime neighbour in The Hague, Carleton's predecessor as England's ambassador-extraordinary to the Dutch Republic, Sir Henry Wotton, that 'an ambassador is an honest man sent to lie abroad for the good of his country'.

The life of even a junior member of an embassy team was not entirely devoid of pleasures, however. Huygens used his leisure time to expand his social and musical network, meeting leading figures such as the composer Nicholas Lanier and the foremost lutenist of the day, Jacques Gautier. He also took the opportunity while in London to improve his knowledge of the sciences, in which he knew he was still comparatively wanting.

He started at the top by seeking out the sixty-one-year-old Sir Francis Bacon, Baron Verulam, whose literary works he had long admired. The prolific lawyer, natural philosopher and historian had risen to the position of Lord Chancellor under King James, but had been recently disgraced by a scandal for which he was briefly imprisoned in the Tower of London. Fortunately, his output as a writer and thinker remained largely unaffected by his travails in public affairs. *Novum Organum*, the work that set out his scientific method based on observation, experiment and inductive reasoning, had been published in 1620, and when Huygens met him he was at work on *New Atlantis*, the utopian novel which would describe a 'Salomon's House' of scientific knowledge – an imaginary prototype of the later Royal Society – and catalogue examples of the marvellous innovations that a new approach to science might bring.

Huygens had 'always possessed a kind of holy awe' for Bacon, even if he felt that the Englishman might have been more modest in the titles he chose for his works. He most admired the astounding breadth of his writing, from histories to the law, science and religion, and essays on many other topics. He quoted approvingly and at length from Bacon's argument in the early sections of *Novum Organum* that science in the early seventeenth century was suffering from a kind of stasis in comparison with its flowering in the classical period. But although in the end Huygens met Bacon several times, the encounters proved a cruel letdown. He found that Bacon was 'not to be outdone in arrogant presumption and affectation'. In Huygens's presence, Bacon adopted a distracting and pompous way of speaking and gesticulating, which the young diplomat observed was not something he did when he was with his friends, among whom he behaved completely naturally.

Although Huygens emerged from these audiences with his respect intact for Bacon's 'exalted learning', such that he helped to arrange the translation of some of his later works into Latin, he could only hope to erase his image of the man himself. 'I do not believe that personal acquaintance has done any great man's name more harm,' he wrote in his youthful autobiography.

Huygens fared better with Cornelis Drebbel, the Alkmaar-born inventor who had lived in London since 1605 except for one long sojourn in Prague. Drebbel impressed the courts of King James and Emperor Rudolf alike with his seemingly impossible inventions, which included an apparent perpetual motion machine, a rudimentary magic lantern capable of projecting images, and even a harpsichord powered by the heat of the sun. His fair, handsome appearance and understated manner were at odds with what people expected to see in a trickster, and they were intrigued as much by the man himself as by the demonstrations he put on. He had little trouble in gaining patronage in both cities, and was set up by his English benefactors with a laboratory at Eltham Palace not far from Greenwich.

Though he was largely without formal education, Drebbel's scientific range was vast. His inventions relied variously on mechanical, hydraulic, thermal and chemical means. He was also a proficient engraver and cartographer, producing a fine map of his home city on copper before he left it for ever. He only ever obtained one or two patents, so it appears that his main interest was in showing off his inventiveness rather than profiting by it, and for all the support he enjoyed he struggled to make a living.

Drebbel's most astonishing coup de théâtre took place on the River Thames, or rather under it, around the time that Huygens arrived in London in 1621, when he demonstrated a passenger-carrying submarine. Contemporary accounts are vague, but the vessel was probably based on an upturned boat sheathed in animal hide and greased to make it waterproof, with leather-jointed holes for rowlocks through which the crew of eight could stick their oars in order to propel themselves underwater. There was supposedly space for sixteen passengers aboard. It seems that the floor of the submarine was open to the water, the craft relying upon the pressure of the air trapped within it and very careful seamanship to maintain its stability.

Although Huygens did not witness the demonstration directly, he later reported:

> One invention to set against all the others is the little boat in which he happily dived under the water such that the King, the court and several thousand Londoners held their breath. When they had not seen him for more than three hours, so the story goes, most thought that he had fallen victim to his own piece of art, until he surfaced again at a great distance from the place where he had dived.

It was said that the vessel could remain submerged for up to twenty-four hours. How the submarine crew were supplied with air for any extended length of time without a tube to the surface has remained a mystery. Robert Boyle, the foremost chemist of the following generation, thought highly of Drebbel – 'that deservedly famous Mechanician and Chymist' – and understood that he had employed 'a chymicall liquor', whose composition he would not disclose, which was able to manufacture 'a certain Quintessence (as Chymists speake) or spiritous part' of air within the craft. Drebbel was well acquainted with the leading alchemists of his day, and knew, for example, that saltpetre could be heated to release the life-sustaining gas that would not be isolated and named as oxygen for another 150 years. It has been widely assumed that this was the reaction he used to maintain a breathable atmosphere aboard. However, there are other chemical possibilities. In 1690, Constantijn Huygens the younger was in London when he was visited by an elderly daughter of Cornelis Drebbel, who told him about her father's famous submarine and explained that he had used a 'pipe with quicksilver' to supply the air. This hints that Drebbel may have employed a different reaction – the thermal decomposition of mercuric oxide into mercury metal and oxygen gas, the very reaction used by Joseph Priestley to isolate 'dephlogisticated air' in 1774.

For a man so excited by scientific spectacle, it is hardly surprising to find that Drebbel was also interested in optics. As he prepared to return to London from Prague in 1613, he hoped to regain the king's favour with a gift, which he claimed to be his own invention, of 'an instrument by which letters can be read at a distance of an English mile' – or five, six or seven miles, or eight or ten, as his effusive description continued. Telescopes were already in widespread circulation in the courts of Europe by this time, but it is likely that Drebbel had effected some improvements of his own, having originally learned to grind lenses in Alkmaar with Jacob Metius. He may have been the first to use a machine to assist in the grinding process to produce standardized lenses. He was also a fine blower of glass, able to make the intricate glass components for his perpetuum mobile. In the early 1620s, Drebbel modified a telescope by introducing two convex lenses, which converted it into a primitive microscope. A different arrangement of lenses enabled him to entertain Londoners with projected images on a wall in early magic-lantern shows.

Constantijn Huygens met Drebbel briefly in the spring of 1621, and was struck by his great knowledge, which he likened to that of Pythagoras and Empedocles. The men were able to meet more frequently when Huygens returned to London for a longer stay the following year, forming both a friendship and a beneficial master–pupil relationship as Drebbel inducted his fellow expatriate into the mysteries of the physical sciences. It is 'very probable' that Huygens learned the techniques of lens-grinding from Drebbel at this time. He also bought from him spectacles for himself and for others, as well as a telescope and a camera obscura, which he took back with him to The Hague at the end of his posting.

They must have made a curious pair, Huygens with his trim dark beard and sharp features, and the strapping, ill-read Drebbel with his shaggy blond hair, careless of social hierarchies. It is odd on the face of it that a junior diplomat always ambitious for promotion should spend time with such a conspicuous and erratic individual. Constantijn's father even wrote to warn his son to stay away from Drebbel, but Constantijn replied: 'I laughed at your last in which you were pleased to warn me about Drebbel's magic, and accused him of being a wizard.' He amusedly reassured his father on that score, and indicated that Drebbel might possess secrets that could prove useful in the war that was expected to resume in the summer, and promised to bring back a camera obscura for their neighbour, the painter Jacques de Gheyn II, 'which is certainly one of the masterpieces of his sorcery'.

The camera obscura is essentially a dark space with a lens or pinhole in one side through which a bright outside scene is projected – reversed and inverted – as an image on to a two-dimensional surface. Leonardo da Vinci gave one of the first accurate descriptions of such a design. Drebbel's version was a 'lightly constructed instrument' designed for portability, with a closed chamber in which the image appeared, thanks to an additional mirror, in its correct orientation on a white translucent screen. Drebbel must have used a lens of exceptional clarity, for Huygens was deeply impressed by the quality of the images he saw. 'It is impossible for me to put the beauty into words for you,' he told his parents. 'All painting is dead in consequence, because here is life itself, or something higher.'

He was equally swept away by Drebbel's microscope. 'Nothing will exhort us more strongly to the worship of the infinite wisdom and power of the Creator than to enter into this other treasure-house of nature,' he wrote. He longed to know what artists, such as de Gheyn, who had died by the time he was writing, would have done with such an instrument. 'If de Gheyn sr. [II] had lived longer, he would presumably have made it his task to delineate the smallest objects and insects correctly with a very fine brush. I had already begun to push him that direction and he had a mind to it. He would collect them in a book and give it the title "The New World".'

How serious de Gheyn himself ever was about these possibilities cannot be known. What is clear is that Huygens immediately intuited – with his topical allusion to the Americas then being opened to colonial exploitation by the Dutch and others – that the *terra nova* of the microscopic realm was above all a treasure-house for the eye. It is notable that Huygens regarded both devices in terms of the uses that artists might find for them rather than as tools for the scientific investigation of nature. While the camera obscura was indeed more widely – and less secretively – employed by Dutch painters after Huygens's demonstration of it in The Hague, the scenes revealed under the microscope remained for a long time too murky, and perhaps too alien and distasteful to Calvinist propriety, for artists to consider them as proper subjects.

3

ENCOUNTERS WITH GENIUS

On these early diplomatic journeys, Constantijn Huygens wrote furiously whenever he could. He wrote verse at sea, in bed, on army camp and even on horseback – two-line epigrams, of which he was to write hundreds, were ideal in such circumstances. When he was concentrating on learning technique from the classical poets, he wrote mainly in Latin, but now, encouraged by a meeting with the poet Jacob Cats and perhaps more conscious of his country's independence, he increasingly chose Dutch.

Towards the end of 1621, between voyages to England, he sent Cats a long work in Dutch, with alternative Dutch and Latin titles, '*t Voorhout* or *Batava Tempe*. The Voorhout was (and is) a pleasant public space in The Hague, too broad to be called an avenue, too irregular to be a square, where the Huygens family lived and under whose lime trees the beau monde gathered to gossip and flirt. With the fondness for a place that often arises from being absent, Huygens boldly reimagines his urban stamping ground as the Vale of Tempe, the rural idyll of Apollo to the Greek poets. Piercing through its 'roof of leaves', the evening sun – 'Thief betrayer, lenses' friend' – reveals a surprisingly modern, secular scene, teeming with the city's youth, as the poet eavesdrops guiltlessly on their amorous exchanges. Hard on its heels came a second, more sharply satirical poem, '*t Costelick Mall*, mocking fashion madness, which Huygens told Cats (in Middelburg) 'may be read just as well in The Hague'.

Huygens was already acquainted with other leading poets through Roemer Visscher, an Amsterdam grain merchant, through whom he had met one of the greatest Dutch men of letters, the historian and poet Pieter Corneliszoon Hooft. These two figures were the leaders of the Muiden Circle, an informal group of writers and artists who met occasionally over a period of more than thirty years. Apart from Huygens, members included the tragedian Joost van den Vondel, Caspar Barlaeus, the professor of philosophy at Leiden, and the architect Jacob van Campen, as well as a noted cartographer and the glass-maker and playwright, Jan Vos, who wasn't above using his verse in an effort to win commercial contracts.

The group's name was taken from the turreted castle at Muiden, not far from Amsterdam, where Hooft had the right of residence in his capacity as the town's magistrate. Hooft modernized the fort for use as a summer home and laid out gardens where parties were held and plays put on. While he conducted his public duties in a fearsome reception hall, he chose a small room in one of the turrets with a panoramic view of the Zuiderzee in which to write his verse. In truth, members of the Muiden Circle met so seldom that the sobriquet – a product of later cultural myth-making – was never justified. Some of the leading members may have coincided with one another, either at Muiden or at Visscher's house in Amsterdam, only once or twice over the decades. Huygens himself was an infrequent visitor because of his diplomatic duties, and most of the others had similar professional pressures.

The 'circle' may have been more virtual than real, but it was nevertheless more than just a literary rubric. The writers collectively felt there was an opportunity 'to mould manners and emotions, in reaction to the prevailing uncouthness', and they hoped to do this by exploiting the latent potential of the Dutch language. Their own brisk poetic sparring did much to realize their goal. Even when obliged to resort to letters, an exchange of verse offered a more direct and concise manner of communication, bypassing the elaborate politesse required in conventional correspondence. When Roemer Visscher died in 1620, his position in the group was taken over by his two beautiful daughters, Anna and Maria Tesselschade. The women served as muses to the men and as frequent dedicatees of their verses, but they were also more than capable of responding to chivalrous overtures with verses of their own, which might lead their admirers on or mockingly put them in their place, as the mood took them. It had been Anna who introduced Huygens to Hooft; Huygens afterwards wrote a verse to Hooft characterizing her as 'wijze Anna' ('wise Anna') and her younger sister as 'schone Tessel' ('beautiful Tessel'), to which Hooft responded, flattering Huygens by comparing him with Achilles. The pair continued to exchange sonnets, cleverly using the same rhyming words in reply each time.

From what little pictorial evidence survives, both women were highly attractive, and both were certainly unusually well educated in literature and languages. Their verses exchanged with Hooft and Huygens matched the men's for poetic references and allusions. They were, too, skilled in drawing, calligraphy and modelling in clay as well as the more expected music and embroidery. They could ride, and had even learned to swim in the Geldersekade canal that ran by the family home in Amsterdam.

Anna's speciality was diamond-point engraving, a painstaking Venetian technique of inscribing designs on glass that demands a steady nerve as well as hand. She engraved rummers (the Dutch for this kind of glass is *roemer*, and she was Roemer's daughter) with ornate calligraphy, and was one of the first artists to incorporate motifs from nature in this medium. Huygens commissioned

4. Maria Tesselschade Visscher, poet and member of the Muiden Circle for whom Constantijn Huygens translated the verse of John Donne.

one of these glasses from Anna, which shows a dragonfly and various flowers, while another, engraved 'AEN Constantinus Huygens' ('TO Constantinus Huygens'), carries her own verse implying that she turned to diamond-point work when writer's block struck – an idea that must have appealed to the multitalented Huygens.

Maria was ten years younger than her sister, and had the fatal attraction of her byname, Tesselschade. Three months before she was born, on Christmas Eve 1593, a great storm had struck a large fleet of ships sheltering in the sea-roads by the island of Texel. Roemer Visscher was among the merchants who sustained losses that day, which he remembered when his daughter was christened; Tesselschade means Texel-loss or Texel-damage.

Huygens addressed many poems to his two 'Amstel-nymphen', and especially to Tesselschade, who remained an active associate of the Muiden Circle even after she married. When he returned to London in 1622, he wrote a long poem to her celebrating the ordinary sights of home: 'O Scheveningen dune, O Hague butter-meadows . . .'. But in a string of poems to mark her wedding in 1623, and which continued after it, Huygens wrote:

> Foe-friendly hand,
> From the first you unlocked the chest
> To bare my fragile breast,
> From the first you set the axe
> That has cut down my freedom

Even today, it seems surprising to find material of such passion addressed to a newly married woman. Lines like these are indicative of the social licence that the members of the Muiden Circle allowed themselves. In later years, Huygens's torment over Tesselschade's conversion to the Roman Catholic faith would lead him to produce 'his best prose and poetry'.

Compared to fellow poets such as Hooft and Cats, Huygens's verse was generally thought 'difficult'. It required effort to extract meaning and pleasure from it, though most readers felt ultimately that the effort was rewarded. His poetic language was dense but rich, occasionally pompous, but also lightened by self-awareness. He had a weakness for homophones and other wordplay, especially applied to the names of people and places. A journey to Zierikzee in 1618 was excuse enough to invoke the sorceress Circe from Greek mythology, for example. He favoured the iambic metre, which lent itself to Dutch pronunciation with its often unstressed word endings, while his peers aimed for less artificial and more musical rhythms. His imagery and vocabulary drew on his exceptionally broad range of interests, his familiarity with many languages, and the things he had seen on his travels. Acquaintance with scientific topics gave his verse a spice that others lacked, and his rare skill of being able to reveal and describe the wonders of nature was greatly admired by his peers. By his association with the Muiden Circle, Huygens found a way to express his thoughts in poems and plays with a gratuitous wit and an emotional honesty that must have provided a welcome corrective to his life of diplomatic communiqués. He also gained a greater facility and quickness from the teasing cut-and-thrust with other members, and in exchange he left the group, and Dutch literary culture generally, with a poetic language greatly enriched by the new ingredients he had brought.

In 1630, when they were both respectively married with young families, Huygens sent Tesselschade some verses of John Donne which he had translated. Huygens may have heard the English metaphysical poet when he visited The Hague and preached there in December 1619, but he was certainly introduced to him during his postings in London from 1621, when Donne was instituted as the Dean of St Paul's Cathedral. He heard Donne deliver sermons there on more than one occasion, and praised the quality of his oratory. Donne's poetry was not published in his lifetime, but it circulated in manuscript form among his supporters. Tesselschade must have known that Huygens had copies. Although he was busy working on his own long poem, *Dagh-werck* ('The day's work', a kind of *symphonia domestica* to his contented married life), Huygens picked out two elegies and two songs, including Donne's well-known amatory complaint, 'The Sunne Rising' ('Must to thy motions lovers' seasons run?'). A few years later, he made translations of a further fifteen poems, including 'The Flea', but he omitted other famous pieces, such as the salacious 'To his mistress going to bed'.

Huygens valued Donne's poetry for its darkness and complexity, but these very qualities made translation exceptionally difficult, and he was never entirely satisfied with his efforts. In a short verse of his own accompanying the second batch of translations in 1633, he sought to manage Tesselschade's expectations:

> Translations fall as short of untranslated verse,
> As shadows of the life; and shadows are the night:
> But let not your discretion these despite;
> They are noble maidens, they are daughters of the light.

Hooft's verdict was that Huygens had successfully 'preserved the English fruit in Holland's honey'. But Vondel teased them all:

> The British Donne,
> That dark sun,
> Shines not for all eyes,
> Says Huygens without lies.
> The language-learnèd Hagenaar.
> The epicure of caviar,
> Of snuff-taking and smoking,
> That set raw brains a-cooking;
> But this is a rare compost,
> 'tis a banquet for the Drost,*
> And for our little comrade,
> The sweet Tesselschade.

---

* Hooft's official title in Muiden.

Vondel's characterization of Huygens may be something of a caricature, but it does suggest that he was more given to the pleasures of the senses than his own writing tends to indicate.

The technical challenges of translation shed light on Huygens's own priorities as a poet. He was fluent in English and accomplished as a poet in Dutch by the time he approached the task. He chose some of Donne's more accessible verses that remain among his best known today, and he avoided ones that were deeply involved with theological argument and those containing cruel or vulgar imagery. If Huygens had hoped to find English poetry different mainly in degree from his own, he found it instead more different in kind, possessing a particular intensity of poetic expression. Donne's emotional spontaneity was something any translator would find it difficult to replicate, and Huygens, more given to calculated cleverness with words, was perhaps not the ideal man for the job.

Donne's lines are complicated enough in English. Translation into Dutch required decisions over the verse form, line lengths and rhyme scheme as well as individual words as Huygens strove to remain faithful to the overall meaning. His translation of 'The Flea', for example, follows Donne's sense line for line, although Huygens's alexandrines demand twelve syllables in place of Donne's alternating eight and ten. The rhyming couplets of the original are kept, but literal accuracy ultimately triumphs over wit. Huygens's version of 'The Sunne Rising' suffers still more, losing the palpable irritation of Donne's interrupted lover. Donne's opening lines – 'Busy old fool, unruly sun, / Why dost thou thus, / Through windows and through curtains call on us?' – emerge when back-translated as hardly more than a gentle enquiry: 'Busy old fool; what let you shine on us / To rouse from bed through windows and curtains?' Much of the bad temper of the accusation made towards the 'unruly sun' is gone; the passion has been drained away.

Huygens admitted these wayside losses. To Hooft, he wrote that Donne's 'fabric was so good that it remains pleasing translated even without poetic form'. His efforts were but 'shadows of beautiful bodies in, which is worse, feeble sunlight, because of the weak rays I project owing to the pressure of many other concerns'. However, if his aim was to communicate the literal sense of Donne's lines, he succeeded admirably. All the material content is conveyed in only a few more syllables than the original. As working translations, intended to give Tesselschade an accurate sense of what Donne had written, his translations do the job. From here, Tesselschade could employ her own poetic sensibility to infer some of the deeper connections made by Donne. Huygens in any case had had enough. In his covering poem to Tesselschade, he complained: 'How solid and how square, how white, how hot, how heavy / is this English dish'.

For some reason, it had to go. The old man's head. His ruddy face turned down to the floor, his bald pate shining, his grey beard spread across his chest like a napkin. He had lots of those. Rembrandt overpainted the largish panel and produced instead the image of an old woman – his mother is assumed to have been the model – in a black velvet hood with an embroidered lining. She stares at us with her grey eyes, tight-lipped and tough, her leathery face etched with the years. The work is tender but not sentimental. The woman is not sick or defeated, and she is certainly not seeking our pity. She is simply old; the painting is an honest, unflinching record of time's work. Perhaps this was the work that Constantijn Huygens admired most when he first stepped into the peeling studio in Leiden that the twenty-two-year-old Rembrandt Harmenszoon van Rijn may have shared with his

younger but already more reputed friend Jan Lievens, for it was
soon packaged up and sent to London as a present from Stadholder
Frederik Hendrik to King Charles I of England.

The year was 1628, and Huygens had been looking for a
Dutch Peter Paul Rubens, a painter who could equal the epic
sweep and expressive touch of the Flemish genius, whom he
called a 'prince among painters, one of the seven wonders of
the world', for his master, the stadholder. And now he had found
two. He was dumbstruck by their talent, the more so since they
were low-born, Rembrandt being the son of a miller, Lievens
an embroiderer's son. 'If I say that they are the only ones who
can go up against the absolute geniuses among the many earlier
great names, even then I do not do sufficient justice to the
merits of these two,' he wrote in his autobiography a couple of
years later. 'And when I say that they will soon surpass those
geniuses, then I am merely interpreting the expectations that
their best connoisseurs entertain based on their astonishing
debut.'

As a more than competent draughtsman and painter himself,
Huygens was perfectly able to make up his own mind on matters
of artistic preference, but on this occasion he was happy to be
found in accord with those other connoisseurs. Clearly, Huygens
did not 'discover' Rembrandt and Lievens, as is sometimes claimed,
but he was in an ideal position to transform their prospects and
so to speed them on their way to the success that surely lay ahead
for them. The way to do that – and to please his master into the
bargain – would be to secure their services with some significant
commissions for the walls of the stadholder's palaces.

First, though, the beardless wonders, each 'more boy than
young man to judge by their faces', needed to be licked into shape.
The 'old woman' painting was a mere token of their promise.
Huygens found Rembrandt superior in 'accuracy and liveliness of

the emotions'. Lievens scored for his 'grandeur of invention and daring of subjects and figures', and showed the greater ultimate potential, he felt. But neither, in his opinion, had been properly taught. With schooling in the techniques of Italian masters such as Raphael and Michelangelo, *then* they might really amount to something. But the two had other ideas. They had no wish to waste time in Italy, they were bursting to get down to work. 'These men, born to drive art to the pinnacle, knew themselves better!'

For all his carefully calibrated assessment of the two painters, Huygens's record of this early encounter cannot disguise that it is in fact Rembrandt whose work truly sets his pulse racing – more than he realizes, or is willing to admit to himself, perhaps. His epiphany came when he saw the artist's large panel painting of a biblical scene, *Repentant Judas Returning the Pieces of Silver* (1629), which could be 'set against all Italy, yes, and everything that has lasted there of the wondrous beauties from earliest antiquity'. A single painted gesture in it was enough to unleash from him an extraordinary torrent of descriptive prose:

The gesture of that Judas collapsed into despair (to say nothing of all the other impressive figures in this one painting), that lone demented Judas crying out, begging for the forgiveness he cannot ever expect, and in whose face all traces of hope have been erased; the visage haggard, the hair pulled out, the clothes torn, the arms twisted, the hands pressed together till they bleed, fallen to his knees in one unseeing dash, his whole body writhing in pitiful hideous remorse; all this I set against the beauty of the ages.

And all done by a 'Hollander who has yet hardly ventured beyond the walls of the city where he was born'!

Huygens's uncontrolled emotional response – far overstating what was actually painted (there is no torn clothing, there are no bleeding hands) – is surely just what Rembrandt was seeking. It is the first known description of the artist's work to single out his exceptional capacity for depicting 'movements of the soul'. Rembrandt would have been flattered indeed if Huygens, a man whose whole professional bearing bespoke discretion and reserve, had been so unguarded when he examined the painting in the artist's presence.

It is probable, though, that Huygens remained cautious. He must have worried what this ungovernable talent might go on to produce, and whether he himself could overcome his own some-what conservative tastes as well as the bias of the court in favour of Flemish and Italian styles of painting to persuade the stadholder to give the young Dutchman a try. The diplomat in him knew that his protégés would find life easier – and so would he – if they painted what was wanted. Although Huygens's advice was sound and well intentioned, and although both artists knew that he was uniquely positioned to be able to give them the entrée they sought, they persisted in refusing to help themselves. Not only did they shun his recommendation to study Italian painting, but also each impishly chose to paint subjects he had been told the other was better at, with Lievens producing history paintings and Rembrandt, portraits.

However, Lievens did want to paint one portrait: Huygens's own. Huygens agreed to sit if Lievens came to stay with him in The Hague, which he hurried to do only a few days later, sleep-less with excitement about the commission. Because Huygens was busy as always, and the winter days were short, Lievens stayed just long enough to do the clothes and hands before returning home, preferring to work from the imagination to do the face, although he did return in the spring to finish the work. Huygens was greatly pleased with the result, and countered friends who said that the 'pensive expression' Lievens had given him belied his 'geniality' by saying that he had been weighed down at the time with family matters (Susanna gave birth to Christiaan in the spring of 1629).

At length, Huygens appears to have exerted some influence over the young artists, as he would do later over his own sons. Lievens was packed off to seek his fortune in London, doubtless with introductions from Huygens. Rembrandt, dishevelled and bareheaded in 1628, was by 1631 at least wearing a hat and able to project a presentable image of himself to prospective patrons. Perhaps a man so aware of the power of his own image would not need much tutelage, but it is easily believable that the consummate courtier was able to impart some useful advice to Rembrandt about the performative aspect of human appearance.

In 1632 Huygens visited Rembrandt again and asked him to paint a pair of small panels – one of his brother Maurits and the other of his friend's son, Jacques de Gheyn III. If it was a test, it was a stern one for Rembrandt to produce likenesses of two men so well known to Huygens. He passed it – though perhaps only just, as we shall see – and shortly began work on a portrait of Amalia van Solms, the wife of Frederik Hendrik. But royal subjects were perhaps not the best for Rembrandt's truth-telling brush. While his finished portrait of Amalia was quietly swapped for one by a more conventional artist, he was given a very different challenge to produce a pair of paintings of the elevation of the Cross and Christ's descent from the Cross, which were probably completed in 1633. These proved more acceptable to the prince, and led to a further commission for three new Passion paintings. An Ascension was finished in 1636. A few years later, Rembrandt completed two more paintings on the same large scale, an Entombment of Christ and a Resurrection. The five works were each nearly a metre high with rounded tops, presumably to fit in a series of arched spaces.

Rembrandt's only surviving correspondence is a set of seven letters that he wrote to Huygens in 1636 and 1639 concerning these paintings, strongly implying that it was Huygens who was responsible for getting him the commissions. Thanks to the diplomat's efficient filing system, we have the only written insight into Rembrandt's priorities as an artist and his situation as a supplier in relation to his client. As Huygens's side of the correspondence has not been kept, the full nature of the relationship between the two men must be divined from Rembrandt's words alone.

The letters began with high hopes on both sides, but were to end in 'mutual dissatisfaction'. Rembrandt was fortunate to win a job that held the possibility of much more work to come. However, he was aware that he would have to follow the example of Rubens if he was to live up to the high expectations set for him. The artist wrote to Huygens in February 1636 to inform him that he was working 'very diligently' and 'skilfully' on the three new paintings. One of them, the Ascension, was already done, while the other two were 'more than half done', which as Simon Schama has pointed out may, in the language of the procrastinating freelance, mean no more than 'nine-sixteenths' done. Then, perhaps in an attempt to deflect attention from his tardiness, Rembrandt threw the ball back into Huygens's court, seeking to know whether the prince wanted all three paintings together, or to take delivery of them one by one as they were finished.

Huygens must have replied that he wanted to see the one finished canvas, for Rembrandt's next letter two weeks later announced that he himself would be following along shortly to see how it looked hung alongside the Elevation and the Descent from the Cross finished in 1633. He requested a fee of £200, 'but I should be content with whatever his Excellency settles upon'. Allowing for the overabundance of customary courtesies that frame seventeenth-century letters, it is clear that Rembrandt employed a minimal level of politeness in his correspondence with Huygens, whom he clearly felt was prepared to put up with what he calls, without apology, his 'impertinence'.

It was another three years before Rembrandt completed the two remaining pieces in this second royal commission. On 12 January 1639, he finally wrote to Huygens again, pleased with what he had achieved, itching to deliver the finished paintings for Huygens to see, 'because these two are the ones where the greatest and the most natural mobility has been observed, which is also the main reason that I have been working on them for so long'. He also promised Huygens handsome personal compensation for his patience, in the form of 'a piece 10 feet long by 8 feet high'. (This single work, if it materialized, would be almost twice the area of all five of the Passion paintings put together.)

The effect that Rembrandt was drawing to Huygens's attention was his expression of motion and emotion combined. A sense of action and feeling is vibrantly communicated by the highly theatrical illumination. Light is pooled towards the centre of each painting, an effect that would be naturally accentuated when they were seen installed in their alcoves. Rembrandt's light does not fall or radiate like natural light, but acts like a vector of the story in each painting: in the Elevation, it slants upward following the Cross as it is raised; in the Descent, it slumps downward with weight of Christ's body. It slips sideways in the Entombment, it comes down with the angel of the Lord descending from Heaven in the Resurrection, and it rises again in the final Ascension. Displayed in the correct narrative sequence, these different thrusts bring an animating rhythm and an overall symmetry to the series.

Huygens attempted to decline Rembrandt's excessive gift, but sent word to have the two final Passion paintings brought to him. Rembrandt sent the paintings, along with a demand for 'no less than a thousand guilders' apiece. This must have come as a shock to Huygens, since for the first two paintings in the series, completed six years before, Rembrandt had been paid a total of 1,200 guilders. However, the artist had that month signed the contract to buy an extravagant house on Breestraat in Amsterdam for 13,000 guilders, and needed the cash to pay the deposit.

Huygens ignored the claim for a few days until an admirer of the artist in Amsterdam stepped in with a solution to advance 1,200 guilders (the undisputed 600 guilders per painting) directly, before the stadholder's accounts office in The Hague was ready to process the payment. Preferring some money soon to the doubtful prospect of more money later, Rembrandt speedily, if somewhat gracelessly, accepted the offer for the paintings, 'since His Highness in goodwill is not to be moved to a higher price, even though they are manifestly worth it, . . . provided that I might be compensated for my outlay on the 2 ebony frames and the crate, which together come to 44 guilders'. Payment of 1,244 Carolus guilders was duly made, and Rembrandt's relations with Huygens came to an abrupt end.

Rembrandt always remained grateful to Huygens. His letters had repeatedly invoked their 'friendship', and in his last communication, he wrote: 'I shall always seek to repay such to your lordship with reverence, service and friendly regard.' Huygens may have been disappointed that his great Dutch hope did not appear to care too much for court patronage (although he did paint two more canvases several years later), and he was surely dismayed at the way their personal friendship ended, even though a resolution had been achieved in their business dealings. However, Rembrandt's art was developing in directions that were no longer in line with Huygens's own tastes, and so it seems inevitable that a split should have come sooner or later. The diplomat could console himself with the fact that it was not hard to find oneself on the wrong side of Rembrandt. The wealthy Amsterdam official Jan Six, perhaps Rembrandt's only other important client in terms of the personal interest he took, also fell out with him in the 1650s, and once again it was the artist's expensive house that was the source of their friction.

But there is a twist in the tale – and a barb at the end of it, too.

Lacking his reply to Rembrandt's offer of a large painting 'that will do honour to my lord in his house', we cannot, of course, tell whether Huygens was truly seeking to deflect the gift, or simply employing excessive politesse, which the artist then misconstrued. Rembrandt sent it anyway, 'against my lord's wishes', with the recommendation that he hang it in a strong light where one could stand away from it, and where 'it will best sparkle'.

This 'first memento', as Rembrandt put it, was also the last he was to give Huygens. What would make a suitable souvenir? What was the subject of this vast canvas? The date, 1636, and the dimensions correspond to *The Blinding of Samson*, an astonishing work where the harsh light of day seeps from the back of the painting into a cave in which a scene of appalling violence is taking place as the betrayed Samson is held down and stabbed in the eye by one Philistine soldier while another stands by with his dark spear dramatically silhouetted against Samson's gleaming chest, ready to take him into servitude.

Would Rembrandt have been so insensitive – or so waspishly cruel – as to offer this particular subject to his bespectacled client

who fretted often about blindness in himself and others? Many
sources accept this version, although the painting was not in the
inventory when Huygens's paintings were sold off nearly a century
after his death. The dimensions correspond with those Rembrandt
gives, but it is known from the existence of near-contemporary
copies that the canvas was once even larger, and Huygens was not
entirely averse to gruesome subjects, admiring a friend's copy of
the famously gory *Head of Medusa* by his beloved Rubens, for
example. Huygens would also have been sensitive to the full story
of the painting, in which Samson is shortly to repent of earlier sins
– in other words, to be restored in his *moral* vision, which was a
theme of his own poems about blindness. Others believe that
Rembrandt must have rewarded Huygens with another painting,
such as the one now known as *Danaë*, also painted in 1636, which
was similarly cut down from a larger canvas and is, now at least, a
little smaller than the work Rembrandt had promised. But perhaps
Huygens's Calvinist mind would have still preferred the message
of the biblical story to the soft pornography of this Greek myth.

Huygens kept his misgivings about Rembrandt's art to himself.
In 1632, while the artist was producing his first test pieces in the
hope of gaining work for the court, Huygens quietly amused
himself by jotting down a number of mocking couplets about the
portrait Rembrandt had made of Jacques de Gheyn III. A typical
example runs:

> Were this at all de Gheyn's visage,
> It would be de Gheyn's true image.

In other words, it was not his face, and no likeness. Huygens was
too discreet to publish the epigrams while he was professionally
involved with Rembrandt. But five years after their working
relationship was over, he did gather seven of the little verses for

inclusion in a collection of verse published in 1644, *Momenta Desultoria*. An eighth couplet that actually named the painter was omitted.

It would have been painfully obvious to Rembrandt that he was the object of the squibs. Others close to the Huygens family, a circle that included very many poets, painters and courtiers, would have known or guessed the fact. Although Huygens carefully prefaced the set with a heading, 'On Jacques de Gheyn's portrait, which is quite unlike him, in jest', and littered them with his usual wordplay, it would be understandable if Rembrandt had been embarrassed and angered by their publication.

It seems that he must have read them immediately, for that same year he made a small pen-and-ink drawing known as his

5. Rembrandt, *Satire on Art Criticism*, 1644. The figure sitting astride the barrel with his glasses left at his feet may be Constantijn Huygens.

Satire on Art Criticism. The sketch depicts an artist squatting on the ground as a succession of his paintings is paraded before a connoisseur arrogantly seated astride a barrel. Various great and good in robes and hats and chains of office hang on the expert's every word as he pronounces on the works one by one, jabbing his pipe in their direction. Rembrandt's scabrous intent is apparent from the huge ass's ears that poke through the critic's hat, and from the fact that the painter is looking round wickedly at us while wiping his arse after defecating.

The timeless image represents the artist's frustrating lack of power in relation to all kinds of influential but often ignorant figures – critics, patrons, pursers, connoisseurs, idle followers of fashion. But it is not only a certain resemblance in the angular face that makes it possible that the figure Rembrandt is portraying is a specific one, namely his former champion Huygens. For lying neglected on the ground at the connoisseur's feet is a pair of spectacles – naturally, he is not using them to examine the work, because he is vain, and because a foolish critic does not truly look.

However disrespectful he was, Rembrandt must have soon regained favour in the Huygens household. For one summer's day in 1645, just a year after Constantijn Huygens had published his verses on de Gheyn's likeness, his sixteen-year-old son Christiaan, then studying at Leiden, found himself in a dry-colour painting class staring at Rembrandt's stock old man, the same one that lay buried beneath the paint of the old woman's portrait that his father had seen in 1628. How was Christiaan able to copy a painting that had been painted over before he was even born? Perhaps copies of the original were made and circulated in Leiden, when Rembrandt was working there, or afterwards. Or perhaps

Constantijn was given a copy, which would then be known to Christiaan, following that first studio visit. Christiaan's highly competent rendering of the painting in graphite and red-and-white chalk reproduces the shine of the old man's domed forehead and bestows an enlivening fleck of red on one eyelid. By his own account, the little sketch cost him a great deal of time and effort, and the result must have almost been the equal of Rembrandt's achievement in oils, since he wrote to his brother Lodewijk, 'you can hardly see the difference'.

In the spring of 1632 Constantijn Huygens travelled to Leiden, where he visited Jacob Golius, the scholar of Arabic who had been appointed as the professor of mathematics at the university in succession to Willebrord Snel. There, he met the thirty-six-year-old René Descartes. The three men discussed the optical property of refraction. Huygens was greatly impressed by the Frenchman, and his ability to offer convincing explanations of physical phenomena. He felt exalted that his own interpretations were rendered superfluous on the spot. After they parted, he felt he was still being shadowed by this 'wonderful Gaul'.

From that moment, Huygens became an ardent admirer of Descartes's physics and philosophy. Already a believer in the power of Baconian scientific reasoning and experiment, he now began to familiarize himself with the Frenchman's more conceptual approach. Together, the contrasting strategies had the potential to give him a foundation for the pursuit of natural philosophy in both its theoretical and its practical aspects as complete as anybody's in Europe.

Descartes was no less effusive in his praise for Huygens. He wrote to Golius: 'There are qualities which occasion one to esteem those who possess them without causing us to love them, and others which cause us to love them without raising our esteem; but I find that he possesses both of these together in perfection.'

Descartes came to the Dutch Republic first in 1618 with a hope of joining the army. He was billeted in Breda, on the southern part of the Dutch ring of defences, but saw no military action as the long truce in the Eighty Years War was still in effect. There was time instead for him to indulge' his taste in gambling and other mathematical challenges. Puzzles were sometimes put up on posters around the city for anybody to attempt. Descartes fell in with Isaac Beeckman, who translated the notices for him, and a friendship developed between them based on their shared interest in science.

Born in Touraine, Descartes was sent to a new but already renowned Jesuit college in the region, where he obtained a sound liberal education in literature and mathematics. One Père François, who sought to refute claims of occult happenings by exposing them as effects achieved by mirrors and distorting glasses, did much to kindle his interest in optics. His debunking of astrology and magic in similar style gave the young Descartes an enduring taste for rigorous methods of enquiry.

After college, Descartes sampled life in Paris, but found the social scene tedious. He tried the city again after travelling abroad, but nothing had changed, and he made a permanent withdrawal to the Dutch Republic in 1629. Although the attraction of such a move might be thought to be its religious toleration and greater freedom of thought, the Republic was for Descartes above all a boring place, where, as he told friends, they were only interested in trade, and he would be free to sleep easy and long. He settled first in the university city of Franeker in the province of Friesland, but soon moved to Amsterdam, which is where he was living when he met Huygens. Later though, he was to retreat from the city once more, to the tiny coastal village of Egmond-Binnen in North Holland, where he would turn from specific problems in physics and begin to consider the universal principles underlying them, and write his first important philosophical works, *Discourse on Method* and the three treatises, on meteorology, optics and geometry, to which it formed an extended introduction.

Descartes lived in the Dutch Republic for twenty years, until he was lured away to the court of Queen Christina of Sweden in 1649. He died in Stockholm the following year, unable to bear the cold and the early risings necessary to make his tutorial appointments with the queen. Shortly before he left for Sweden, Frans Hals painted the portrait of him that has become our standard image of the philosopher. He is looking grimly out of the frame at us with a broad mouth and set jaw, and lank brown hair flowing over his collar. An apparently disembodied hand just creeping into the bottom corner of the painting, holding his hat, could almost be a joke at the expense of the philosopher's ideas of mind–body separation.

Discourse on Method set out the philosophical approach that came to be known as Cartesian doubt. In it, Descartes's starting point is to doubt everything that can be doubted; except that, in order to be able to do this, the one thing we cannot doubt is our own existence. This is the origin of his dictum *cogito ergo sum*: I think, therefore I am. In order for the world to be known by the human mind, it first must be sensed. However, doubt enters here, too, for what our senses tell us may not be true. Objects may be no more than sense perceptions, and our senses may not perceive accurately. Since sight is the most important of our senses, as well as the sense that seemed during Descartes's time the most susceptible to scientific investigation, it is possible to see why optics became an important part of Descartes's project, even leading him to perform his own dissection of bulls' eyes.

Consider the puzzle posed by the production of optical images – images of the sort that Drebbel produced with his camera obscura that so entertained Constantijn Huygens in London perhaps, or images like those which Descartes's teacher Père François was able to prove were not the devil's work. How is the appearance of an object seen in one place transmitted to another place? How do the two object-images differ? What sort of reality does each possess? The question need not even involve mediation by a manmade instrument. Every now and again, through a ruffled surface of water, you might glimpse a fragmented image – a reflection of a sunlit branch that suddenly appears as if snapped in two, say, or a stick emerging from the water that appears to kink abruptly just at the water's surface. How does this discontinuous image arrive at the eye? Is it sent out whole and perfect only to become broken in transit, as it were? Or is it disassembled in some way and then imperfectly reassembled at the end of its journey, if indeed it is right to describe its translation as a journey? Optical reflections were, of course, familiar, and could be said to be understood at a basic level. However, the refraction that causes the stick in the water to appear kinked defied human intuition (and the work of Harriot and Snel was not widely known).

Now, imagine the viewer of a planet through a telescope or of a microbe through a microscope. This person sees an image made large, brought close to the eye, as if they could reach in through the eyepiece of the instrument to grasp it in their fingers. How is this apparition made at this new size? Certainly the glass lenses through which the viewer has had to look have been carefully fashioned to a particular shape, and carefully positioned along the line of sight. It is clear that these simple glass shapes alone must perform the trick. But before the ray diagrams that could at least begin to explain the change of scale between object and image, it was not uncommon for people to believe that the optics actually generated a kind of replica of the object.

These questions had exercised ancient Greek and Arab philosophers such as Empedocles and al-Ḥasan ibn al-Haytham (often known as Al-Hazen). But it was the invention of the telescope, which had not required any fundamental theory of optics to bring it into being, that gave them new impetus. Descartes was distressed that such an innovation could be made without reference to an underlying scientific principle at all. With the benefit of this missing guidance, he thought, it might be possible to construct telescopes with limitless powers of magnification. However, Descartes was not satisfied simply with establishing an empirical rule of refraction, the so called 'ratio of sines' (of the two angles made by light bending at the interface between one medium and another), which was useful for building optical instruments. He also, characteristically, wanted to trace the ultimate *cause* of light bending in this way.

Descartes's practical engagement with optics was guided by his friendship with the versatile Middelburg engineer Isaac Beeckman. Their relationship was one of bright apprentice to accomplished master as Descartes brought ideas about the theory of lenses into contact with Beeckman's matchless expertise in making them. At the same time, Descartes also hoped to bring to Amsterdam a Parisian artisan, Jean Ferrier, whom he had engaged to build a machine that he hoped would be able to grind more perfect lenses. He warned Ferrier that it would be hard graft, but there would be great rewards: 'if you were to take a year or two to adjust yourself to all that is necessary, I would dare hope that we would see, by your means, whether there are animals on the moon'.

The difficulty in obtaining optical perfection appeared to lie with the precise curvature of the lenses. Despite their apparent geometric ideality, spherical lenses (whose surface or surfaces, convex or concave, follow the curvature of a sphere) do not bring rays of light that pass through them close to the rim to the same focal point as more central rays. In addition to this spherical aberration, they also produce chromatic aberration, owing to the glass of the lens acting like a prism and separating the light passing through it into colours. To a geometer and theoretician such as Descartes, the knowledge that a spherical lens does not produce a single focal point immediately set the challenge of finding a form that did. The presumption was that such a form, even if not spherical, would still have its basis in ideal mathematics. Influenced by his knowledge of the geometry of conic sections, Descartes reasoned that another mathematically generated perfect curve, the hyperbola, might be the answer.

Descartes soon fell out with both Beeckman and Ferrier, but he persisted in his wish to grind a hyperbolic lens. His initial encounter with Huygens was timely, therefore. Huygens was not able to contribute theoretically since his mathematical knowledge was largely confined to accountancy. But he did have first-hand experience of lens-grinding from Drebbel, and many useful connections. Unfortunately, he was only able to devote time to the project in 1635, upon his return from that summer's army campaign. He assured Descartes that his enthusiasm for helping him had not cooled, and told him that he had found a turner in Amsterdam who could do the job. Between them, they agreed the dimensions and the desired optical properties of the lens, which was to be flat on one side and convex on the other, with a focal length of fourteen thumbs. Huygens then made a tracing of the hyperbolic curve they had sketched together for the turner. After a few weeks, Huygens was able to write to Descartes, proudly enclosing the lens made 'out of my Hyperbola', punning that 'not without real hyperbole, for the first attempt, I think it is well done'. But Descartes found the lens came to a focus at many different points and was therefore useless. Not wishing to hurt his new patron's feelings, perhaps, he suggested that the turner might not have followed Huygens's drawing; the turner blamed the drawing.

The two men issued new instructions, this time using a hyperbola drawn by Descartes, but this lens, too, proved unsatisfactory. For their third attempt, a year later, Huygens enlisted the mathematician Frans van Schooten, a young acolyte of Descartes, to draw the troublesome curve. The figure was mathematically accurate, but once again the resulting lens did not produced a clear image. After two years of effort, Descartes and Huygens had failed to match the craftsmanship of Ferrier a decade earlier, and their collaboration in practical science came to an end, with Descartes remaining convinced that the artisanal hand was the source of error, and hoping to find a mechanical way to accomplish the task.

Huygens's eagerness to be involved at the forefront of new thinking is well displayed in this unhappy episode, but so too is his naivety in thinking that such a fundamental problem in optics could be resolved so expediently. However, there were more important ways in which Huygens was able to make himself useful to the French philosopher. Descartes had held back from publishing previous work in 1633 when he heard that Galileo had been put on trial for heresy in Rome in the wake of the controversy that had arisen following the publication of his *Dialogue Concerning the Two Chief World Systems*. His reluctance seems surprising since he would have run into no such difficulties in the Netherlands, where his ideas, as well as those of Galileo, were not considered dangerous.

Huygens was not a man to pit theology and science against each other. He regarded Descartes as a beacon of rationalism, and was prepared to help his friend in any way that he could when the *Discourse* was ready to be published. He wrote to him full of praise for the work: 'I devoured your *discours de la Méthode*, which is truly the best, the most daring, and, as I think the Italians keenly express it, *la più saporita* [the tastiest], piece that I have ever seen. If it matters that you should know my opinion, I protest that it satisfies me in every way.' He persuaded van Schooten to comment critically on the manuscript, and to provide the figures for the scientific and mathematical sections of the work. The *Discourse* was published anonymously in June 1637 by Jan Maire of Leiden. Huygens was able to use his diplomatic privilege to smooth its path to publication in France, too, by forwarding Descartes's letters to Paris colleagues via the Dutch embassy in that city, and by negotiating guarantees that the Dutch publisher's interest in the work would be respected there. Thus, a Dutchman acted as the midwife to one of the most important of all works of French philosophy in its own country.

Later, when his ideas did come under attack in the Dutch Republic (though never to the level of existential threat faced by Galileo in Italy), Descartes was able to turn to Huygens to ask for the stadholder's intercession to guarantee his safety from his intellectual adversaries. However, the Frenchman may not always have grasped quite how robustly the machinery of Dutch tolerance turned. It is true that his views were sometimes subject to legal strictures and local bans, but he was wrong to interpret these actions as personal assaults. Rather than caving in to these furious condemnations, he might have done better to treat them as an exhortation to rejoin the fray with greater force. For all the difficulties that he faced, he always remained free to work and to publish in Holland, and the Cartesian way of thinking steadily gained ground there during the middle years of the seventeenth century.

In the Dutch Republic, Descartes found himself the beneficiary of a set of conditions unique in Europe. True freedom of thought was evidenced in the constant eruption of conflicting ideas and the sense that almost anything was thinkable. The soporific country he had sought out that wished only to trade, wished to trade in ideas too. Science was not confined to established universities where it could be easily smothered by orthodoxy, as it was in the Italian cities or in Paris. Instead, following the pattern set by Simon Stevin, it arose most vigorously in the civic setting, with the expectation that it might offer practical benefits, whether these were for the defence of cities, the creation of new agricultural land or the accelerated movement of people and goods for increased trade. The sight of new canals and polders of farmland claimed from the marshes stretching in grids to the horizon demonstrated a potential for the control of nature that surely excited and sustained Descartes during his Dutch years, while the cities harboured a cultured citizenry ready to assist in the dissemination of new ideas, and among whom Descartes had found a veritable paragon in Constantijn Huygens.

4

AT HOME

The family that Constantijn Huygens was to build around him would grow up to become a formidable generation, working for the Dutch Republic, the House of Orange and always for each other. In later life, his four sons and their sister were to function together with him in a network that stretched across Europe, as they travelled here and there on a variety of missions. Christiaan was the principal beneficiary of this familial arrangement, depending not only on his father's financial support, personal contacts and diplomatic advice, but also on his brothers' practical engagement in his work with telescopes and his sister's ever-present concern for his welfare.

The engraved glass that Constantijn Huygens commissioned from Anna Roemers Visscher in 1621 was a gift for Dorothea van Dorp, his neighbour on the Voorhout and a close friend since they met in 1614, when he was eighteen years old and she was already in her early twenties. She seems to have taken the initiative in the relationship that developed. He wrote her poems and called her 'Song' and 'Songetje'. When he went away to Leiden University in 1616 they exchanged rings, but the relationship soon faltered amid mutual mistrust. Huygens assured her of his constancy, and seemed to accuse her in a quatrain monogrammed with their interlocking initials: 'Although the D is broken, the C is yet whole'. But his idea of faithfulness may have been more casual than hers. Their ardour cooled into an occasionally troubled friendship. Dorothea never married, while Huygens was left poetically protesting his dislike of women.

He protested too much. In July 1622 Susanna van Baerle visited the Huygenses' house with her sisters. She was beautiful, accomplished, twenty-three years old and in possession of a fortune from her father's Amsterdam trading business, having lost both parents by the time she was eighteen. Constantijn's father Christiaen had it in mind that she should marry his elder son Maurits. However, Susanna refused Maurits, who soon married someone else, and Constantijn was free to try his suit. Since Susanna wrote verse and drew and painted (her subjects included birds, flowers and insects), she had far more in common with the younger man. Constantijn found that he was also able to discuss political and scientific subjects with Susanna in a way that had not been possible with Dorothea. The recently widowed Hooft courted her, too, but Constantijn eventually won her over with a barrage of verse, including a smart, satirical love poem, 'Anatomy', full of bodily parts and functions, not unlike Shakespeare's sonnet, 'My mistress' eyes are nothing like the sun':

> Walk into your garden, and you will see me touch
> Rubies more beautiful and sweeter than your flesh,
> These cherries are they, these strawberries and currants,

There was ample time for wooing. After his return from England, Huygens waited for his next diplomatic assignment, passing the time by experimenting with verse forms and learning Spanish, the language of the enemy, which was sure to come in useful in state negotiations. In 1625 Frederik Hendrik, the youngest son of William the Silent, succeeded his half-brother as stadholder. The old stadholder's secretary died very shortly afterwards. Huygens's encomium – 'Jan could read and write untiring, / Jan could reckon up the score, / Jan was loyal and never petty, / Jan was loved by high and low. / Jan had once been all in all things' – showed that he knew what the work entailed; it was in effect his application for the job, which he got.

His modest salary of 500 guilders would be generously augmented by travel expenses and gifts for favours done in line with his duties. But he had few illusions about the reality of the work or the paradoxes that it engendered. The stadholder was by now a person of real power as well as a symbol of the state. The fact that Frederik Hendrik was also the new Prince of Orange, among sundry other titles, was an uncomfortable hangover from the time when the dynastic rule of the Habsburgs was yet to be challenged in the Low Countries. Huygens's position was thus analogous to that of a courtier in a monarchy, yet the stadholder was the leader of a republic, and answerable to the States General. Huygens had a finely tuned understanding of the contradictions inherent in the role, and therefore in his own role too. He was not an aristocrat, but his family had been close enough to the Oranges for many years that he knew instinctively how to behave in both 'royal' and civic settings. He learned to read the whims and tempers of the stadholder, and to bear it when he was interrupted in his secretarial work, as he often was. Nevertheless, Frederik Hendrik was on the whole a careful, reasonable man who, like his father before him, sought compromise whenever he could. He was a soldier but also a thinker and, with Huygens at his side, would become a significant patron of the arts.

Constantijn proposed to Susanna in September 1626, and, as he recorded it in verse, her hard-as-diamond heart finally crumbled to crystal dust at the sound of his lute the following January. As a gift, she sent him a brooch set with diamonds in the form of an S. The woman he would always call Sterre – 'Star' – was his. She brought 80,000 guilders to the marriage; he brought relatively little except his good prospects.

Constantijn's growing confidence in his professional position is apparent in the portrait painted of him at this time by the leading portrait painter in Amsterdam, Thomas de Keyser. Huygens appears very young, smooth-skinned, with his fine bones and pointed face accentuated by a trim goatee. His eyes are large and prominent. His hands are delicate, the right held out to receive the folded letter brought in his by his clerk, the left ungloved and resting on a table loaded with signs of learning and culture picked out in the cold, slanting light – pen and ink, a compass, a watch, globes, architectural plans and books and an extravagant lute. Huygens is smartly dressed in a black hat, white ruff and a brown matching cloak and tunic with light gold embroidering. The calfskin tops of his boots are fashionably folded down to reveal the lace trimmings of his breeches. As he sat for the picture, he happily prattled on about the impending wedding.

Susanna and Constantijn were married in Amsterdam on 6 April 1627. Later that year, they made their married home on Lange Houtstraat in The Hague. The domestic routine demanded all of the administrative skills that Susanna had built up as a merchant heiress. Well educated and self-assured, she had a powerfully logical, even mathematical mind. When they were together, she and Constantijn made important family decisions jointly. But Constantijn was often called away from home during the summer months on campaign with Frederik Hendrik's army, and at these times Susanna's businesslike independence was essential to the running of the home.

In their first months of married life, Constantijn had his first real experience of battle. The resumption of hostilities in the Eighty Years War following the Twelve Years Truce that lasted from 1609 to 1621 had seen reversals for the Dutch Republic, culminating in the surrender of the important garrison city of Breda in 1625, just a few weeks after Frederik Hendrik's accession. The stadholder moved to calm religious tensions in the cities and strengthened the army. In the summer of 1627, Constantijn was with Frederik Hendrik in eastern Gelderland, where the capture of Groenlo marked an important step in breaking through the Spanish encirclement of the Republic. He found himself immersed in a strange blend of medieval and modern military technology, hearing bugle signals, drums and cannon-fire, and the clash of pikes and halberds, as well as muskets and new explosive mines set off by means of a tripwire and a sparking device. But there were longueurs during which he had time to write to Frederik Hendrik's wife, Amalia, as she had requested, to assure her of her husband's safety, and to exchange poems with Hooft about the progress of the campaign. He consoled himself with thoughts of home, writing love poems to Susanna and beginning to outline his major verse work based on a day in their idealized domestic existence.

Susanna and Constantijn had five children during the next ten years. Their first son was born on 10 March 1628. The godparents wanted the boy to be named Christiaen after his grandfather, but Susanna preferred to name him for her husband, and he was baptized Constantijn in the Kloosterkerk on the Voorhout.

Christiaan was born just over a year later, on 14 April 1629. The boys played together from infancy and later shared tutors. Being similar in aptitudes and interests, as well as so close in age, they formed a close bond that would lead them to work together on many occasions and to correspond frequently when they were apart.

Lodewijk, born on 13 March 1631, was the next closest in age to Christiaan. He cried the least and laughed the most of all the children, and grew healthy and bold, his father noted. He was also, 'so somebody said, the most handsome of our children'. His more physical and less academic inclinations meant that his education would take a slightly different path from that of his brothers. Nevertheless, he too later became a useful member of the family alliance when he travelled abroad, although he was to bring shame on the Huygens name.

The fourth son, Philips, was born after another interval of two years, on 12 October 1633. During her labour, Susanna had hoped she would give birth to a daughter this time, but the pains were so great, and she had never felt herself so close to death, that she promptly stopped wishing for a daughter, lest the poor girl one day experience the same agony. Philips was the only one of his children whose birth Huygens was unable to attend, as he was away on army service; he received the news a week later, and returned home in time for the baptism. Four years later, returning once more from the field, he found the infant transformed into a healthy lad always ready to launch a surprising question or a funny remark upon anybody who happened to be in the house. Philips was the only one of Constantijn and Susanna's children who did not live to old age; he died aged only twenty-three while abroad on an ambassadorial mission in 1657.

Susanna did finally give birth to a daughter, also called Susanna, on 13 March 1637. Constantijn remained at home for long enough to observe the infant begin to take notice and to laugh and play, and to see her grey eyes turn brown. But when summer came, he was called away once more to the field. The girl would later become an important lynchpin at home, coordinating the comings and goings of her brothers and father, ensuring that they were kept abreast of family and city news when they were away and that they always had clean laundry. She enjoyed keeping up with the fashions of the places they visited and ordering fine goods for them to send on to her. She married her cousin Philips Doublet in 1660 and long outlived her brothers, dying in 1725.

As his own father had done before him, Constantijn carefully described the birth and early life of his children. In each case, he named the godparents and the wet-nurse, and itemized the christening gifts, giving their weight when they were silver. Unsurprisingly, the most comprehensive account is given over to his first-born, Constantijn. As a new father, he lovingly described each little development, detailing month by month the trials and sicknesses of infancy, noting how it felt to handle the baby and observing each new behaviour with the curiosity of one brought before an unfamiliar wonder of nature.

But the events of the day are such that it is the birth of Christiaan, thirteen months later, that is more revealing, both in the immediacy of the account – it is the only time that Constantijn records the infant's weight, for example – and for what it reveals of the unusual nature of the father's life:

Christiaan our second child came into the world anno XVI.c twenty-nine on the fourteenth of April, being the Saturday before Easter, in the night, just on two o'clock, being the beginning of the aforementioned day, in the same house and Room in which Constantijn was born. From the morning of the previous day the mother began to be aware that her time was approaching, so that I, having been out at an Anatomy from 7 o'clock, was called back to the house at 9 o'clock.

99

Constantijn's hunger for knowledge in art and science, especially where they might overlap, led him to the anatomy demonstrations that had begun to be held irregularly for the instruction of aspiring physicians, curiosos of all sorts and any ghoulish citizen who was prepared to pay the entrance for the show. The demonstrator on the occasion to which Huygens refers was his scholarly friend Christiaan Romph, who had performed the autopsy on Prince Maurits. Anatomy lessons typically took place in the winter – this one was late in the season – and began early in the day when it was still cold in order to minimize the stench rising from the decaying body parts. Huygens picked up his account of the day when he returned home:

> Nevertheless, it appeared to pass, and at midday my wife ate with us at the Table. In the evening at around 10 o'clock the pains first truly came upon her, and from 11 she began to endure a very hard labour, certainly harder than the first, such that this child was found to be larger, measuring, on the day of baptism, 9 pounds in weight . . . He came into the world without any injury or deformity, even though my wife had feared the contrary, because she had been frightened by a poor boy passing by in the street with a gross, misshapen cheek, whose appearance was monstrous to behold.

The baptism was held on 22 April, as soon as Susanna had recovered from a postnatal fever. Two weeks later, on 3 May, Constantijn had to leave the house to join the army at 's-Hertogenbosch, where the stadholder was preparing for a massive assault on the Spanish-held city.

Frederik Hendrik came to appreciate the true measure of his secretary at 's-Hertogenbosch, as Huygens worked through the night to translate coded Spanish letters smuggled out of the besieged city. When the stadholder marvelled at his ability, Huygens replied that it was mere donkey work, and that he would rather spend a week turning millwheels. But he was proud of his contribution and pleased to be thus acknowledged. The success of the siege proved that it was now the Dutch who had the upper hand in the war, and it made Frederik Hendrik's name as a military leader.

The following year, Huygens bought the estate of Zuilichem in Gelderland on the south bank of the River Waal for 41,820 guilders. With it came a coat of arms, a moated castle, lands, church goods, the local magistracy and the right to be called Lord of Zuilichem. A fine sketch of the place done many years later by the younger Constantijn shows a bend in the broad Waal with a sailing barge moored up to the dyke path that meanders close by the many-chimneyed castle. Huygens had the Zuilichem arms altered to show the outstretched branch of a tree bearing three oranges in acknowledgement of his debt to Frederik Hendrik. He acquired further lands along the riverbank in 1638 and 1642, but there were continual difficulties with rent collection and church disbursements, and frequent legal disputes, and the estate ultimately brought him little financial advantage.

As a true connoisseur, Huygens naturally chose to commemorate his elevation with more art. In addition to his wedding portrait by de Keyser, and the picture that he ordered when he discovered the miraculous talent of Jan Lievens a year later, he sat for Antoon van Dyck, when the artist paid a visit to The Hague in January 1632. This work, now lost, was perhaps a grisaille, intended to be used as the basis for a copper engraving to be included in van Dyck's *Iconography*, a long-running project to produce an illustrated compendium of eminent European contemporaries that eventually included a hundred nobles, statesmen, scholars and artists. A copy engraving shows Huygens grandly robed with his hair beginning to thin across the forehead. He is facing towards us, his tired eyes apparently drawn to some object lying on a table beside the artist.

Although land may have been a safer investment than tulips or the Indies companies' voyages, Huygens did not do well out of other property that came his way, either.* Through his marriage, he had come into possession of 150 'measures' (about 70 acres) at Hatfield Chace near Doncaster in England, where King Charles I had employed the Dutch engineer Cornelis Vermuyden to drain the marshy parts of his hunting estate. Huygens later bought more acres here, but the English courts imposed heavy taxes on the improved land, and had an annoying habit of upholding the ancient rights of those who lived there. When his mother died in 1633, his inheritance included other polder land in Zeeland and Brabant, as well as the majority share of the ancestral family home in Antwerp, along with crimson wall hangings, damask sheets and family portraits. His father Christiaen had drained the land in 1617, but it had been flooded as a defensive measure when the war recommenced in 1621. Constantijn sought state compensation for the damage, but without success. Later, he was awarded another substantial estate of seventy-two houses at Zeelhem (not far from Hasselt in modern Belgium) as a favour from the stadholder, which gave him a further noble title. Perhaps his most unusual investment was in a project to build a canal to link Lake Neuchâtel and Lake Geneva as part of a bold scheme to create a route for

---

* The apocryphal advice to 'buy land, they're not making it any more', often attributed to Mark Twain, did not apply in the Dutch Republic. During the first half of the seventeenth century, the area of land available for farming within the borders of the province of Holland, for example, increased by nearly a third thanks to poldering (Helmers and Janssen 35).

shipping between the North Sea and the Mediterranean that would bypass enemy waters around the Iberian peninsula. Huygens's three per cent stake purchased in 1637 only began to produce a yield thirty years later. In all, although they may not have made his fortune, these holdings were not frivolous investments, and they gave Huygens something he coveted more than wealth, which was social status.

When Constantijn returned to The Hague from Venice in August 1620, he brought back more than his formal appointment to the Dutch embassy and the gold chain that came with it. His head was filled with visions of the new classical architecture of northern Italy, and in particular the work of Andrea Palladio. In Venice itself, he would have seen Palladio's magnificent churches. But he was drawn especially to the more innovative secular buildings. In Vicenza, the playwright in Huygens appreciated the witty drama of Palladio's final work, the Teatro Olimpico, completed in 1585, five years after his death. He found it a 'modern building, but truly such that there is not anything more beautiful to be seen in Europe'. Huygens carefully measured its false perspectives, 'a wonderful thing to see, which could fool the eye of the most alert, especially by candlelight'. He sailed in a *burchiello* along the Brenta river, which he observed was about as wide as the canal between Delft and Rotterdam. The houses along the banks made 'a continuous neighbourhood of the most elevated palaces and villas that one could imagine, so many that I lost the will to take notes'.

Another Palladian model was Inigo Jones's Banqueting House in Whitehall, which opened in March 1621 during Huygens's first visit to London as an official diplomat – a great double-cube room flooded with light from two storeys of windows.* These magnificent buildings, at once restrained and grand, perfectly proportioned in every detail yet retaining an essential simplicity, now inspired the ambitious Calvinist to build his own house in The Hague.

He needed a larger home for his growing family. More important, though, was the fact that building his own house in such a prestigious location would flaunt his skill as an architectural taste-maker and conspicuously enhance his social status. The opportunity arose when a new square, known now simply as Het Plein ('The Square'), was laid out in the centre of the city, close to the Orange court in the Binnenhof, in 1633. Frederik Hendrik made available a long plot, 360 feet by 90 feet, along the west side of the square to Huygens, while an adjoining plot went to the Count of Nassau-Siegen, Johan Maurits, the governor-general of Dutch Brazil, a cousin of Frederik Hendrik's.

Huygens the *kenner* had, of course, studied architecture, which meant that he had acquired a gentlemanly understanding of the classical orders as laid down by the Roman architect Vitruvius. However, he had neither the time nor the practical skills to manage his own project. Fortunately, though, he knew the Haarlem painter and architect Jacob van Campen, an outer member of the Muiden Circle, whom he had met at Tesselschade's wedding in 1623 or shortly after. The resumption of the Eighty Years War had put an end to the extravagance of the early years of the century, and van Campen had found his niche when he inaugurated a version of the new classicism he had seen while travelling as an artist in

---

* Following the accession of Charles I in 1625, Rubens was commissioned to paint the ceiling of the building with patriotic allegories, which were greatly admired by Huygens on a later visit, and became an inspiration for him when painters were commissioned for the Oranjezaal, the memorial room in the palace outside The Hague created on the orders of Amalia van Solms after the death of her husband in 1647.

Italy, tempered by an austere northern rigour. The style could hardly have been better attuned to the expressive needs of the Dutch state, and Huygens's own house would be its first great advertisement.

The two friends worked together on the design, with Huygens closely involved in aspects of the detailing, down to the level of what cornice sections should be employed. They even talked about making Dutch translations of Vitruvius's and Palladio's canonical books of architecture. The job would be doubly rewarding for van Campen if it came off well, for Huygens was in a good position to feed him new commissions. As construction began, a second architect from Haarlem, Pieter Post, was engaged to implement van Campen's intentions on site. Huygens would later praise van Campen and Post in verse, writing that they had lifted the dirty Gothic scales from the eyes of blind Dutch 'mis-builders', while in a letter to Rubens – in which he assured the artist that he was working to resolve the matter of a refused passport to England – he bragged of his hope that the house might do a little to revive classical architecture in the Low Countries.*

6. Huygens's house and walled garden on the Plein in
The Hague shown shortly after its completion. The gates to the
Mauritshuis lie on the extreme right.

---

* Ironically, the house was demolished in 1876, and replaced by the neo-Gothic Ministry of Justice building, which still stands.

Built of brick, the main house was rectangular and symmetrical, with two narrow wings projecting forward to create a courtyard. The interiors were laid out largely by Constantijn and Susanna themselves, with separate apartments leading off to each side of the house. External details such as the capitals of the pilasters were executed in the style of Palladio's protégé Vincenzo Scamozzi, Huygens having laboriously compiled tables comparing versions of the classical orders according to Vitruvius, Serlio, Palladio, Scamozzi and Henry Wotton. The statues on the pediment personified the Vitruvian architectural virtues of Firmitas, Utilitas and Venustas (famously parsed into English in 1624 by Wotton, Huygens's one-time ambassadorial neighbour on the Voorhout, as 'Commodity, Firmness and Delight'). Johan Maurits's house, though larger overall, had no statues, and comprised a mere seven bays across the front to Huygens's nine.

In practice, it was Susanna, now the mother of four children, who was the day-to-day client. She oversaw the works and was often on site during Constantijn's long periods of absence. She held meetings, negotiated prices and prepared the bills. One 'bill', forged by the seven-year-old Constantijn in his mother's handwriting, was presented to his father and fooled him completely. A draughtswoman and painter in her own right – better than he, in her husband's view – Susanna also made aesthetic decisions. Huygens, meanwhile, paid regular visits to van Campen at his own country house, or received him when he was off on army campaign in a convenient part of the country. In October 1635, when plague broke out in The Hague, the whole family joined Huygens near Amersfoort, where he met again with van Campen. At home, all he could do was to complain about the workmen, who were 'lazier than sleeping sickness and slower than syrup', and sympathize with Susanna, who had to deal with it all.

Despite taking the economical step of ensuring that both

houses made use of the same materials, progress was slow because of shortages owing to the war. For example, oak from Hesse required an import permit from the Spanish governor in Brussels, which in turn necessitated diplomatic intervention from Huygens. Bluestone for the cornices, lintels and sills was held up in transit. The builder became ill and then died. In the end, though, the cost of construction was entirely covered by the proceeds from the sale of the Huygenses' former house on Lange Houtstraat.

During 1634, as construction began to slow, van Campen filled the time by painting a remarkable portrait of his clients. Constantijn appears in profile, sat staring straight ahead. His fine black hair is uncovered and a trim goatee tucks under his chin. Too vain to wear his customary glasses, his eye bulges and his eyelid droops. In his hand, he holds a sheet of music – a reference to marital harmony – but he is peering right over the top of it. Susanna, sitting on Constantijn's right, and behind him from the viewer's perspective, leans forward past her husband to look directly out from the canvas with beady dark eyes, her mouth pursed, as if issuing a challenge to the painter, her architect: is this really how you want me?

Constantijn responded archly to the unusual optics of van Campen's composition:

> Blessed are the faithful rays
> That light the way for man and wife
> Through the joy and through the strife
> Looking out in double ways:
> But holier still is this conception;
> Man and wife see one direction.

By the end of 1636 the house was still not ready, and Susanna was six months pregnant. The need now became urgent. The sale of the old house was agreed in February 1637, with a moving date of 1 May, a few weeks after the baby was due. At the begin-ning of March young Constantijn suddenly had to be taken to Utrecht for an operation on his neck. Then, on the thirteenth, the baby was born – the longed-for daughter. A little over two weeks later, Susanna was suddenly taken ill with an eruption of mouth ulcers. Her condition worsened as the deadline for moving approached, until, with three days to go, and now seeming to be in mortal danger, she was removed to be cared for in the quieter surroundings of her sister's house. Huygens began to panic about how he would cope with the children, and how he would explain the worst when his eldest came home. He observed how the eight-year-old Christiaan would not leave his mother's bedside. Susanna cuddled her infant girl for the last time, calling her 'soet mockeltje' ('sweet chickling'), and made her will. She died quietly – 'that beautiful death', according to her grief-stricken husband – late in the afternoon of 10 May. The move had gone ahead as planned. The day after Susanna was buried, Huygens wrote: 'I take possession of my new dwelling, but, alas, without my turtle-dove.'

For Huygens, the house on the Plein would always be a memorial to his beloved Susanna, and for him the best way to remember her was to maintain it as a family home for his boys and their new sister. 'My beloved has dedicated this house to me as the last reminder of her love and as a token of herself,' he wrote later. 'I have taken it over, together with my own little consequence, in tears, deprived of her, my better half, for whom I shall eternally mourn like a lonely dove.' Fifty years later, in his last will, Huygens repeated that the house was never to be sold, but was to remain within the family.

For the present, too, the house was indispensable in a political sense, because of its proximity to the centre of stadholderate power, the Binnenhof. It made a public display of the bond between the secretary and his stadholder, especially when it was pressed into service for royal entertainments, as it was, for example, in 1638 almost before the mortar had set, during a five-day ring-tilting contest arranged for the wedding of the Count of Brederode.

Its value was hardly less as a cultural symbol. One of the first important designs by van Campen and Post, Huygens's house signalled a new turn in Dutch architecture, and before long in Dutch art, too. Between them, the two architects went on to create many of the emblematic buildings of the 'Golden Age'. Post designed the sylvan Huis ten Bosch as a royal retreat in the woods of The Hague. Van Campen created the stadholder's palace, Noordeinde, adopting some of the features of Huygens's house. Later came the city hall of Amsterdam, the largest secular building

7. Jacob van Campen, the architect of Huygens's house, Amsterdam City Hall and other landmarks of the Dutch baroque, shown in his later years.

in the world at the time of its completion, and 'the single most imposing architectural venture ever undertaken in the Republic'. This vast edifice, roughly imposed on the chaotic medieval network of streets and canals, provides the clearest expression of the architect's intention to merge the ideals of classical architecture learned from Italy with the strictures of Calvinism to create an authentic Dutch baroque style. For all its carved swags of fruit, the building remains cold and forbidding, an effect achieved by two tiers of severe pilasters running around its full perimeter like a rank of sentries.

All this makes a heavy load for one house to bear. Perhaps it is not surprising that Huygens later commissioned a second house, different in almost every significant respect from the one on the Plein, where he spent the greater amount of his time in later years, finding poetic inspiration and respite from city life.

It was a pun, of course. Hofwijck: *hof* being 'the court' or merely a court or garden; *wijk* being an area, a quarter or a district, also a walk or one's regular round or beat, as well as a retreat or even a flight from somewhere else. So, Hofwijck was his 'courtly quarter', or 'garden circuit', or equally his 'escape from the court'. Except that he was far from withdrawn from the life of the court. The name was a self-mocking jest from a poet who loved to play with words.

He bought the land near the 'pretty village, or, rather, little town' of Voorburg on the bank of the Vliet, the old canal linking the major cities of Holland, in December 1639. To the south, across the Vliet, lay water meadows. (Much of the area bounded by the cities of Delft, Leiden, Gouda and Rotterdam was once a large freshwater lake, the Zoetermeer, which was gradually turned into polder land during Huygens's lifetime.) In other directions, Huygens could see the walls of Delft, many windmills and the church tower in The Hague. He was a 'weathercock' uncertain where to turn for the best view.

Huygens swiftly acquired adjoining slivers of land that enabled him to expand the estate, and directed the planting of trees to furnish his arcadia. Building work began the following year and the house was ready by late 1641. His architects were those who had worked for him in the city, the house conceived by the 'reason-rich mind' of van Campen, and 'midwifed' into the world by the pen of Post, as Huygens put it. Once again, Huygens himself played a major part in the conception of the overall design. As on the Plein, axial symmetry was a guiding principle, but the mathematics of this ideal villa were purer, a simple brick cube apparently afloat on a square lake. There was no place for 'crooked corners' or 'imparity' in a house modelled after God's perfect creation, man. Doubled windows made eyes, ears and nostrils in the facade.

It was a true retreat in the sense that it was too small to entertain staying visitors or to live at all grandly. The entrance opened directly into the main room, which was often used for musical gatherings. A small library space led off to one side. Downstairs lay a kitchen where the family usually ate without formality. Upstairs were a couple of small bedrooms, and above that an attic space that Christiaan would later adopt for his scientific work. Just as he had helped to introduce a new Dutch baroque style for city architecture, Huygens was setting a trend in the countryside, too. Few wealthy Dutch citizens possessed a country seat in the 1630s, but within a generation nearly half of them did, and by the end of the seventeenth century four out of five of these 'pseudo-aristocrats' owned new estates. The Vliet in particular was soon lined with grand villas, transforming it into a northern equivalent of the Brenta that Huygens had admired on his visit to Venice.

The Vliet had become part of a growing public transport or *trekvaart* network in 1636; a spur was added into The Hague in 1638, just before Huygens made his purchase. By the mid seventeenth century, there was a more or less hourly barge service linking Rotterdam, Delft, The Hague, Leiden, Haarlem and Amsterdam, and by 1665 there were 400 miles of inter-city canals. The barges were pulled by men at first, and later by horses. When the canals froze, the towpaths still provided convenient connections by foot. Huygens erected a landing stage at Hofwijck from which it took forty minutes to reach The Hague. On his journeys, he enjoyed the banter of the skippers, listening for dialect terms which he might use to enliven the dialogue in his plays.

It was in the poet's paradise garden where he was able to demonstrate most fully the Vitruvian axiom that ideal planning should take the human body as its template, an idea perhaps inspired by the elongated proportions of the site. The house, with its window eyes, was the head; the bridge across the moat, the neck. From there, arms branched off down each side of the garden. In the middle was the 'stomach', an orchard where the family planted apple and cherry trees and grew melons. The waist was made by a public right of way, the Westeinde road, that cut across the land Huygens had bought. Beyond this, the Vitruvian man's legs stretched out along paths and water-filled ditches, the latter suggesting to the scientifically minded Huygens the newly discovered circulation of the blood and the regulation of body (and house) temperature. The entire estate was a compact 35 rods at its widest and 110 rods in length (125 × 410 metres).*

---

* Hofwijck's garden survives only above the waist. The legs were amputated by the arrival of the railway in the nineteenth century, and a later motorway makes further encroachments. It is ironic that good transport links were one

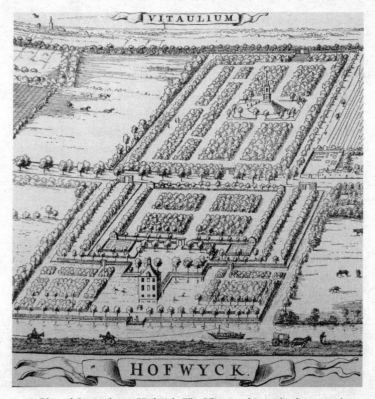

8. Plan of the garden at Hofwijck. The Vliet canal is in the foreground.
The house is surrounded by a moat, while the planting and paths are
laid out following Vitruvian ideals like a human body.

Trees were planted according to a detailed plan such that each
one would play a role in the whole composition. From Frederik
Hendrik, Huygens received a present of tall pines and other
conifers. Birches stood around 'like tapers in church', but oaks
suffered in the sandy soil. The plot was long enough that the

of Huygens's reasons for acquiring the site.

many varieties of tree would at first obscure any view of the house. The walk out from the house and through these woods thus made a transition from the material world to one more spiritual, 'a tame wilderness from savage civilities', as Huygens put it in one of his characteristic antitheses. Concerned about the cost and disruption of removing the soil dug out to create the ditches, Huygens even had the thought to build a central mound of surplus earth in the middle of the garden. From the summit, it would be possible to see The Hague, 'the white wall of dunes at Scheveningen', and the sea beyond. He crowned the hillock with an obelisk. 'The key to my heart,' Huygens wrote, 'is the one to this garden.'

Huygens's new retreat inspired his longest and perhaps most charming poem, which he finished in December 1651. In 2,824 densely packed lines of alexandrine verse, he presented his vision of Hofwijck as a place of miraculous organic fulfilment, like an overnight 'mushroom revealed in the light'. *Hofwijck* is in effect an early example of the country-house poem, a genre of works that began to be produced by many poets in appreciation of a visit to the estate of a wealthy friend or patron. These poems typically blended admiration for the house and the good taste of its owner with praise for the rustic way of life and the civilized pursuits made possible in the country, from arboriculture to star-gazing.*

---

* Huygens's *Hofwijck* bears sufficient similarities with Andrew Marvell's 'Upon Appleton House', written in the same year, for some to believe that the two men must have compared notes (Huygens tr. Davidson and van der Weel 208). It is true that both poems use the idea of a walk through woods to the house, and both invoke Vitruvian man, although Marvell is obliged to place this conception in tension with his subject, an older house, a former priory, that lacked the requisite architectural ideality. Marvell also offers a lengthy history of Appleton House (impossible in the case of the new Hofwijck) and an account of the heroic deeds of its owner (which it would have been immodest for Huygens to include). It is possible that Huygens conducted Marvell around Hofwijck when the Englishman visited Holland in the 1640s.

Because Huygens was both the poet and the lord of the manor, he was able to offer rather more. He unfolded not only the house and its garden but also much of his philosophy of life. He confirmed Hofwijck's importance as a refuge, with a list of things banned, which began: 'I ban the whole of The Hague / With all its backbiting. I ban the filthy plague . . .'. With the Vliet barges passing by and the road cutting through the estate, it was nevertheless a populated refuge, and he gladly pictured his splendid surroundings as part of a community. It was his conceit that his lofty trees would shield the townsfolk of Voorburg from harsh winds. But above all, Hofwijck was a place for Huygens to discourse with friends or to contemplate the world alone outdoors while sitting upon its grassy banks, or to be indoors with his musical instruments and books – he feared for the strain the latter would place on the timbers of the house. He imagined guiding his reader round the place in a future time 'as if our yesterday were an age ago', when his grandchildren might be living there. His cherished saplings he visualized now in their maturity as an inheritance that these descendants too should nurture. He realized without false modesty that *Hofwijck* the poem might outlast Hofwijck the house:

> Hofwijck as it is, I would have the stranger see,
> Hofwijck as it will be, the Hollander must read.
> So feeble is man's work, it lasts less than paper.

Meanwhile, the reflections in the still water seemed to double his riches in an illusion of 'opulent alchemy, or I never knew any'. In winter, the lawns flooded, the waters froze, and out came the

---

However, there is no evidence that the two men ever met, either in Holland or in England.

skaters, carving patterns in the ice. In summer, nightingales nested in the garden, reminding him of the sweet-voiced woman – one of his musical companions, Utricia Ogle – who once sang there.

Yet this whole idyll was presented with an aching sense of loss. This is the effect of Huygens's imagined standpoint in the distant future, which serves to display his proper humility as well as his satisfaction in his creation. He even foresaw Hofwijck left upturned in ruins following some divine cataclysm, with the memory of his happy marriage lying among the fallen stones:

> . . . the lowest thrust up high,
> The highest brought to ground: the name-plate far off
> seen,
> Lay sadly toppled down, SUSANN and CONSTANTIN:
> But never sundered yet, as e'er they must remain,
> Their souls, I now do mean, as were their bodies twain.

In fact, he enjoyed many summers at Hofwijck, especially during the Stadholderless Period beginning in 1650, when his diplomatic duties became lighter, before it passed to his children and eventually became Christiaan's less happy home.

5

ALMOST A PRODIGY

The view from the top of Maurits's tower in the Binnenhof reveals the Huygenses' world. At the foot of the tower lies the Hofvijver lake with the stadholder's court stretching along its bank to the right. Beyond it, van Campen's grand houses, the Mauritshuis on the far corner of the lake, and just to its right, Huygens's on the Plein. A little way to the north, the saplings of the Voorhout, where Constantijn Huygens spent his youth, wave their leaves above the rooftops. To the west rises the hexagonal brick tower of the city's great church, Sint-Jacobskerk. You hardly need take up Lipperhey's telescope to see the rest: three miles to the south-east, the little town of Voorburg and the Vliet canal with Constantijn's retreat, Hofwijck, on its bank; to the north, a great band of dunes, with the spire of Scheveningen church beyond, and behind that only the sea.

Imagine the house on the Plein not as a mausoleum or a museum – though it was almost two-thirds the size of the next-door Mauritshuis, the picture gallery which now holds some of the most important paintings of the Huygens family – but with four small boys tearing through the hallways and up and down the stairs, and the infant Susanna adding her gurgles to the din. When he was at home, Constantijn Huygens ran the house along modern, bourgeois lines rather than in emulation of the stilted manners that he saw in operation at court. The family took their meals together even when the children were still very young. The father showed an attentive concern for their development, observing their different characters and finding for each of them activities suited to their individual temperaments, as well as moulding them in the way he thought best. Punishments were fines taken out of pocket money rather than beatings.

But often, he was absent, either serving in the field with the stadholder, or called away on diplomatic errands. Then, there were only his serious portraits looking down from the walls of the great house to remind the children of their only parent's existence. At these times, the youngest had the household staff to turn to, while the older boys must have developed a strong degree of self-reliance and found their own ways to keep themselves amused.

Constantijn's commitment to family life shines out from a portrait by Adriaen Hanneman, completed in 1640, nearly three years after Sterre's death. It shows the five children, the boys' ages ranging from eleven to six and Susanna not yet three years old, in medallions closely encircling their father. A possible early sketch by the artist showing a central married couple with only four medallions spaced around them for the children suggests that the painting may have been planned before the arrival of their last child. If so, then the fact that Huygens went ahead with the commission after Susanna's death reveals a stoical acceptance of the disaster that has befallen him.

In the finished painting, the boys appear remarkably alike, despite their range of ages, each with similar tumbling auburn hair and pretty face. Young Constantijn looks out of the canvas with a watchful expression and the barest trace of amusement on his lips, as if he is taking in everything the painter is doing with his big brown eyes. Christiaan's rounder cheeks, retroussé nose and pursed lips make him look the most girlish. He was, his father noted, 'the image of his mother', and before the boys were breeched he was indeed sometimes mistaken for a girl. Unlike his older brother, he is not inspecting the artist, but looking past him into some greater distance. Lodewijk, aged eight, looks hardly younger than Christiaan, and was in fact sometimes taken to be his twin; he is obviously more comfortable with the idea of sitting for the artist. Six-year-old Philips is wearing a plumed cap and a gold-embroidered cloak that makes him look older than he is, closing the gap with his brothers. Perhaps conscious of his extravagant attire, he appears to be suppressing a smirk into the serious expression that the artist requires. The toddler Susanna, in the topmost medallion, has fairer hair than her brothers and is wearing a white lace bonnet with a floral attachment and clutching an apple tightly in both hands.

A few weeks after his wife's death, Huygens arranged for a cousin, Catharina Zuerius (or Sweerts), to join the household. By all accounts, she was unable to fulfil many of the maternal functions – an impossible thing to expect. Huygens soon came to dislike her, and in time so too did his sons, although she remained in the family's service for thirty years. Years later, writing letters that criss-crossed Europe, the brothers would still bring each other up to date on her latest antics, and treat her as a yardstick against which other women could only be favourably compared. When she died in October 1680, twelve years after her retirement from her thankless labours, Huygens wrote a brutal memorial verse:

> Here lies Auntie Catharine: what more can there be writ?
> Because, dare it be said,
> In eighty years and three what deeds did she befit,
> Than that she haggled and nagged and domineered and
>   died.

Although the boys were indeed close enough in appearance to be mistaken for one another, Christiaan was less physically robust than his elder brother and would always remain in respectful awe of him. But he soon showed himself to be his equal in other respects, and Tien and Tiaen, as they were affectionately abbreviated, spent a lot of their time together. They were not sent to school as their father judged that this would be a waste of their time and the syllabus would be too restricted. The far broader and more ambitious education that they received at home had the consequence of making the two still more alike. The teaching was modelled on their father's upbringing, and included classical and modern languages, mathematics, theology, logic and philosophy, as well as classes in horsemanship, fencing, dancing, music and drawing and painting. Christiaan early on showed himself to be the better musician, able to sing and play several instruments, while Constantijn was the superior draughtsman.

Young Constantijn was the quieter and more serious of the two boys, and perhaps felt the paternal burden of expectation more keenly. Christiaan, on the other hand, generally bubbled with enthusiasm, cheerfully chattering and singing, and following his father 'dog-like' around the house to pester him with questions. It is possible that his true intellectual potential was not identified as early as it might have been because he was the younger.

Lodewijk was always the most difficult son, less bright and less well behaved than Tien and Tiaen, always seeking attention by playing the fool. His predominant childhood passions were horses and fencing, and Huygens feared that he would become a soldier. Philips was also less academic than the older boys, and his father followed his progress with less interest. Susanna, on the other hand, was clearly intelligent. Nearly four years younger than the youngest of the boys, she was schooled separately. She learned sewing and embroidery like any girl of her social status.

She also learned French – the language of the court – but not Latin or Greek. In addition, she proved to be highly musical, which greatly pleased her father, who taught her at the keyboard. As she grew up, Huygens recognized her potential to become like the intellectual women he knew in the Muiden Circle, but he did not want this life for his daughter.

Huygens was proud of all his children. It is an indication of his closeness to them, as well as of his familiar relationship with the Prince of Orange, that sometimes when he was at army camp, he would show him the letters they had sent him. Another time, in The Hague, Huygens took little Susanna along when he had a meeting with him. Once back home, Susanna declared that she had seen no prince, but only 'a worn old man'. Huygens repeated this afterwards to Frederik Hendrik, who roared with laughter.

The file heading has been amended in Christiaan's own hand: *Juvenilia pleraque*, it reads. He has crossed out the word *Puerilia*: 'Mainly juvenilia', not 'Mainly boyhood'. He does not wish to be thought of as a *child* genius. A scrap of paper among the loose sheaf in the file illustrates his youthful range. On one side is a sketch of a tree and some bushes, rapidly executed in charcoal and red chalk, with expressively upward-reaching branches and grass in the foreground suggested by confident flicks; on the other is a scattering of diagrams, one showing a device for drawing perfect ellipses, another of a circular board cut with perpendicular slots, probably for lens-grinding, together with some geometric doodles in the margins – tangents and parabolas and a catenary curve.

Their father was able to offer instruction in arithmetic and music, but soon the older Huygens boys began to see a procession of tutors to their new home. The first to arrive, in July 1637, was Abraham Mirkinius, who taught them Latin. A year later came Hendrik Bruno, only eighteen years old himself, who had studied theology at Leiden University and was an aspiring poet in neo-Latin. Unfortunately, his first gift to the boys was scabies, and he had trouble keeping them under control. But he must have soon established a good working relationship for he stayed on for nine years.

Teaching both boys together over such a long period, Bruno naturally began to notice their particular aptitudes. Christiaan had such ability that 'he might almost be called a prodigy', he wrote in one report to the father – if only he wouldn't spend so much time on 'devices of his own invention, constructions and machines, which, though they might be ingenious, are but distractions that will always break down'. These elaborate toys – little contraptions with wheels and gears resembling model mills and lathes – concerned Christiaan's father, too, who discouraged such unscholarly activity. He had regarded the third and fourth years of his boys' lives as dull because they were concerned entirely with play of a kind in which he had no contribution to make. Now that they were a little older, everything had to be directed towards learning, and the rare games at home always had a moral lesson. As Christiaan grew older, such manual recreations raised an unthinkable new fear in his father's mind that his son might be content merely to pursue a handicraft trade.

Other tutors came. Jan Stampioen, who had studied with the mathematician Frans van Schooten, moved in during the spring of 1644 to give the fifteen-year-old Christiaan a more solid grounding in the subjects for which it was now clear that he showed the greatest flair: logic, philosophy, physics, astronomy, optics and geometry. His reading list for Christiaan included Descartes on lens-grinding, Stevin on perspective, Lansbergen on astronomy, Al-Hazen's astrology and Scamozzi's books of architecture.

Stampioen had been the professor of mathematics at the illustrious college in Rotterdam and then a tutor to Frederik Hendrik's son, the future William II of Orange. A specialist in geometry and trigonometry, he had earned a degree of notoriety for engaging in public exchanges of mathematical problems and solutions, even going so far as to announce anonymously solutions to problems he himself had set in order to publicize his talents. One challenge nearly went too far. When Stampioen set Descartes a problem whose solution required the use of an unstated quartic equation, Descartes responded by giving the equation, but failed to use it to solve the problem. Stampioen judged that the answer was therefore incomplete, which provoked Descartes to an angry response. With a wager of 600 guilders at stake, this soon led to a duel being arranged between the disputants. As one who knew both men, it was the Huygens boys' father who reluctantly agreed to step in and defuse the row.

This makes the appointment of Stampioen as his son's tutor seem somewhat surprising. Stampioen's track record with the royal family, as well as Huygens's awareness of Descartes's short fuse, ultimately may have counted in his favour. Or, he may have simply matured by the time of this appointment. Hendrik Bruno observed his startling impact on Christiaan in mathematics, and gave a critical description of his methods, noting that he required 'first a good intelligence, secondly constant application, and finally a perfect desire to achieve; if all these conditions are met, this understanding can be acquired not all at once, but little by little, each piece in its turn, by prolonged study'.

On 11 May 1646 the two brothers matriculated at Leiden University, with their father's detailed academic and moral instructions for how they were to conduct themselves in their pockets. They were encouraged to write home frequently and always to employ proper forms of address. Constantijn adapted better to this regime than Christiaan, who was inclined to write only as necessary, for example when, two months after arriving at university, he found he had run out of money. Both were quiet, diligent students who tended to avoid the more boisterous side of student life. On one occasion Constantijn was shocked to find that he had incurred the wrath of the violent landlord with whom they were lodging, and wrote home about the incident. His father felt it necessary to intervene in his son's interests, influential as always.

From its beginning, the Dutch Revolt had seen a great flowering in education. Leiden was chosen in 1575 as the location of the first university in the Netherlands, with the object of preparing the young men who would run the new country and its Reformed church. By the 1640s, with more than 500 students, half of them drawn from abroad, it was the largest university in the Protestant world. Although natural philosophy did not feature explicitly on the curriculum, the foundation boasted one of the first botanic gardens and the first anatomy theatre in northern Europe. In 1633, Leiden became the first university anywhere to have its own astronomical observatory.

Encouraged by their father, who thought good things might come of it if their rivalry were constructively harnessed, Constantijn and Christiaan took several classes together, including law with Arnold Vinnius and mathematics with Frans van Schooten. Constantijn perhaps gained more from the former, a celebrated jurist, but for Christiaan, it was undoubtedly van Schooten who most greatly inspired him among his professors at Leiden.

Frans van Schooten (the younger) was a notable Protestant, a

freethinker and – of greatest significance for Christiaan's education
– a convinced Cartesian. A fine mathematician specializing in
analytical geometry, he had known Descartes personally for ten
years, and defended his work at Leiden, where his philosophical
ideas could be discussed only with the greatest circumspection,
and would soon be banned altogether because they could be taken
as sympathetic to atheism. Other talented students taught by van
Schooten would also go on to notable achievements in mathematics
and in other fields. Chief among them, a year above Christiaan,
was Johan de Witt, the Holland councillor whose spell as the de
facto national leader during the long period from 1650 to 1672
when the Dutch provinces were without an overall stadholder did
not stop him also publishing his own treatise on geometry. The
two did not overlap directly, however, and it seems that Christiaan
perhaps did not find many like-minded souls with whom to forge
the sort of lifelong friendships that university often provides.

Van Schooten had a rare ability to take the geometries found
in the practical world and distil their pure essence. Like his father,
he had been schooled in Stevin's mathematics of military engineering. This training left him with a 'kinematic' feel for geometric
curves; he was able to sense, as it were, the meaning behind their
shape. For example, he showed that the 'gardener's ellipse' – a
method of generating ellipses empirically by using a board with
two pegs and a rope in order to set out plant beds – was the same
as the pure ellipse in the family of conic sections. This sense of
a deep connection between abstract geometry and the physical
world was something that he transmitted to Christiaan Huygens,
who began to tackle the problem of the catenary – the 'natural'
curve followed by a chain freely hanging between two points,
which had resisted all attempts to reveal its mathematical secret
– during his year with van Schooten. As well as enjoying close
relations with Descartes, van Schooten had travelled to Paris a

few years earlier, and had made the acquaintance of the leading mathematicians there. Many of these men would later welcome Christiaan Huygens into their midst.

Christiaan's father was also busy building connections. On 12 September 1646 he wrote to the French mathematician and theologian Marin Mersenne, with whom he corresponded regularly on musical matters and scientific curiosities such as the magnet, drawing attention to his teenage sons' mathematical prowess. They were, he improbably claimed, 'most eager about your quadrature of the hyperbola and your centre of percussion'. The boys' own enclosed letters demonstrated that this was indeed the case. Mersenne was so impressed that he sent back a new mathematical puzzle to test the boys. Christiaan returned something promising on catenaries, Mersenne responded by setting another problem about pendulums, and so the correspondence blossomed. After a few months of such exchanges, Mersenne wrote to Christiaan's father: 'I don't doubt that if he continues he will one day surpass Archimedes.' Thereafter, both men habitually spoke of Christiaan as their 'little Archimedes' or the 'Dutch Archimedes'. Mindful perhaps that such a badge could weigh heavily on the young student, Mersenne gave Christiaan an old man's advice: 'do not worry too much, for you have so many years remaining, that if you were to make one demonstration each year, as beautiful as that of this chord [i.e. the catenary], you would have enough to rank at the top end of all the nobility'. Mersenne soon spread word of Christiaan's unique facility to other French mathematicians such as Blaise Pascal, who also began to share their problems with him. But Christiaan's correspondence came to an abrupt end with Mersenne's death in 1648, and the eruption of civil war in France in that year put paid to plans for an excursion to Paris.

What was the power of curves? Why did these shapes engage such able minds at this time? Some were surely drawn by the promise that greater understanding would help them to address practical challenges in construction or ballistics. But for others it must have been the sheer abstraction that appealed. With the discovery of perspective, the artists of the Renaissance had tamed the straight line. It was a line that pointed to where it was headed and went directly there. Its intention, if it could be said to have one, was clear. But curved lines were another matter. Some, such as the circumference of a circle, clearly followed a simple rule. But others, like the line taken by a quill producing an extravagant signature, could clearly go anywhere. What lay in between? Some curves, such as the conic sections, those exposed by slicing at angles through a cone – the ellipse, parabola and hyperbola, seemed almost as fundamental as the circle. Could these be represented mathematically like the circle? And what of other curves found in nature, in the growth of plants or the waves on the sea?

Then there was the question of quadrature. Quadrature, or squaring, a precursor of integral calculus, was the general method of calculating areas bounded by lines, using pure algebra rather than direct measurement. But curved lines resisted squaring, the circle famously so. Indeed, Christiaan, who busied himself with these matters after his year with van Schooten, was to become caught up in a ten-year dispute with French and Flemish rivals over their methods of performing this particular calculation.

The catenary seemed to be another fundamental curve. Like the parabolic path followed by a rising and falling cannonball, it is found in the real world: Huygens uses the Latin word *catenaria*, derived from *catena*, meaning 'chain' in his notes and correspondence, because a catenary is the curve made by a rope or chain of uniform heaviness hanging under gravity. But efforts to generate and manipulate this curve using geometrical methods had failed. Huygens approached the problem in a physical way, drawing upon his investigation of the centre of gravity of beams hung with weights. This was the problem of 'centres of percussion' set by Mersenne, the centre of percussion being the point along a beam suspended by its end where it should be struck in order to make it swing like a pendulum without producing a reaction at the pivot. From there, Huygens began to explore where the centre of gravity of a suspended flexible cord would lie when weights were hung at various points along its length. The first weight forces the cord to adopt a path of two straight lines angled at the point where the weight is hanging. Then, by adding more weights along the length of the cord, Huygens was able to approach the condition of a uniformly heavy chain. In this way, he was able to prove that the catenary was not the same as the parabola, which it closely resembles, and as Stevin and Galileo had thought, but a new kind of curve not found among the conic sections.

Early on in their correspondence in 1646, Mersenne drew the young Huygens's attention to another special curve that had thwarted efforts to understand it. This was the cycloid, the path traced by a point on the circumference of a wheel as it rolls along a flat surface. Such a curve might almost have been designed to demonstrate the link that would become so important for Huygens between pure mathematics and the mechanical world. Although Huygens did not get to grips with the problem at this time, it was not to be many years before the cycloid returned to occupy a central place in his life.

Constantijn left Leiden after only a few months in order to join his father in secretarial service to the stadholder. Christiaan left in March 1647 to move on to the House of Orange college at Breda, established following the Dutch recapture of the city in 1637 as a training academy for the diplomatic service. Although his father was one of the governors, the school had struggled, with only around sixty students by the time Christiaan arrived there, and moral scandals among the staff. Christiaan wrote to Constantijn about the girls and the music he found in Breda, but he attended only a few lectures of the unpopular mathematics professor. The classes in law were little better, but he was able to defend his thesis in the subject. Christiaan's more boisterous brother Lodewijk was clearly better suited to the place and had already been there for two years when Christiaan joined him. When Lodewijk became involved in a duel, however, their father finally realized it was not right for either young man and pulled them out.

It was a new kind of world into which the Huygens brothers were now stepping. For the whole of their father's life, and for the whole adult life of *his* father, the country had been at war. Now, five years of negotiations between overextended Spanish forces and the tired and sick stadholder were drawing to a close. Frederik Hendrik died in March 1647 and was succeeded by his son William. The Peace of Münster, ending the Eighty Years War and finally confirming the independent sovereignty of the Dutch Republic, was signed in April 1648. Feasts and celebrations were held in many Dutch cities. From The Hague, Constantijn wrote to Christiaan, who was still in Breda: 'Today we burn the Victory and the artillery that has been brought to the Denneweg will be discharged nine times.' Perhaps as a personal gesture of rapprochement, their father wrote his first piece of verse in Spanish, later to be followed by translations of Spanish proverbs no doubt gleaned from his negotiating sessions during the long years of the war.

Did Christiaan remember the man with the strange accent who surely must have called on his father at the old house when he was small? Although many accounts suggest that the boy must have met Descartes, it seems unlikely, or at least that, if he did, he did not recall it as a special occasion in a household accustomed to seeing many illustrious visitors. Constantijn's close involvement with Descartes ended in 1637, when Christiaan was just eight years old, although the two men maintained a lively correspondence for another ten years. Circumstantial evidence leans the other way. Christiaan offered no anecdote of an encounter when he wrote in later years about the deep impression Descartes made on him as a young man. And in the autumn of 1649, after leaving the college at Breda, Christiaan set off on a diplomatic mission to Holstein and Denmark, with hopes of travelling on to see Descartes at the court of the Queen Christina in Sweden – a quest that might have lacked urgency if he had met him properly during his student years.

What is certain is that Christiaan did absorb Descartes's ideas at a young age, thanks to his father's assistance with publication of his works, as well as to Stampioen's and van Schooten's teaching. Before long, he too was a Cartesian. He read *Principia Philosophiae*, in which Descartes outlined his scientific theories in detail, on its publication in 1644, when he was no more than sixteen years old. It is clear that he found the logic of Descartes's arguments highly persuasive. The Cartesian pattern of thought, which was logical and unpedantic, stayed with him all his life, and his own descriptions of physical phenomena in optics and mechanics and other fields, though informed by more thorough observation and experiment, always followed Descartes's lucid example.

Descartes's philosophy was exciting in its radicalism and in its ambition to construct an entirely new understanding of nature. The Frenchman believed that all matter extended through space and time in a manner that could be quantified. Geometry, and in particular Descartes's introduction of algebra into geometry, was an important tool for revealing this universal framework. The innovation attributed to him (though in fact due to Huygens's teacher van Schooten) of what are now known as Cartesian coordinates broke down dimensional space in such a way that it might be manipulated in terms of algebra, which provided a powerful new weapon for analysing spatial events such as the motion of bodies.

In mechanics, Descartes accepted that moving bodies tend to continue moving in a straight line (later known as Newton's first law of motion), but he did not believe in the existence of atoms, the vacuum, or action produced at a distance. Gravitational attraction he interpreted, like all physical interactions, as the result of impacts between bodies. Where bodies did not make actual physical contact, as in the case of the sun holding the planets of the solar system within its orbit, the action was explained by the presence of 'vortices' around each body. Light, for example, was transmitted as a form of luminous pressure acting from vortex to vortex through the transparent medium between the light source and the receiver.

Despite his conviction that everything could be understood in terms of numbers, Descartes himself did not have the facility to make much practical headway. Huygens was struck by the fact that his work on optics, *La dioptrique*, for example, contained no mathematical proofs. Spying an opportunity, Huygens set about trying to understand the geometry of all curves in terms of spherical curves only. His underlying practical hope was that it might become easier to grind telescope lenses with the complex shapes required to bring an image to a perfect focus if they could be ground in a sequence of operations using moulds shaped to different spherical radii. He did not achieve this goal, but the work did enable him to produce a general theory of the properties of spherical lenses that took spherical aberration into account, and yielded a set of rules for calculating the focal distance of such lenses.

That Huygens came to be known as a leading Cartesian now reads as something of a slur. This has little to do with Huygens, and much more to do with the changing reputation of Descartes during the latter half of the seventeenth century and after. The term 'Cartesian' quickly gained a pejorative edge owing to the philosopher's dogmatic insistence on the separation of body and soul, his disagreeable assertion that animals are unfeeling machines, and the fear that 'Cartesian doubt' might be a cover for atheism – none of which conceptions was central to Huygens's concerns. While Descartes's reputation as a philosopher has prospered since these troubled beginnings, his standing as a scientist has sharply declined. This is in part a relative decline that must be set against the rising standard of scientific investigation during this period, which ushered in a clearer understanding of experimental method and emerging protocols for the reporting of results – developments to which, ironically perhaps, Huygens was a principal contributor. In fact, Huygens was never an irrational defender of Descartes, and when his observations or calculations had the unintended effect of challenging the philosopher's beliefs, it was his own results that he tended to trust more. Nevertheless, Huygens remained faithful all his life to Descartes's broad philosophy of nature, and very much shared his serious ambition to *know*.

Once Newton's physics had exposed the errors in Descartes, it would become easy to criticize Huygens for being a Cartesian. But in 1650, it would have been astonishing if he had been anything else.

6

REVERSALS AND COLLISIONS

News spread slowly that Descartes had died of pneumonia during his first winter in Stockholm. In April 1650, two months after the event, Christiaan Huygens passed on to his brother Constantijn that he had seen it reported in the Antwerp gazette 'That in Sweden a madman had died who said that he could live as long as he wanted'. Christiaan added: 'Note that this is M. des Cartes.' Constantijn found this 'quite funny'.

The philosopher's death moved Christiaan to write a short verse epitaph, the only time he showed any sign of following his father's literary path. The last of its four stanzas runs:

> Nature, take to mourning, come foremost to lament
> The great Descartes, and demonstrate your woe;
> When he let go the day, you were lost to light,
> It is only by that torch we could imagine you.

Earlier that winter, the same harsh weather had forced Christiaan to abort his plan to travel on from Denmark in order to see his idol in Sweden. His elder brother, meanwhile, was enjoying his own travels, proceeding 'galliardement' through Switzerland and Italy, where he enjoyed ices 'powerfully flavoured with cherry, strawberry, lemon, amber, cinnamon &c', which he discovered were made in little bottles placed in snow mixed with salt and saltpetre. He marvelled at the 'great peaks of the Alps, which I can see from my room, terrible to behold', and complained that Christiaan in return was telling him nothing of what he was seeing on his trip. He hoped that he, too, was keeping a journal so that they could read them together when they returned home. In fact, Christiaan was holding little back. He replied sourly of his scenery: 'the greatest I have seen [in Denmark] is less like the Alps and more like the mountain of Voorburg which my father has had made [in the garden at Hofwijck].'

Later that year, on 6 November, came a more shocking death, and one that was to have a severe impact on the Huygens family. The stadholder, Prince William II of Orange, succumbed to smallpox at the age of twenty-four, after just three years in the post. Eight days later, his wife Mary Henrietta Stuart gave birth to a son, William III of Orange, who would eventually succeed him as stadholder and become king of England, Scotland and Ireland.

During his brief and tempestuous reign, William II had been more bullish in pushing back the Spanish than his predecessor, Frederik Hendrik, and had aggressively promoted the Protestant Reformation in areas formerly controlled by Spain. But his progress came at some cost to internal unity in the Republic as squabbles arose over the sovereignty rights of individual provinces. On 30 July, in an attempt to impose overall control, William had staged a coup designed to bring Holland (the most powerful province, with its capital in Amsterdam) into line with the other provinces, and to define a peacetime role for the House of Orange. This bold action appeared to resolve the sovereignty issue, and, by agreeing that the titular figure of the stadholder was not a monarchical rank, and was therefore not incompatible with the existence of the republic, William promised a new political order. His untimely death immediately undid much of this work, and the United Provinces entered a long period of stadholderless government. In The Hague, there was such heavy snowfall that William's funeral service had to be postponed because the mourners were unable to reach the city. The meltwater that spring combined with storm surges to produce Holland's worst flooding in eighty years. These were omens that the Dutch knew too well how to read.

For Constantijn, three decades of service as a secretary to various stadholders was at an end, and with it his influence over statecraft. However, he retained his involvement with the House of Orange, continuing to serve Frederik Hendrik's demanding widow Amalia, not least in the creation of a magnificent memorial to her husband at the Huis ten Bosch palace in The Hague. This was a great hall of paintings by the best Dutch and Flemish artists, commissioned to commemorate the victories he had gained for the Dutch Republic.

The sudden reversal in the Huygens family fortunes put paid to any lingering expectation that Christiaan might follow in his father's (and his grandfather's) footsteps by joining the secretariat of an Orange stadholder, and he gratefully seized the opportunity to give himself completely to mathematics and physics. From now on, although he would typically be pursuing a variety of investigations at the same time, all would have their basis in a fundamental wish to learn more about the physical world. Christiaan Huygens was beginning to mould himself in the image of a creature that did not yet exist, the professional scientist. The projects he took on at first were both practical and theoretical in nature, and included improving the design of his telescopes, a consideration of the nature of fluids, and problems more easily described than solved, such as how solid shapes float on water.* An extended

---

* The full theory behind this decidedly Archimedean question was not

analysis of the geometry of curves included a new method of calculating the length of given sections of the circumference of an ellipse. When applied to a bisected circle (a special case of an ellipse), this yielded the mathematical constant π accurate to nine decimal places by the first new method since Archimedes. The work greatly pleased his former teacher van Schooten, with whom Christiaan corresponded frequently at this time, and it made his name in mathematics.

Christiaan's achievement vindicated Marin Mersenne's early championing of his friend Constantijn Huygens's young 'Archimedes'. Before he died, Mersenne had done much to spread the young Dutchman's name in Paris and beyond, making France's greatest mathematical minds, Pascal and Pierre de Fermat, aware of his talent. When Lodewijk Huygens travelled to London in December 1651, he discovered that the philosopher and would-be mathematician Thomas Hobbes, too, knew all about his brother's abilities from Mersenne, and indeed claimed to have been perusing his quadrature of the parabola and hyperbola only a few days before. 'He praised it abundantly and said that in all probability he would be among the greatest mathematicians of the century if he continued in this field,' Lodewijk reported.

Huygens began to make even greater progress in the field of mechanics as he noticed flaws in Descartes's laws of motion. He first raised his concerns with Gérard van Gutschoven, the professor of mathematics at Leuven University and a former associate of Descartes, in January 1652. Descartes had enumerated laws

developed until the nineteenth century, building on the principle established by Huygens that, whatever its shape, the centre of gravity of a floating body will tend to lie as low as it can (OC11 83–92).

governing collisions between bodies of various masses moving at various speeds. These included the statement that a smaller moving body colliding with a larger stationary body would not cause the stationary body to move, but would only rebound itself. This was clearly wrong, as a moment's play with two pebbles would have shown him. But Descartes was not a man given to experimentation, and he may have felt that, since he was discussing *ideal* bodies rather than anything that existed in nature, such demonstrations would have no bearing.

Despite his Cartesian indoctrination, Huygens was always mindful of the physical world. His re-examination of the Frenchman's theory of collisions led him to the realization that it was not one body or the other, but the centre of gravity of the whole system – in this example, the two unequal bodies taken together – that was significant. This centre has the same velocity and direction after the impact as before, or as Huygens put it, the 'quantity of movement' is conserved. This new theory of collisions, in which he reluctantly overturned Descartes's idea, came close to describing a concept of force that was only fully worked out by Newton in the following decade.

Van Schooten, however, was not keen on where his protégé's train of thought was leading him. He insisted that Descartes must be right, and may have persuaded Huygens not to publish his analysis for that reason. The work only appeared in a paper in the French *Journal des Sçavans* in 1669, a few years after Newton's laws of motion, and then in a longer treatise, *De Motu Corporum ex Percussione*, which was not published until after Huygens's death.

An engraving made for *De Motu* shows two men, one standing in a passing barge, the other closely facing him standing on the water's bank. Perhaps the idea came to Huygens from watching the ferrymen plying the Vliet waterway that runs past the garden at Hofwijck. In his outstretched arms, the man on the bank is

9. A diagram illustrating relative motion from *De Motu Corporum ex Percussione*, Huygens's treatise on the collision of bodies, which was not published until 1703.

holding two identical balls suspended on strings. The man in the barge has his arms outstretched, too, so we can imagine that the strings with their equal weights might shortly be passed to him.

Setting the balls swinging like pendulums, the man on the bank finds that they exchange their velocities when they collide; in other words, the one in his right hand that was swinging to the left until the impact now moves to the right, and vice versa, and the speed of their separation is the same as the speed of their original approach. When the man on the moving barge follows this action, the situation appears to him exactly the same. However, the man observing him from the bank now sees the two *initially unequal* relative velocities of the balls exchanged after collision because of the additional component of velocity due to the passage of the barge. Proceeding in Archimedean fashion by using geometric propositions rather than the algebra we would expect to see today, Huygens was able to extend his analysis of this problem to include bodies of unequal mass, leading him to a more complete theory of collisions that admitted the concept of relative motion. Huygens believed that the visual image of the two men and the moving barge would convince even the greatest sceptic of the rightness of this principle of relativity.*

Further consideration of the mechanics of these reversible collisions in 1652 led Huygens to the discovery that the total kinetic energy of the system is conserved. He stated this proposition as follows: 'In the case of two bodies which meet, the quantity obtained by taking the sum of their masses multiplied by the squares of their velocities will be found to be equal before and after the collision.' This time he did find it more profitable to proceed using algebra, and his expression of this rule in algebraic symbols marks the first time that a mathematical formula was used to describe a relationship in physics.

Aristotle had denoted simple quantities such as lengths by the use of letters in geometric problems. Galileo understood that mathematics could be used to describe certain laws of nature, although he used ratios and sequences of numbers to do this rather than algebraic relationships. But Huygens appears to have been the first to designate a function, in this case velocity or speed, which is a function of time, using a single symbol, and to manipulate it with other quantities in mathematical equations. In his nomenclature, a and b represent the two masses seen in the engraving of the men and the barge, and their respective velocities are denoted x and y. The equations for their kinetic energies before and after collision then take the form $axx + byy$ and so on. These equations occur in his notes, but Huygens did not persist with the

---

* The editors of Christiaan Huygens's *Oeuvres Complètes* add: 'one must attach [Huygens's] name, rather than any other, to distinguish it from the more general Principle of the same name developed in recent times by Einstein' (OC16 27).

use of algebra in the final text of *De Motu*, or on other occasions when it might have proved helpful, such as describing the properties of the pendulum. He remained loyal to the geometric methods of the classical mathematicians, even after the calculus of Leibniz and Newton had shown a more powerful way forward, and in this respect truly lived up to the nickname given him by his father and Mersenne: the 'Dutch Archimedes'.

We are accustomed to the idea that many of the laws of science are best – which is to say, most concisely and powerfully, and least ambiguously – expressed in mathematical formulae. The symbolic representation and manipulation of real physical quantities is now regarded as an essential tool of science. But in the seventeenth century it was still usual to try to explain the physical world using lines and angles to represent variable quantities. This was no primitive method, however. 'It requires ingenuity to draw the correct triangles, to notice about the areas, and to figure out how to do this,' according to the twentieth-century physicist Richard Feynman, in a discussion of the respective merits of geometry and algebra in his book *The Character of Physical Law*. The reward was a visual and quantitative representation of events such as the acceleration of an object falling under gravity that before could be observed and described only in qualitative terms. Geometry used in this way is an attempt to show rather than tell, which mathematical formulae cannot hope to emulate. For Huygens, it surely also fell within a Dutch tradition that included surveyors such as Stevin and Snel, cartographers such as Gerard Mercator and Willem Blaeu, and the many painters demonstrating their mastery of perspective, all reliant upon the geometry of triangles, and inspired by the land lying like a sheet of paper before them.

Later in 1652, Christiaan Huygens began to think about lenses. Like Descartes, his approach to the topic was theoretical and mathematical at first. Descartes had tried to show that elliptical or hyperbolic lenses would not suffer from spherical aberration – the tendency of a spherical lens to produce a focused image only through the central part, while rays of light passing through the edges of the lens converge at other focal lengths. The problem, as Christiaan's father had found from his unsuccessful exercises working with lens-grinders, was that lenses of this form were very hard to make. However, Christiaan believed that optical images might be made free of aberration if a second spherically ground lens were to be placed in the path of the light. As with Descartes's hyperbolas, though, his mathematical demonstration of this possibility could not be converted into practical devices. But the defeat did awaken Huygens's interest in the theory of refraction.

In his workshop in the attic of the family house, he now measured the refractive index of various glasses and waters as accurately as he could. Although it was the practicalities of telescope design that most concerned him, he treated the optical property of refraction with a high degree of mathematical rigour, aiming to provide a solid theory of optics for the instrument, which had now been in widespread use without one for nearly half a century. In a series of geometric propositions and demonstrations making use of the sine law of refraction, he analysed the propagation and refraction of light through increasingly complex lenses. Geometry was a persuasive tool here, revealing at a glance the index of refraction in the respective lengths of lines representing the incident and refracted rays of light. Huygens's mathematical analysis even extended to an examination of the rainbow, which yielded angles of refraction, but no clue to the origin of its colours. The work was valuable nevertheless. Huygens's geometrical figures were often also optical diagrams, accurately representing the path of light at a time when it was still not known that light possessed a direction of travel, and most drawings of optical instruments were notably vague about what happened as it passed through the mysterious space between their lenses.

Like so much of Huygens's work, however, this 'theory of the telescope' was not published in his lifetime, despite van Schooten's support this time. If he had published, it would have ensured his priority in explaining some aspects of refraction over later works on optics by the English mathematician Isaac Barrow as well as by Newton. Huygens continued to make new optical discoveries, which were eventually compiled in his major treatise on light (*Traité de la Lumière*, 1690) and in a posthumous work dedicated to the topic of refraction. These early forays into mathematical optics were perhaps most important to him as mental preparation as he set about making his own lenses for the first time. The lucky invention of the telescope had embarrassed natural philosophers into developing a theory of optics; now theory stood ready to inform practice.

One of the best descriptions of how to grind lenses comes from the journal kept by the Middelburg Latin teacher and natural philosopher Isaac Beeckman. Towards the end of his life, still only in his forties, in ill health and often appearing morose and dishevelled, Beeckman turned to grinding his own lenses, having failed to find a supplier of ready-made telescopes to his liking. His diary for the year 1634 draws on all that he was able to read about the subject and glean directly from the best lens-grinders in Middelburg and beyond, and records his struggle to equal them in their art in such detail that it came to serve as an instruction manual for others.

Having studied the literature, Beeckman bought a grinding basin and the other necessary equipment, and took some preliminary lessons from Johannes Sachariassen, the man who had claimed that his father had invented the telescope. Although Sachariassen was happy to use 'collet' – recycled glass – as his raw material, Beeckman soon came to prefer *cristallo* or 'Venice crystal', glass of very high optical clarity, which was available from Middelburg's superior glassworks. Rock crystal, or pure silica, which is even more transparent, he disliked, however, because of its highly reflective surfaces. First, the basic form of the lens is cut from a plane sheet of glass using hot pincers. Then the shaping begins. For a convex lens, the grinder uses a basin of cast iron or other metal with a very slightly greater radius of curvature than the desired lens. For a concave lens, a rounded stone is employed, either held and used freehand, or suspended on a rope to scribe a constant radius.

The flat piece of glass is ground into shape by means of ever finer abrasives. The grinder starts with coarse sand, which is moistened to make a paste. He then repeats the procedure several times using finer sands, prepared, as Beeckman describes it, by sifting through successively finer weaves of cloth and then washing in rainwater. The finest sand must be moistened with distilled rainwater so that the abrasive medium remains free of organic contamination as well as larger grains. Grooves scored into the grinding basin help to retain the sand in position. Sometimes a layer of leather or cloth is introduced between the glass and the basin, too, in order to even out the pressure that is applied.

Much is left to the skill and care of the lens-grinder. Each stage takes longer than the last, as the finer new compound must erase the scratches left in the glass by the coarser medium used in the previous stage. It requires close attention to avoid pits and wells forming in the glass, and concentration to 'hold the centre' of the future lens. 'One must grind lightly for sufficiently long, otherwise one begins to press before the sand is greatly broken up, which makes scratches,' Beeckman wrote. In his diary entries, he compared the methods not only of professional lens-grinders but also of artisans in other fields, searching widely for possible improvements in technique. He observed, for example, that oil is not suitable for mixing with the sand, noting that painters grind their pigments with water first before mixing them with oil for application to the canvas.

Even a careless final wipe of the glass could be enough to undo hours of effort if a single outsize grain of abrasive remained. There were other precautions to be observed, too. For instance, an English lens-grinder in Amsterdam whom he visited for further instruction warned Beeckman that he should take care his hands were not greasy, because that might damage the leather he was using.

Beeckman laboured diligently through the spring and summer at the task he had set himself, but occasionally his frustration got the better of him, and he felt he was getting nowhere:

> Up until this 25th of July, I have polished no glass on the basin except by accident, and can do it no longer. It is possible that I am wiping off the glass too heavily, and that there remains always some dust hanging to it. Going at it too carelessly: stuff sticks to the cloths, on my hands, arms, etc., which I set down here and there as I wash the glass, not shaking it off well enough, not washing my hands, touching my dusty clothes with my arms etc.

But his persistence had its reward in the end. In the spring of 1635 an impromptu contest was held at a friend's house in Middelburg, where Beeckman's lenses were compared and found to be superior to those of Sachariassen. So far as is known, however, Beeckman never went on to make a working telescope, and he died two years later, leaving his work unpublished.

In November 1652, having likewise failed to find good enough telescopes for sale, Christiaan Huygens set to work much in the same way as Beeckman, proceeding by trial and error and seeking advice on best practice. He wrote to Gutschoven asking what mould shapes he should use, how fine to sift his sand, and what sort of glue to use (a strong adhesive was needed to temporarily bond the glass to a handheld wooden form). Gutschoven responded helpfully, and reminded Huygens of the importance of maintaining the axis of symmetry while the glass was being ground so that the lens would have a central focus.

Huygens copied many of Beeckman's techniques, but also made some decisions for himself. He found, for example, that very clear white glass had a tendency to 'sweat', or lose its high lustre, and so he preferred other glasses even though they might have a slight colour. Huygens's surviving early lenses have noticeable tints of grey and green, but are remarkably free of bubbles and other defects. He also experimented with counterweights and a lathe to turn the basin against which the glass was ground. These were labour-saving innovations, although it is probable that an element of Cartesian idealism was also present in his enthusiasm for automation.* As he learned more, Huygens also made notes – keep the sand wet to prolong its life in the grinding basin, but not too wet or else the glass was liable to jolt, and so

---

* A very different philosopher, Baruch Spinoza, always preferred to grind lenses manually. See Chapter 12.

on. Stopping for an occasional rest was feasible, but while working, intense manual concentration and fine motor control were essential. He reminded himself: '*Always think about keeping an equal pressure*. Often paused and brought the hand to bear again with an equal pressure. It is preferable to be alone.'

In fact, Christiaan often worked with his brother Constantijn when grinding lenses, and found joy in their collaboration. The addition of the manual lathe demanded the presence of a second person to provide the motive power. But there was clearly also something highly congenial about sharing such a demanding and monotonous job with familial company. Late in life, when he was back in The Hague after his glorious career in Paris, and Constantijn was away accompanying William III on the invasion of England in 1688, he pined for these placid times.

Although it has obvious visual possibilities, and although the trade was reasonably widespread, the lens-grinder at his table never became a popular subject among Dutch painters. It has the same sense of quiet creative purpose as letter-writers and music-makers, and even comes ready with a fitting moral theme of honest labour directed towards achieving greater clarity of (spiritual) vision. But it was perhaps too specialized an activity to make a saleable picture.

Which of the brothers cranked the handle that drove the rotating platform? And which held the glass to be ground? It is likely that they took turns, if only to alleviate the tedium of the work, although there is reason to think that it was the more artistic Constantijn who was better at the skilled task of offering up the glass. Even with the lathe rumbling away, it was a contemplative business. The grinding compounds are so fine that little sound arises from the abrasion. There is no fire or smoke and little smell. Only the gentle wooden rhythm of the pulleys was there to break the silence, unless one of them spoke. The continuous labour could be interrupted from time to time in order to wash off or add more compound. But apart from that, there was a kind of meditative frugality about the task.* Because of the dust, it was best not to wear good clothes. Very modest quantities of the abrasive materials were all that was required, and the best results were obtained with very little – but even! – pressure applied by the operator, whose main actions otherwise were to make slight adjustments in the lateral and rotational position of the glass on its form with respect to the basin platform underneath. The room in which the work was done had to be scrupulously arranged and kept clean to minimize the risk of cross-contamination of the fine compounds by grains of the coarser ones used in the early stages of the process. In short, lens-grinding might almost have been designed as an activity with little other purpose than to satisfy Calvinist piety.

The Stadholderless Period, as the years from 1650 to 1672 are known in Dutch history, began inauspiciously, with much jostling for influence among the provinces and between the nobility and military, loyal to the House of Orange, and the civic interests of the merchants and traders who generated the country's consider-able wealth. Any mood of republican idealism was soon shattered by the reality of riots in the cities and the rise of fresh tensions abroad.

In October 1651, the English Parliament passed the Navigation

---

* It is interesting to compare the Huygens brothers' lens-making with that of two other siblings, William and Caroline Herschel, polishing the mirrors for their reflecting telescope nearly a century later, which was 'dirty, exhausting and monotonous work, for which they wore rough clothes and ignored ordin-ary household routines and niceties' (quoted in Holmes 86).

Act in an effort to protect its seafarers' commercial interests against the more efficient Dutch maritime trade. Growing harassment of Dutch ships at sea precipitated the first of three Anglo-Dutch wars the following year. At the same time, Johan de Witt amassed support at home by promising the provinces greater sovereignty in a federated republic, and at the age of twenty-seven was duly confirmed in office as Grand Pensionary of Holland. Victory at sea swiftly followed.

The difficult relations between England and the Dutch Republic at this time – allies in their antimonarchical aspirations, but rivals in trade and international expansion – are entertainingly glossed by Andrew Marvell in his satirical poem 'The Character of Holland'. Marvell was well acquainted with the Dutch, had learned to speak the language and may even have visited the Huygenses in The Hague when he travelled to the country as a young man on a five-year tour of Europe during the English Civil War in the 1640s. His poem is riven with contradictions as he struggles to turn his former admiration for the Dutch Republic to the purposes of propaganda.

He begins by disparaging the country 'as but the off-scouring of the British sand; . . . This indigested vomit of the Sea / Fell to the Dutch by just propriety.' It is not just the low-lying terrain that is found wanting, in Marvell's opinion, but the polyglot population that has had the temerity to raise 'their watery Babel' upon it. He is discomfited, too, by the religious pluralism of the place, and especially its relative tolerance of Roman Catholic worship alongside Protestantism – 'Faith, that never could Twins conceive before, / Never so fertile, spawned upon this shore:' – and by the fact that the constant threat of flooding has engendered a spirit of cooperation that has led in turn to the emergence of a true republic, a condition that England was at that moment seeking to emulate. In a reference to The Hague, which did not gain city status until 1806, he adds: 'Nor can civility there want for tillage, / Where wisely for their court they chose a village.'

Marvell's verse served its patriotic purpose – and its author – well, and would be reissued twice during the subsequent Anglo-Dutch wars of the 1660s and 1670s. Meanwhile, the English poet's love–hate relationship with Holland was answered in the play that Dutch writers often made of the words *Engelsen* and *engelen*, characterizing their English former allies as fallen angels. A few polemics were less ambivalent. Vondel, for example, likened London to a 'new Carthage', implying Holland was the civilized Rome.

These were testing times for all the Huygenses, who had to strike a fine balance between old loyalties and the new regime, especially when tensions flared between the Orangists and de Witt's republican government as the war looked like being lost. It must have helped to ensure smooth relations that Christiaan at least was able to maintain personal contact with de Witt because of their shared passion for mathematics. Christiaan thought it advantageous to dedicate his first major treatise on clocks – with its promise of advances to come in marine chronometry and greater glory for the Dutch at sea – to de Witt in 1658, and de Witt, for his part, the following year asked Huygens to read his treatise on the geometry of curves for errors and suggestions.

Christiaan's brother Lodewijk, meanwhile, had been the most closely involved in international politicking. His youthful encounter with Hobbes had come when his father secured him a place as an aide on a special embassy to London led by the poet and former Holland Grand Pensionary Jacob Cats in an ill-fated effort to avert the war. On 12 February 1652, Lodewijk wrote in his journal that he had paid a visit to the renowned philosopher Hobbius who, upon having been exiled from France for the strange notions in the book which he entitled *Leviathan*, has come back to live here again. He is a man of rather over than under sixty years of age and sickly most of the time. He was still dressed in the French manner, however, in trousers with points and boots with white buttons and fashionable tops, wearing besides a long dressing-gown.

They talked happily for more than an hour, Hobbes speaking only English, with Lodewijk interjecting in Latin. Doubtless they discussed the state of their respective nations, but they ranged, too, over the physics of motion and gravity, with Hobbes full of admiration for Christiaan's mathematics.

10. Lodewijk, the third Huygens son. Though less scientifically inclined than his older brothers, he was still able to assist them by sending news of the astronomers and philosophers he met on his missions abroad.

Although the diplomatic effort ended in failure, the trip may have been a personal success for the twenty-year-old Lodewijk, whose father had hoped that it would shock him out of his immaturity and give him a chance to learn English. As a member of a prominent Orangist family who could be presumed to have royalist sympathies, Lodewijk was in a potentially awkward position in Oliver Cromwell's England. He would have had to take care who he spoke to and think carefully about what he said – lessons he was in greater need of than his more thoughtful and responsible brothers.

And the father? How was he to pass the time now that there was no stadholder to serve? There was poetry, of course. He completed his great verse of retreat, *Hofwijck*, at this time, and, assisted by Christiaan, began to organize the contents of a major edition of his poems under the title *Koren-bloemen* ('Cornflowers'). This was one of the first collections of verse to be published in Dutch and it eventually (an enlarged edition was published in 1672) included *Hofwijck*, his other long verses, *Dagh-werck* and *Zee-straet* (concerning his proposal for a grand avenue stretching from The Hague to the sea at Scheveningen, which was later built). Also featured was *Trijntje Cornelis*, 'one of the most uproarious farces of the seventeenth century', about the misadventures of a Holland bargeman's wife when she finds herself in the alien world of Catholic Antwerp. The play was notable for its use of local dialect to comic effect, and was definitely not the sort of thing approved of by the dourer sort of Calvinist.* Bringing the work to more

---

* Huygens suggested that his 'rejected nestling' be reserved for private performance only, and for three centuries it was, until 1950, when its public premiere was given, appropriately in Antwerp, by the National Theatre of

than 1,300 pages were several thousand *sneldichten* – epigrammatic 'quick poems' of four lines or so in a variety of languages.

Although it was his eldest son, named after him, who most closely followed his professional path, it was always Christiaan whose company Constantijn longed for the most. Christiaan was clearly literary enough to be useful in preparing his father's verses for publication, and they enjoyed making music together and could discuss painting and the arts. They were also able to converse on scientific matters, especially optics, which began to revive in Constantijn's interest as he found himself with time on his hands. However, father and son necessarily had different conceptions of what science (then still called natural philosophy) actually was. Their generations fall conveniently either side of the beginning of what is sometimes called the Scientific Revolution, and so it is worth dwelling briefly on the two men's contrasting approaches. For Christiaan, as we have seen, science already had the status of a vocation or profession, to be undertaken with rigour and dedication, even if he did pursue many topics at once. Constantijn, on the other hand, remained 'a dilettante in the best sense of the word'. He had a highly developed curiosity about the physical world, and some sense of which fields it might be rewarding to explore (optics and medicine interested him, but he was properly sceptical about alchemy and astrology). But for the most part he lacked the tools – experimental design, painstaking observation, accurate measurement, mathematical analysis – to answer his own questions, as is revealed in a remarkable episode involving one of the greatest intellectual women of the age.

Belgium (Huygens ed. Hermkens 28, 21).

One of the longer poems that Constantijn Huygens included in *Koren-bloemen* is called *Ooghentroost*, and is addressed to a family friend who had lost the sight of one eye owing to cataracts. The word is translated literally as 'eyes' comfort', but it is also the Dutch name for the plant eyebright or euphrasia, traditionally favoured by herbalists for the relief of eye irritations. This was not the only occasion on which Huygens used this title, and his choice reflects his ever-present preoccupation with his own and others' eyesight. In his case, it was probably a thyroid condition called exophthalmus, which causes tissue to build up behind the eyeballs, that left him with his 'wide-open, large and bulging eyes', and necessitated his wearing spectacles from boyhood. Perhaps it also sparked his interest in optics.

Huygens puts his own poor eyesight on a par with that of his more seriously impaired friend as he itemizes the many kinds of people who might be counted (morally) 'more blind . . . than we': the virtuous and the sinful, misers and prodigals, the happy and the sad, the ambitious and the powerful, and even – more surprisingly and self-mockingly – poets and 'the whole Court'.

Another category of the blind are the learned who see only through their books. This thought leads Huygens into a brief survey of contemporary scientific controversies. After all, one faction at least must be blind when there are both those who believe the Earth goes round the sun and those who believe the opposite. Alchemy, the circulation of the blood, and the mathematical conundrum of squaring the circle (or should one 'round the square'?) get the same treatment. Even the science of vision has its controversial opposition, with some believing 'our eyes are bows / That shoot out rays of light' – a 'vulgar lie', Huygens adds.

Huygens reserves a little of his fire for another surprising group in *Ooghentroost*: 'Painters call I blind . . . / . . . They see but through the palette, / And erect a Nature . . . sweet and pleasant: but do you think to read there / How grandmother Nature really is?' When he wrote this, he was in fact more closely involved with painters than at any other time in his life, as councillor to the House of Orange and as the confidant of Amalia van Solms, who had announced her wish to commemorate her husband Frederik Hendrik with a grand hall of paintings in the Huis ten Bosch.

The room itself, now known as the Oranjezaal, was planned by Jacob van Campen, who had been the architect of Huygens's own house. Among those contributing heroic scenes from Frederik Hendrik's life were Gerard van Honthorst, Jacob Jordaens and Thomas Bosschaert, as well as Adriaen Hanneman, who had painted the Huygens family group. Huygens's 'discovery', Jan Lievens, produced a painting of the Muses, but Rembrandt did not receive a commission. In general, Amalia and van Campen selected the painters on the basis of their stylistic closeness to Rubens, a somewhat old-fashioned preference that neglected the legion of Dutch artists working in new ways, but which was in tune with Huygens's own taste, and more importantly with that of Frederik Hendrik, whose memorial this was.

Other than performing his secretarial function, Huygens himself played little part in furnishing the Oranjezaal. Perhaps it was just as well: he must have quailed when Amalia reappointed Gonzales Coques, the artist who had disgraced himself when he was caught out having subcontracted a previous commission for Frederik Hendrik to another painter.

Christiaan's father often sought distraction in female companionship. The list of Constantijn's women friends in the desolate years after the death of his Sterre would have made a veritable directory of Low Countries pioneers of feminism, had there been call for such a document in the seventeenth century.

The chief object of his poetic attention in the Muiden Circle, Tesselschade Visscher, died in June 1649. He had once marvelled when she came to stay with him in The Hague and slept in the room above his own, kept apart only by 'my cold ceiling, and her cool honour'. But she caused him bitter disappointment when she turned to Catholicism and, although they stayed in touch until the end, he grieved wittily for the spiritual loss of 'Beroemde, maer, eilaes! be Roomde Tesselscha' ('Famed, but, alas, be-Romed Tesselschade').

His favourite singer, Utricia Ogle, the daughter of the English governor of Utrecht, Sir John Ogle, was another lifelong friend (her unusual name is a reference to the Dutch city of her birth). He called her his 'bewitching little bird', playing on the rhyme of the Dutch word for bird, *vogel*, with her maiden name. He delighted at the opportunity for further wordplay in the same vein when she married and acquired the name of Swann, and he became a friend of her husband too.

But unquestionably the most intellectual of Huygens's women friends was Cologne-born Anna Maria van Schurman. Her father had taken the unusual decision that she should be educated alongside her brothers, and she became proficient in more than a dozen languages, as well as in theology, history, geography and mathematics. She also mastered many crafts, including paper-cutting, drawing, wood-carving and embroidery, and achieved particular distinction in the art of engraving on copper. In 1636 she was invited to write a Latin verse for the inauguration of the University of Utrecht, in which she made a point of recording that she herself was nevertheless disqualified from actually studying there. Thereafter, she was permitted to attend some lectures, though only if she sat screened from the gaze of the male students.

11. Anna Maria van Schurman engraved this self-portrait in 1632.
An artist and scholar, proficient in more than a dozen languages,
she was the first woman to study at a Dutch university.

Huygens's acquaintance with van Schurman dates from 1633, when she was twenty-six years old and becoming known for her verse in the leading poetic circles. She sent him a finely detailed gravure self-portrait in which she pretends to look bashfully aside as if seeking approval, while a suppressed smile knowingly suggests she needs no such thing. In front of her, she holds up a cartouche, which reads that, if she has not here made a good job of rendering her own features 'in everlasting copper', then she will not take on the 'more important task' of portraying anybody else. It was in effect a business card. Huygens replied with a dozen lines, teasing 'the handless maid' for hiding from view the instruments of her talent behind the cartouche.

Theirs was a meeting of minds above all. Huygens praised van Schurman's polyglottism – which exceeded even his own – as 'mannelijk', or manly, the same complimentary term that he had once applied to his wife's skills of project management. He asked her to review *Dagh-werck*, his long poem in celebration of domestic life, because he wanted a poet's view as well as a woman's. Huygens did not employ with Anna Maria the familiar, and sometimes risqué, language that he adopted with some of his other women friends. Instead, he addressed her in his Latin correspondence as 'Nobilissima virginum' and 'N virgo', or 'Nobilissima domina, Amica nobilissima'. He valued her great learning and her broad network of contact with leading thinkers across Europe. Her declared celibacy intrigued him, but if he thought this was the price she paid for creative freedom (it was usual for women to give up poetry and other intellectual pursuits upon marriage), then it was one that he ultimately felt was too high. He believed that his own daughter, Susanna, had an intelligence such that 'one could make a Schurman out of her', but he avoided following Anna Maria's father's example when it came to her education.

A number of younger women went some way towards satisfying longings that Anna Maria van Schurman certainly did not. In 1652 Huygens was introduced to Béatrix de Cusance, the well-connected Duchess of Lorraine, who maintained a wide social network that included various royal families. She was another musician, and Huygens wrote keyboard pieces for her to play as well as a number of poems, often rich in double entendre. The most notorious of these is a riddle describing the removal of a corset, in which Huygens contrived the same rhyme to every line in the original French as he described his feelings – 'I love it more than / Amber, civet and musk' – leading up to the closing line's invitation to 'Guess whose is the busk'. Taking a chance, Huygens sent the poem off to her, and got no reaction at first, although later, Béatrix, who was ill-treated by her husband, told him she had been pleased to receive his attention.*

In their various ways, Béatrix, Utricia and Anna Maria were all untouchable ideals. Maria Casembroot, on the other hand, twenty-five years younger than Huygens, lively and beautiful, was able to brighten daily living in more immediate ways. She was the closest Huygens came to finding a true partner after the death of Susanna. Although she, too, was an able musician, Huygens accentuates her more down-to-earth qualities in verses where he plays on her name as *kaas-en-brood*, or bread-and-cheese.

Another young woman friend was Maria van Oosterwijck, the daughter of a church minister who moved to Voorburg when she was six years old. Huygens must have known of her early on, and was doubtless attentive when she began to show promise as a painter. She became extremely successful, and a number of courts in Europe acquired examples of her precise flower still-lifes. Her portrait painted when she was forty-one years old shows her as very beautiful still, with a broad mouth and face and fine black hair, holding her palette and brushes with an open book on her lap.

Whatever the nature of his relationships with these women, Huygens clearly adored female company, and never gave up trying to find it. In 1682, at the age of eighty-six and still going about on the business of the prince (the restored stadholder William III), he rode out one day, like Don Quixote, to Nijenrode Castle near Utrecht in a desperate courtship of one Maria Magdalena

---

* In a further example of this seventeenth-century 'sexting', Huygens wrote barely a month later to Utricia Swann, who had gone to take the waters at a spa. His letter in English rather ungallantly expresses the hope that she will return with 'a paire of gelegentheitjes fatt and plumpe, and such as I suppose they were a dozen yeares since'. The word that translates literally as 'little opportunities' clearly indicates Utricia's breasts (quoted in Joby 231).

Pergens. Sadly for him, beautiful Leen, or Leentje, married another a few months later.

Though his badinage often appears to err on the side of bawdiness – when the Princess of Hohenzollern, Maria Elisabeth, could not play the lute for him because, she said, of the wide collar she was wearing, he parsed her refusal into the thought that 'the princess never played better than when she was undressed' – Huygens was always sincere in his personal interest, and the subjects of his attention understood this. He may have come across as an old goat at times, but he was also a true charmer, and none of his female associates ever severed her connection with him. His attitude of openness and generosity is apparent from the way that these connections were established in the first place. It was Utricia Ogle who introduced him to Anna Maria van Schurman; he introduced Maria Casembroot to van Schurman; he mentions Casembroot admiringly in letters to Utricia and Béatrix; in with an amorous letter to Utricia was also a note for 'her Sibyl', van Schurman; and so on. None of these intersections would have been likely if he had entertained a serious romantic interest in any one individual.

One more woman deserves notice here, for with her Huygens found common ground in the discussion of a scientific matter of intense contemporary interest. On 15 September 1653 Constantijn wrote to his musical friend Utricia Swann in raptures, having been introduced to Margaret Cavendish, the Duchess of Newcastle, whose husband was her cousin: 'I am fallen upon this Lady, by the late lecture of her wonderful Book, whose extravagant atoms kept me from sleeping a great part of last night.' The book was her newly published *Poems and Fancies*, which contained a long section setting out her ideas about atomic theory.

As a young woman, Margaret Lucas served as a lady-in-waiting to Henrietta Maria, the consort of King Charles I, at first in Oxford and then in exile in France during the English Civil War. In Paris, she married the widower William Cavendish, thirty years her senior, and like her from a resolutely royalist family. Now she found herself playing hostess to the city's leading scholars. She met Descartes and mathematicians such as Pierre Gassendi and Gilles de Roberval (who would later become associates of Christiaan Huygens). Although her early education had furnished her with nothing more than the usual ladies' accomplishments, her lively, inquisitive nature clearly appealed to such men. Through her correspondence with Mersenne, she became aware of other intellectuals across Europe, including Evangelista Torricelli, Hobbes and Constantijn Huygens.

Unable to return to England following the execution of Charles I, William and Margaret settled in Antwerp, where they lived in some style in Rubens's old house, holding musical soirées to which other exiles and their new Dutch neighbours came. The sojourn on the continent must have greatly assisted Margaret's development as a writer and thinker. She was still in her twenties in a land where women such as Huygens's friends Anna Maria van Schurman and Tesselschade Visscher were at least not discouraged from learning. Emulating these women of letters, Margaret overcame her lack of formal education and dyslexia, and eventually wrote (with the assistance of a scrivener) many plays, poems, stories and – exceptionally for a woman at the time – several treatises on natural philosophy. She placed on record her regret that it 'was not a woman that invented perspective glasses to pierce into the Moon', and that there was no woman to equal Paracelsus, Galen or Vitruvius. Her most famous work, *The Description of a New World, Called The Blazing-World*, is an early example of utopian science fiction which imaginatively incorporates aspects of her autobiography, and has been acclaimed as a proto-feminist polemic.

Cavendish found it advantageous to cultivate a striking appearance in order to promote her work and became a well-known personage wherever she went. To this end, she designed her own extravagant and sometimes revealing clothes. Her later English nickname, 'mad Madge', may have some connection with the redoubtable battle-axe known as 'dulle Griet' or 'mad Meg' in a sixteenth-century Antwerp play (and subsequent Bruegel painting). This bravado was perhaps a means of compensating for shyness, but it was more likely a strategy to warn those whom she met to be ready to engage with an equally vibrant intellect. Some – especially men – were just distracted, but others successfully decoded the message. Dorothy Osborne, the wife of Sir William Temple, a future English ambassador to the Dutch Republic, wrote of *Poems and Fancies*: 'this book is ten times more extravagant than her dresses'.

Huygens met Cavendish at least twice in Antwerp in 1657 and 1658, and they corresponded long afterwards. As a poet and playwright himself, and a keen follower of new science and philosophy, he became her leading admirer, happy to engage with the range of her ideas. Huygens was able to offer his new friend more than just moral support. Cavendish knew that her ideas needed to be communicated in Latin if they were to reach the greatest audience, but she had no language skills herself. Huygens, who was, of course, accustomed to reading and writing in Latin, became a vital adjutant, producing a Latin index of her work so that scholars in all countries might have access to it. His committed interest is evidence that Cavendish's scientific work was taken seriously in its day, which some modern historians have assumed was not the case.

Their principal scientific dialogue concerned their mutual fascination with the phenomenon that came to be known in

English as Prince Rupert's drops. These were teardrop-shaped glass beads a couple of inches long which exhibited apparently paradoxical properties not seen in other forms of glass. The head of each tadpole-like bead was extremely strong, able to resist a sledgehammer blow. But if the tip of the tail was snapped off, the whole bead would explode, leaving behind only a scattering of sugar-like granules. It made for a dramatic opposition: the fragile beauty that glass objects always possess set against the potential for violent destruction.

Although the earliest drops date from 1625, they became more generally known around 1650 in Holland and Germany, where they were called *Bataafse tranen*, 'Batavian tears' or 'Holland tears'. Their strange behaviour was demonstrated at Henri Louis Habert de Montmor's circle of savants in Paris in 1656 and they quickly spread as a topic of speculation of natural philosophers across Europe. They became a widespread fad, however, and gained their English name when Prince Rupert of the Rhine brought some specimens to his cousin King Charles II and they were subjected to the experiments of the Royal Society. Their momentary fame was captured in the satirical 'Ballad of Gresham College' in 1663:

> And that which makes their Fame ring louder,
> With much adoe they shew'd the King
> To make glasse Buttons turn to powder,
> If off them their tayles you doe but wring.
> How this was donne by soe small force
> Did cost the Colledg a Month's discourse.

On 12 March 1657, Huygens sought to discover Cavendish's opinion concerning 'the natural reason of these wonderful glasses, which, as I told you, Madam, will fly into powder, if one breaks but the least top of their tailes, whereas without that way they are hardly to be broken by any waight or strength'. He can have done no harm to her self-esteem when he added that neither the king of France nor 'the best philosophers of Paris' can explain it. Presuming that she would resort to a practical experiment, he advised her how to proceed to work safely with 'these little innoxious Gunns': 'a servant may hold them close in his fists, and yourselfe can break the little end of their taile without the least danger. But, as I was bold to tell your Ex.^{cie}, I should bee loth to beleeve, any female feare should reigne amongst so much over-masculine wisdom as the world doth admire.'

Cavendish responded with alacrity a week later, offering a detailed argument as to the possible cause of the phenomenon: 'to myne outward sense these glasses doe appeare to have on the head, body or belly a liquid and oyly substance, which may be the oyly spirrits or essences of sulpher'. Sealed inside, the volatile liquid might escape with the force of an explosion when the tail is broken off. She arrived at this conclusion not only from direct observation of her own glass drops, which must have appeared hollow to her, but also, as she candidly explained, from a feminine familiarity with the glass orbs of earrings into which shreds of brightly coloured silk were often placed.

Huygens was not satisfied with this, however. Surely, he argued, if sulphurous liquor is encapsulated in the glass, a flame brought near will cause it to ignite. He had tried this, but it did not happen:

> Madam, I found myself so farre short of my opinion, that firing one of these bottels to the reddest hight of heat, I have not onely seene it without any effect, but also being cooled againe, I have wondred to see all his vertue spent and spoiled, so that I could breake of the whole taile by peeces even to the belly, without any motion more then you would see in an ordinarie peece of glass.

Cavendish replied again, reasoning – less plausibly this time – that the contained liquor might have quenched the fire or leaked away as a vapour.

> Thus, Sir, you may perceive by my argueings, I strive to make my former opinion or sense good, although I doe not binde myselfe to opinions, but truth; and the truth is that though I cannot finde out the truth of the glasses, yet in truth I am
>
> Sr
>> Your humble servant
>> M Newcastle.

This correspondence is notable not so much for its resolution of a scientific puzzle – Cavendish and Huygens do not come close to a full understanding, as we shall see in a moment – but for the ready acquiescence to the Baconian scientific method by two intellectuals who, for all their many accomplishments, can hardly be thought of as 'professional' natural philosophers (as some of their contemporaries, such as Descartes, Robert Boyle and, of course, Christiaan Huygens, certainly were). They switch from hypothesis and experimental test and back like old hands in their search for the truth even when, as Cavendish admits, the truth is that they do not really know what is going on. Cavendish believed, in fact, that the truth cannot be fully known, and that natural philosophers can only speak in terms of probabilities, a position very similar to that which Christiaan would later come to articulate.

Though the Batavian tears were above all a curiosity, there were good reasons for wanting to know why they behaved as they did. Lenses for telescopes and microscopes typically began life as molten glass beads. Here, any irregularity of the glass medium as it solidifies is a cause for concern as it may impair optical quality. Some may have envisaged military uses for the little explosive devices, too. So it was not unnatural that members of the Royal Society among others should have wished to learn more about them.

The presentation given at Gresham College in March 1661 was written up by the Royal Society president, Sir Robert Moray, who had himself lived in exile during the Commonwealth years in the Dutch Republic and was familiar with the country's glassworks. He closely observed five of the comma-shaped drops that Prince Rupert had given to the king, and which the king, out of admirable self-restraint or sheer ennui, had resisted the urge to tweak. Two of them appeared to contain liquid while the remaining three appeared solid. Although those who assembled for the demonstration had no special knowledge of how they had been made, an experimental assistant apparently had little difficulty in forming some additional drops, which were found to behave in the same way as the king's gifts. The exhaustive series of tests included 'detonating' the drops underwater, surrounded by cement, in a fire, and in the vacuum of Boyle's air pump.

It was Robert Hooke who best intuited the principle of their operation. He explained that as the glass suddenly cools upon being dropped into water, its surface forms a hard shell while the interior, slower to solidify, tries to contract, producing a great build-up of tension. (When Constantijn Huygens heated one of the drops during his exchange with Margaret Cavendish and found that it lost its explosive property, he had without knowing it converted the glass from a 'tempered' form like that used today in car windscreens to an annealed form without tension.) Differing rates of thermal contraction of the interior would account for the fact that some drops turn out hollow while others are solid. Hooke made a comparison with an arch to explain the distribution of forces within the curved surface of the drop, and in *Micrographia* included a drawing complete with fracture lines in the glass, which he was able to capture and sketch by triggering a drop to explode while held in place by fish glue.

Some members of the Royal Society did not pause to take in this analysis, or else they simply wished not to spoil a neat party

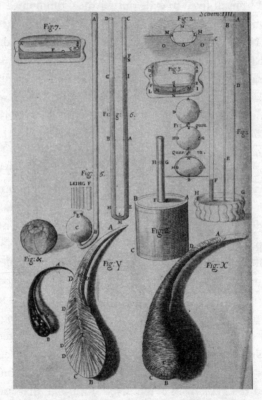

12. Prince Rupert's drops or 'Batavian tears' diagrammed by Robert Hooke in *Micrographia*. Fig. X shows lines of stress captured on the surface of a drop coated in fish glue to prevent it from flying apart, while Fig. Y shows conjectured lines of internal stress.

trick with science.* At a dinner in January 1662, Samuel Pepys witnessed a repeat demonstration of 'the Chymicall glasses, which break all to dust by breaking off the little small end – which is a great mystery to me'.

Pepys was all eyes, however, when Mad Madge paid a visit to the Royal Society a few years later. She was shown experiments related to light, magnetism and chemical solvents. Optics was another topic of discussion, Cavendish having disparaged the efforts of microscopists including Hooke, whose instruments she believed could reveal only a distorted picture of nature. It was the first visit by a woman to the society and exceptionally well attended, for Cavendish was by then a major celebrity. Pepys had recorded earlier in the month how he had seen her spectacular black and silver coach pursued through the park by other coaches and accompanied by '100 boys and girls running looking upon her'. But unlike Constantijn Huygens, he remained quite blind to her merit as a scientific intellectual. On the day of her visit to the Royal Society, 30 May 1667 (OS), he wrote in his diary: 'The Duchesse hath been a good, comely woman; but her dress so antic, and her deportment so unordinary, that I do not like her at all, nor did I hear her say anything that was worth hearing, but that she was full of admiration, all admiration.'

---

* The drops are still the subject of contemporary research. See, for example, H. Aben et al., 'On the Extraordinary Strength of Prince Rupert's Drops', in *Applied Physics Letters*, vol. 109, 231903 (2016).

7

SATURN

In the early spring of 1655, most likely in the garden of the family house on the Plein, Christiaan Huygens set up a telescope he had made with his brother Constantijn – it comprised a twelve-foot tube fitted with a planoconvex objective lens and a simple eyepiece that gave overall fifty-fold magnification – and turned it towards Saturn.

In the seventeenth century, Saturn was the remotest known planet in the solar system, which was unchanged from the one studied by classical astronomers except in one important respect. The identification of four moons orbiting Jupiter by Galileo in 1610 raised the possibility that there was more to be learned. Saturn was the new puzzle, as its size and brightness appeared to vary in a most unplanetlike way, and it now became the focus of attention for many astronomers. It was perhaps all the incentive the Huygens brothers needed to keep on striving to improve their telescopes.

During the frosty nights – the few that were clear – Christiaan carefully observed the planet's changes, making many sketches, some perhaps done with the assistance of magic-lantern projections. On 25 March, he detected what seemed to be a satellite as well as another body close to the planet, which he labelled 'étoile b.'. Surprisingly, he did not measure the exact positions of these bodies, but he did take enough readings to conclude that 'étoile b.', which he was unable to find again after a few days' break owing to cloud cover, was 'errant'. The true satellite, which he observed to lie three minutes of arc away from its parent planet, had a fixed orbital period that he measured to be fifteen days, twenty-two hours and thirty-nine minutes. He continued his observations through April and May in order to confirm that he had indeed identified the first new body in the solar system since Galileo's 'Medicean stars'.

He communicated his news to the English mathematician John Wallis in a letter mostly concerned with his analysis of geometric curves. Referring only in the briefest general terms to his use of the telescope, he adopted the form of an anagram, then a customary means of disclosing that one had made an important discovery while not yet revealing its exact nature.* He formulated his anagram such that it would be virtually impossible to solve by using a quotation from the Roman poet Ovid, and wrote it out in clear capitals along with a jumble of left-over letters: ADMOVERE OCVLIS DISTANTIA SIDERA NOSTRIS, VVVVVVVCC CRRHNBQX. Suitably rearranged, these letters decode as 'Saturno luna sua circunducitur diebus sexdecim, horis quatuor', which would, whenever Huygens chose to explain it, be enough to establish his priority in discovering this first moon of Saturn, which, he now added, having refined his earlier measurements, circles its planet in sixteen days and four hours.

---

* Anagrams were also something of a Huygens family obsession. Christiaan's father sometimes amused himself by working out the number of permutations of the letters of people's names, and finding character-revealing anagrams for them. His demanding charge Amalia van Solms (Amelie de Solms in courtly French) became 'sommeille d'aise' and 'sale de sommeil' ('comfortably dozy'; 'dishevelled from sleep'). His own name, Constantinus Hugenius, he transformed into 'Continuus, haut segnis' ('enduring, not at all slothful') (KA 48 ff.240–1).

1. *The Spectacles Seller*, by Rembrandt, c. 1624. The painting was one of a series of allegories on the five senses, which are his earliest known works.

2. Constantijn Huygens, by Jan Lievens, c. 1628–29. Huygens marvelled at the raw talent he identified in the work of the young Lievens and Rembrandt.

3. *Muiden Castle in Winter*, by Jan van Beerstraaten, 1658.
Pieter Corneliszoon Hooft hosted meetings of the Muiden Circle
of Dutch poets and artists here during the 1630s and 1640s.

4. Jan Lievens self-portrait,
c. 1629–30. He shared a studio
in Leiden with Rembrandt
for a number of years.

5. Rembrandt self-portrait, 1629. He and Lievens ignored Huygens's advice to learn from the Italian masters.

6. Frederik Hendrik, Prince of Orange. Stadholder, 1625–47, he employed the elder Constantijn Huygens as secretary.

7. *The Blinding of Samson*, by Rembrandt, 1636. This painting may have been given to Constantijn Huygens in gratitude for the artist's commissions for the Dutch court.

8. René Descartes, by Frans Hals, c. 1649, shortly before the French philosopher left Holland for Stockholm, where he died the next year.

9. *Constantijn Huygens and his Clerk*, by Thomas de Keyser, 1627. The painting, suggestive of his good prospects, was made just before his marriage to Susanna van Baerle.

10. Constantijn Huygens and his wife painted by the architect of their house, Jacob van Campen, c. 1635. Between them, they hold a sheet of music, symbolizing marital harmony.

11. Hofwijck, the Huygens family's retreat a few miles from The Hague, sketched by Christiaan Huygens, 1658.

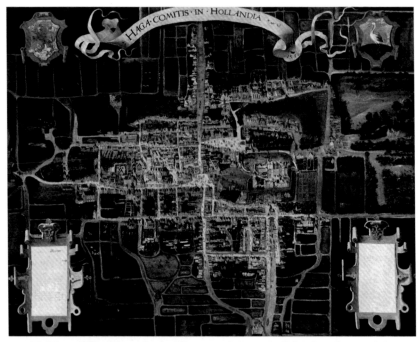

12. Map of The Hague in the Huygenses' time: city life is based around the Binnenhof, with the Hofvijver lake (centre right), and Sint-Jacobskerk (centre left). The Lange Voorhout is the residential area surrounded by trees above the lake. Het Plein, not yet developed, is the tilled field to the right of the Binnenhof complex.

13. The Huygens family, by Adriaen Hanneman, 1640. Surrounding their father clockwise from the top are Susanna, Christiaan, Philips, Lodewijk and the young Constantijn.

14. William II of Orange and Mary Henrietta Stuart. William II succeeded his father Frederik Hendrik as Stadholder for three years until his early death in 1650.

15. Johan de Witt. He led the government of the Dutch Republic during most of the Stadholderless Period, 1650–72.

16. Amalia van Solms, wife of Stadholder Frederik Hendrik. Widowed in 1650, she relied heavily on Constantijn Huygens for his counsel and secretarial services.

17. Margaret Cavendish, Duchess of Newcastle, by Peter Lely, 1665. One of the liveliest minds of the age, she corresponded with Constantijn Huygens on topics of scientific curiosity.

13. Huygens's anagram of March 1655 concealing news
of his discovery of the first moon of Saturn.

Although he did not name the moon – it was called Titan by
John Herschel more than a century later – Christiaan was thrilled
with his discovery. He picked up his diamond-point and along
the glass rim of the objective lens through which he had seen it
he scratched the Ovidian part of the anagram, marking forever
the lens that had 'brought the distant stars to our eyes'.

This was the secret that Christiaan carried with him when, a few
weeks later, he arrived in Paris for the first time, travelling with
Lodewijk on their father's instructions. On 23 July, Christiaan
wrote a short letter to his older brother describing his first impressions of the place, which Constantijn had visited on his Grand
Tour five years before:

It is 9 days that we have been in this city, which time, if you knew how we used it, I guarantee that compared to us Gargantua would seem to you but a lazy-bones. I will not mention the beautiful and magnificent things we have seen, because it would be nothing new for you. What I find most pleasant is the garden of Bagnolet which we saw yesterday coming back from the Bois de Boulogne, I don't know if you have been. Our counsel in ordinary is A. Tassin, who comes to find us regularly each morning. I find him a great braggart and greatly irate, just as you had described him, even to his voice which I think I could put into notes of Music. I have as yet hardly been to find the people of letters or music, and I am just running about the streets with the 2 others of my companions. We went together to Ambassador Boreel's, who holds forth with seriousness. Afterwards at Monsieur Bracet's, and we also saw Mademoiselle his daughter, whom I found highly knowledgeable about everything, even the least, that happens in Holland, and the rest who were a little less beautiful than in the past, but nevertheless in very good humour as always. We have only been to the Cours once, and once to the Comédie, which is hardly any more fine than that of His Highness at The Hague, and costs an écu per head. Everything is expensive here, but mainly I foresee that we shall incur pretty costs in carriages. I have accommodation on my own, in a room that is mostly carpeted: there is an attic above where the rats and mice often come to keep me company. Apart from that the bugs bother me at night so much that I carry the marks on my hands and forehead. Perhaps you would hardly wish to be in my place.

Their diplomat father, no doubt, saw Christiaan and Lodewijk's advent in Paris as the latest thread in his Europe-wide family web. When their elder brother saw the city, he had then travelled on via Orange in Provence (the city and its surroundings being a feudal principality in Dutch hands since William the Silent inherited it in 1544) and Geneva to the Alps and Italy, having had to abort an earlier trip to London when, four days after his arrival there, King Charles I was beheaded. Christiaan had already been north on his disappointing trip to Holstein and Denmark. In 1652 Lodewijk made his more successful tour of England, notwithstanding the outbreak of the first Anglo-Dutch War, and he later joined an embassy to Madrid. The youngest brother, Philips, would later travel east on a similar mission along the Baltic coast to Danzig; after a sudden illness, he died at Malbork on 14 May 1657 at the age of twenty-three.

To the twenty-six-year-old Christiaan, however, the big city was a revelation. He would pay two more short visits there, in 1660 and 1663, before making it his home for sixteen years in 1666. His ever more extended visits would eventually see him spend half his productive career in the French capital.

Wishing to avoid a difficult overland journey though the Spanish-held southern Netherlands, Christiaan and Lodewijk had sailed from Brielle, accompanied by their cousin Philips Doublet and another friend, passing between Dover and Calais, 'which we saw both very distinctly, the coasts being very high on both sides, particularly the English coast which appeared completely white'. The party disembarked at Dieppe and then rode through the beautifully cultivated countryside to Paris, where they found their lodgings with friends of their father in the rue de Seine in Faubourg Saint-Germain, a Protestant area and a popular meeting-place for Flemish, Dutch and Germans. They passed the next few weeks of the summer touring the great houses round Paris.

Fontainebleau provided a highlight, with its terraces and fountains and, indoors, paintings by Titian, a Raphael Nativity 'estimated at 50 thousand écus', and Leonardo da Vinci's *La Gioconda*.

Christiaan stayed on until November. He was interested in everything. He met many of the mathematicians and natural philosophers who would in due course become his correspondents, colleagues and friends. Chief among these were Gilles de Roberval, Ismaël Boulliau, Pierre de Carcavi, Melchisédech Thévenot and Claude Mylon. He was lucky also to meet the older and weightier figure of the mathematician and astronomer Pierre Gassendi, who was the first to observe the transit of a planet (Mercury) across the sun and who died that October while Huygens was still in Paris.

He visited a Parisian lens-maker and had discussions with the Dutch ambassador Willem Boreel, Middelburg-born and an enthusiast for the latest in telescopes and microscopes. It is likely that he visited the leading clock-makers, perhaps including Isaac Thuret, with whom he would later collaborate and then fall out over the design of a more accurate pendulum mechanism. He also tracked down a number of old musical acquaintances of his father, although possibly more memorable for him were encounters with the courtesan and patron of the arts Ninon de l'Enclos and the celebrated Italian soprano Anna Bergerotti.

In September he travelled along the Loire to collect the doctoral degree in law he had purchased from Angers, one of the few French universities still to admit Protestants. (The selling of degrees would later become a scandal, but for now it was standard practice.) Christiaan's older brother, meanwhile, was spending most of his time at home in The Hague in an effort to avoid an epidemic of the plague. The brothers generally corresponded by weekly letters, although not every one arrived, and often one would chide the other for his apparent silence. Their discussion tacked easily between technical matters to do with their scientific observations and the more conventional concerns of young men. On one occasion, for example, Constantijn asked Christiaan for

> some new, light little songs of the kind that girls like most, because it is for this kind of creature that I ask them, since about a month ago I made the acquaintance of some English misses . . . If brother Lodewijk were a man of music I am sure he would be delighted to oblige the English beauties. Joking aside, if you can send me some I would be most grateful.

Constantijn was not completely distracted, though. He was also working on a twenty-foot telescope, and comparing glass samples sent by various opticians which he judged to be no better than 'our convex glasses'. Often in their astronomical endeavours the brothers were to find that the lenses they made for themselves were superior to any that could be supplied. And, besides, the manual labour involved in grinding the glass was companionable work. 'Making the lenses, we will have something to keep us both busy,' Constantijn wrote, anticipating Christiaan's return to Holland.

Christiaan replied to his brother's musical request:

> I have new tunes aplenty, but not those you need . . . You promised me a drawing of Jupiter with its Zones, I beg you to send me it, so that I can confer with Mr. Boulliau. The day before yesterday he showed me his great glass that the Duke of Florence sent him, which seems good enough to me, but exposes too little at a time, as being of 10 feet, having at the eye only a concave lens.

Despite bad weather in Holland, Constantijn had managed to observe the equatorial stripe lying across the Jovian planet, 'in the middle or a little above, a pale line and on each side of it two further black lines which bound it', and promised to send a drawing of his observation for comparison with Christiaan's sketches.

A few weeks later, Christiaan dispatched his own drawing of Jupiter along with words and music for two songs, plus another 'which is completely new'. The (French) lyrics of the latter ran:

> It is true, I confess it, your beauty holds my soul in
>   captivity,
> Alas if I could smack your bottom I would know liberty,
> And if you wished to rest upon my couch to satisfy my
>   desire,
> I would kiss a hundred times your eyes and mouth and
>   die of pleasure.

Christiaan appended that he had made the song 'for la Méneville, the prettiest of the Queen's maids of honour'.

But it was Christiaan's revelatory observations of Saturn that most greatly titillated his new friends in Paris. Although Huygens would not be ready to announce his discovery of its moon in a formal way for another year, he compared notes with the astronomers he met in France, and together they speculated about the mysterious nature of the bright bodies that seemed to hover about the planet. These discussions were sufficient to sow some confusion and jealousy later on, although the precautionary anagram that Huygens had communicated to Wallis, which he also brought to the attention of his Parisian associates, would fortunately be enough to secure his claim to the discovery of the planet's first satellite.

Huygens left Paris with happy memories, many new contacts and a head full of mathematical puzzles. The trip had changed him for ever. On 15 October, knowing that he was shortly being recalled home, he wrote to his father:

> For the rest of this sojourn in Paris, I believe that the
> longer I stay here the more I will gain from conversation
> with the people of this nation. But I do not know how it
> will be taken if I return quite altered from when I left. As
> for me, I do not notice this change, which is perhaps the
> same as when one grows in height.

In addition, he had fallen for one Marie Perriquet, and clearly hoped to keep in touch, sending her gifts and regularly asking that his Parisian correspondents pass his greetings on to her. His interest was encouraged by Valentin Conrart, one of the founders of the Académie Française, who wrote to Huygens in January 1656:

> She speaks to me very often, & most worthily of you, &
> the present that you gave her has greatly satisfied her. If
> you had stayed longer here, you would have seen, in this
> rare person, wonders which it would be difficult to find
> elsewhere, at least in persons of her sex. I do hope the
> curiosity of discovering them might oblige you to make a
> second voyage.

But it gradually became clear that Marie was not quite what she seemed, and Roberval, the most reliable of the friends he had made in Paris, gently advised him that he had been 'led skilfully by the nose' by Conrart, 'who has professed a complete friendship [amitié entiere] with P.', as had a number of others, all 'virtuous old men, long married'.

Of more lasting value was the impression Huygens made on another senior academician, the poet and critic Jean Chapelain, who admired not only the Dutchman's scientific acumen and his fine manners, but also his perfect command of French.

> I was afraid at first to reply to you in my own tongue
> when I read what you had written in that language, in a
> style so pure, so unhindered and so unusual even among
> we French. It seemed that you set me a dangerous test,
> and that were my sentences ever compared with yours, it
> could be who you is taken for the native and me for the
> Foreigner.

A couple of years later, Chapelain begged Huygens to be allowed to be the first to communicate 'the beautiful things that your noble Genius and the great lights of your Studies have given you for the ornament of the World and the instruction of human Kind'. In fact, it was Chapelain who was in a position to grant favours, as Huygens was no doubt aware. Chapelain's sustained flow of flattering compliments throughout the late 1650s must have warmed Huygens with a strong sense of the welcome he would receive if he were to return to Paris. But he also prepared the ground for that day in a practical manner, mentioning Huygens in his absence to Henri Louis Habert de Montmor.

Montmor was a senior aristocrat and government official who had long hosted an intellectual salon at his house on the rue du Temple. Following the death of Gassendi in 1655, he had become a champion of Cartesian thought, and the focus of the meetings turned increasingly to the sciences. Among the regulars at the gatherings were Gilles de Roberval, the physicist Jacques Rohault, the physician Samuel de Sorbière, the astronomer Adrien Auzout

14. Jean Chapelain, poet and one of the founders of the Académie Française, who became Christiaan Huygens's greatest champion in France.

and the engineer Pierre Petit, as well as Huygens's most important Paris correspondent, Chapelain. Montmor displayed the curious instruments and machines that he owned for the delight of his visitors, and together he and his guests discussed all kinds of unexplained phenomena, sometimes making experimental investigations, as on the occasion when Prince Rupert's drops were demonstrated.

Montmor wrote back immediately to Chapelain: 'I have no doubt that your Friend, who has already had the good fortune to have discovered the Moon [of Saturn], still has the advantage of finding the reasons for this Phenomenon which I have great impatience to know.' Montmor spoke for many of the savants in Paris, and his praise made Huygens feel 'most glorious'.

With the longer winter nights back in The Hague, Christiaan was able to give more time to observing Saturn again. His earlier drawing of the planet had shown luminous extensions from either side like long ears. But now these inexplicable features had almost disappeared. (Their disappearance had made the sky around Saturn darker, which had helped Huygens detect its moon.) The true form of the overall planet remained a puzzle. Any complete explanation had to account for this apparently impossible changing morphology.

Many theories had been advanced in the few years since telescopes had revealed the planet's strange truth. Could it be that two large, diametrically opposed moons occupied an orbit very close to the planet? This would explain the elongated shape sometimes observed (when the supposed moons were blurrily visible on either side of the planet) and the reversion to a simple disc (when the moons were eclipsed). Or were there in fact lobes of some kind attached? A drawing that Wallis sent to Huygens showed looping 'handles' bigger than the planet itself. Or was Saturn itself ellipsoidal, which was Gassendi's favoured hypothesis? Roberval suggested the planet might have an equatorial 'torrid Zone', which from time to time ejected giant flares like those that had been seen coming from the sun. Christopher Wren was closest in thinking to Huygens, believing that an attached thin corona might rotate with the planet.

Huygens argued, based on the gradual way that the observed shape changed, that the planet overall must be a symmetrical body, and that its orbit and rotation must be so as to present different aspects at different times to observers on Earth. The symmetrical feature he favoured was a ring tilted slightly from the plane of the ecliptic (the plane containing the sun and the Earth's orbit). Such a ring would appear bright around the planet when most greatly tilted from the ecliptic. But it would virtually disappear from view

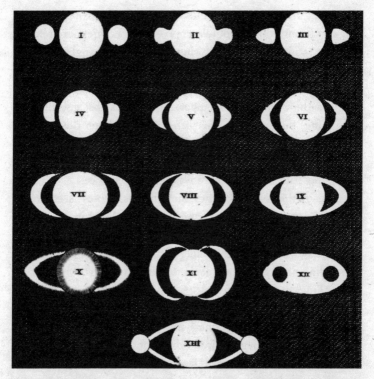

15. Interpretations of the form of Saturn as seen through early telescopes, compiled in Huygens's *Systema Saturnium*. I is Galileo's proposal of 1610; others show later proposals, including those of Gassendi (XII) and Hevelius (IV–VII).

either when it fell in line with this plane or when it fell edge-on to the sun and little light was reflected from it. In March 1656 Huygens communicated his idea, once again in anagram form (an insoluble alphabetical ordering of the letters of the message this time; no Ovid), in a short paper, *De Saturni Luna Observatio Nova*, and sent sketches of the proposed planetary ring to Roberval and other colleagues in Paris. Huygens's working sketches had begun with crescent-like shapes touching each side of the planet like ears or wings, but as the quality of his observations improved – thanks to a new twenty-three-foot telescope – they became more refined, showing Saturn and a discrete ring from different angles. In these sketches, the swollen appearance that had led people to think of ears and lobes was gone, and the shape was refined to what was clearly a thin circular belt (appearing elliptical because of foreshortening) surrounding the planet. Throughout this process, Huygens was careful to remain faithful to what he saw through the telescope, and did not allow this mathematically ideal shape to interpose itself before there was proper evidence.

One of the many sketches in Huygens's notes is especially beautiful. It shows the planet and its ring outlined in ink against the void indicated by a grey wash of watercolour. The annulus is a little less wide than we see in photographs today, and there is no shading or anything about the outline to indicate whether we are looking at the ring from the 'top' or the 'bottom'. Without this perspective, the ring could still be some odd feature attached to the planet rather than one lying around it at a distance, but for the fact that Huygens draws the edges of the ring very carefully as perfect ellipses, which leads the reader to the expectation that a pure geometry must be present.

Huygens's proposition was met with scepticism at first, but his freshly made discovery of Saturn's moon already testified to the superior quality of his observations. Furthermore, a ring that appeared and disappeared according to a regular periodic cycle was consistent not only with recent observations of the planet, but also with sightings going back over decades by Galileo and others, as Huygens judiciously pointed out. Huygens here asserts himself as a model scientist, combining observation and a hypothesis based on the logic of Occam's razor as well as on mathematical principles to produce a finding open to verification (or falsification) at any time in the future.

He was not correct in every detail, however. He estimated that the ring had a thickness about one fortieth of its width and that the edge must be dark in order for it to disappear from view at certain times, although we now know that the ring is simply vanishingly thin. This possibility for some reason Huygens explicitly ruled out: 'That the whole body of the ring is so thin that, although its edge is in fact shining, it remains nevertheless imperceptible to our telescopes because of its too small width? Not so.' Surprisingly, he also speculated on 'the effects that the ring that surrounds them must have on those who inhabit' Saturn. Towards the end of his life, in the 1690s, Huygens was to produce a comprehensive volume on the possibility of extraterrestrial life, which became and remained for long afterwards the most scientifically informed examination of this questionable topic.

Among the astronomers around Europe with whom Huygens communicated at this time was Johannes Hevelius, who had built his own observatory above the banks of the Vistula in Danzig. Here he developed telescopes of extraordinarily long focal length, which enabled him to make a detailed mapping of the Moon. He had rather different ideas about Saturn, which included an ellipsoidal planetary body with crescents at the sides, but this hypothesis was based on observations made using a telescope that Huygens was confident was not even capable of detecting Saturn's moon, and did not explain the observed phase changes of the planet.

Huygens was able to compare notes with Hevelius during the summer of 1656, not only by writing to him but also thanks to his brother Philips, who was in the city on an embassy mission (the Dutch were shortly to send a large fleet to Danzig to contain Swedish ambitions in the Baltic, and to ensure the port remained open to the grain trade on which they depended). In May, Philips reported that he had had an enjoyable meeting with the German astronomer. He was impressed with the professional tidiness of his workshop, but was unable to see anything through Hevelius's telescopes because the weather was overcast. Nevertheless, he was able to communicate important technical intelligence to Christiaan about his methods of lens-grinding: 'he grinds in red copper dishes which are a good 2 times as big as yours and he insisted that the size of the dishes contributes much to the perfection of the lenses . . . He is now busy grinding Hyperbolics and had already made a small one.' He added a familial postscript: 'Get Sus to write me some news. I will bring her a Polish sable or something.'

A few weeks later, Philips moved on to the former Teutonic stronghold of Marienburg or Malbork. His next letter to his brother resumed the description of Hevelius's optical innovations, using Latin for the purpose, as scholars were accustomed to do for classical topics such as astronomy and mathematics, for which there was an established vocabulary in the language. Occasionally, though, words did not exist. Philips was forced to invent a dog-Latin Dutch phrase, 'verrekyckatorum Slypatores', to describe the grinders of telescope lenses. Then, suddenly, in the midst of this optical discussion in Latin, he slips into Dutch for a startling anecdote about a young man of Danzig, 'heer Salomons, a trader: who until around his 17th or 18th year passed for a young woman, so that some others I know there have told me that they have often seen him going to weddings and other occasions always as a young woman, Flora Salomons'. Philips goes on – now in humorous verse – to relate how this Flora developed swellings like a male member, which caused the mother to cry out with shock. Flora Salomons apparently then took a ship to Amsterdam and was treated by Nicolaes Tulp, the illustrious physician and surgeon portrayed in Rembrandt's celebrated early painting *The Anatomy Lesson of Dr Tulp*. 'The same being a great artist, he gave a snip or two and everything came all right.' Trousered from then on, Salomons spent a year or two in Paris before returning to Danzig, where Philips went around a few times 'with him or her': 'Now he's called Floris and has a little beard, is around twenty-eight years old and cannot be kept off the girls. [He] is a very small man and most unabashed. Mais à nos moutons.'

And with that, Philips reverts to Latin once more in order to shepherd his further thoughts about Hevelius's lenses and the woolly shapes people claimed to have seen through them in Danzig. Hevelius had observed Saturn for more than twenty years. In fact it was he who first observed its phases and measured their period (a contribution Huygens would fail to acknowledge when it came time for his own treatise). That June, he rushed out his own paper, containing this data together with his erroneous deduction about the nature of the planet and the observation of a prick of light close by it, which he took to be a star.

It is not known what Christiaan made of Philips's surprising story as no reply survives. The next time his brother is mentioned is when Christiaan writes to an uncle with the news that he was dead. 'We are told that his sickness was a bloody diarrhoea along with a constant fever that snatched him away in the space of 7 days complaining of nothing else than that he would not be able to say a last adieu to his Father and close friends.'

The form and even the existence of Saturn's ring continued to be contested for some years before Huygens's interpretation prevailed. The French astronomer Ismaël Boulliau, who maintained a lively correspondence with Hevelius, called on Huygens in The Hague in 1657. The Dutchman encouraged him to look through his telescope and then begged him 'not to communicate to anyone what you know about the Saturnian world'. Only in

March 1658 did Huygens finally write to his most loyal confidant Jean Chapelain, giving him permission to reveal to Montmor's academy the solution to the anagram he had distributed two years before. By then he had sufficient supplementary observations of Saturn in different aspects to dispel any lingering personal doubts, and he could begin to prepare a detailed treatise on the planet. But the best of Huygens's rivals in astronomy knew immediately that he had cracked the mystery. Wren surrendered graciously in 1658: 'when . . . the Hypothesis of Hugenius was sent over in writing, I confesse I was so fond of the neatenesse of it & the naturall simplicity of the contrivance, agreing so well with the Physicall causes of the heavenly bodies, that I loved the invention beyond my owne'.

All new science happens at the limit of resolution: it is about seeing and understanding what has never been seen and understood before. But discoveries do not come from nothing, nor do they manifest themselves all at once in their entirety.

Today, the most powerful optical telescopes – using reflectors rather than lenses, for various practical reasons – are focused on obtaining visual images of some the first exoplanets, planets in orbit around stars far beyond the edge of the solar system. But in the middle of the seventeenth century, it was Saturn that lay at the limit. When Galileo identified four Jovian satellites in 1610, he naturally turned his telescope on Saturn as well. He saw the ring but could only resolve it into two bright blurs on either side of the planet, each about a third of its diameter. He was puzzled when these features disappeared in 1612, and puzzled again by their reappearance some time later. Other astronomers observed the 'ring' too. But they did not draw the correct conclusion from what they saw. Nevertheless, the 'ears' and 'handles' and 'lobes', the close-lying 'moons' and the 'flares' and 'coronas' that they documented were what they truly believed to be there, and they were being no less honest than Huygens when they committed these shapes to paper.*

Huygens knew from direct experience that a weaker telescope, or one with less perfect lenses, would not reveal the ring. It would show blurry lobes or spots. He had seen these forms many times himself. Yet who was to say his lenses were superior now? The Italians in particular, believing they had the best telescopes, were especially frustrated at their inability to see what Huygens had seen. 'Perhaps the Dutch sky was different,' was their mocking complaint.

Ultimately, Huygens's result would require independent confirmation, which would have to come from telescopes at least as good as his own. In the meantime, he had other important arguments in his favour, in particular the rigour of his theoretical reasoning based on his expertise in mathematics and geometry. According to this logic, whatever it was that accompanied Saturn had to be symmetric about the planet's axis, since no irregularity had been observed in its period of rotation. This immediately ruled out oddities such as handles. Huygens's ring of Saturn was a mathematical deduction as well as a scientific observation.

Huygens's thinking was undoubtedly also guided by his considerable skill as a draughtsman. He had the necessary visual understanding of how light falls in relation to solid forms, and

---

* A clue to the fact that the limit of resolution is always apt to be pushed further lies in my referring here to the 'ring' of Saturn. Nineteen years after Huygens's discovery, better telescopes enabled Giovanni Cassini to observe that the ring was in fact split in two, hence the 'rings of Saturn' we think of today, although close-up images captured by spacecraft now show that this label too is a gross simplification.

the angles and shadows it produces as it does so, to imagine – in effect to deduce – the one geometric shape that could account for what he saw through his telescope. The challenge of working at the optical limit of resolution, according to the art historian and expert on visualization Martin Kemp, is to 'translate the seen patterns of lights and darks into a coherent, three-dimensional image *with reference to known forms*' (italics added). A ring, though unexpected and without precedent in the solar system, is just such a known form. It was this mathematical and visual logic that sustained Huygens's vision until enough other astronomers, using more powerful telescopes, were able to satisfy themselves of its veracity by their own direct observations.

For many, however, there remained powerful reasons to reject Huygens's discovery. First of all, it challenged the accepted idea inherited from Greek philosophers that the solar system consisted exclusively of perfect spherical bodies occupying ideal circular orbits in relation to one another. (The asteroid belt, a further disruption to this surview, was not discovered until much later.) These ideal heavens were also part of Christian doctrine. Second, Huygens interpreted the ring as offering support for the Copernican solar system. Inclined by more than twenty degrees to the ecliptic, it gave its parent planet something important in common with Earth, which is similarly inclined; it made the two planets in a sense equal, with an equivalence that could not prevail if the Earth lay at the centre of the solar system, but only if all the planets were in orbit around the sun. This sort of reasoning was still anathema to Huygens's Catholic rivals such as Eustachio Divini in Rome. Third, Saturn's indistinct and constantly changing ring was suddenly the only hazy object in a solar system whose denizens had always appeared as crisply defined discs and crescents. To some, such as the English clergyman Joseph Glanvill, an avid enthusiast for the new science but a fierce opponent of materialism, acceptance of such a ring based on evidence of the senses was indicative of sloppy thinking that was bound to lead natural philosophers into error, so that they would proceed 'no further towards Science, then to imperfect guesses, and timerous supposals'.*

Huygens published his full analysis of the planet in 1659 under the title *Systema Saturnium*, its appearance slightly delayed by the hope – unfulfilled – of getting some better observations and by his other work on clocks. He dedicated the work to Prince Leopoldo de' Medici, who was the patron of the Accademia del Cimento in Florence, which had supported the work of Huygens's most illustrious forebear, Galileo. Perhaps the Dutchman's choice of dedicatee indicated a wish to see such an organization closer to home. But he also hoped for a more tangible gain – a copy of the drawing of the clock pendulum mechanism that Galileo was supposed to have left with his patron.

Systema Saturnium was printed by the Elzevir family in Leiden, which had printed Galileo's work after it was banned in Rome. Following the fulsome dedication came a prefatory poem by Christiaan's brother Constantijn in tribute to their astronomical work together. In his own introduction, Christiaan provocatively set the scene for his planetary discoveries by claiming the invention of the telescope for the Low Countries, a claim that could realistically be contested only by Italy.

The work's chief glory lay in Huygens's drawings and diagrams

---

* The anomaly of Saturn soon came to be regarded more positively. Following William Herschel's discovery of the next outer planet, Uranus, in 1781, Joseph Banks wrote in excited anticipation of 'what new rings, new satellites, or what other nameless and numberless phenomena' it might reveal (quoted in Holmes 105).

of Saturn. In one diagram in particular, Huygens showed the planet at sixteen different positions in its orbit around the sun, revealing its appearance from 'above', with part of the ring hidden behind the planet above the equator and the visible part crossing the lower hemisphere, from 'below', with the hidden part of the ring below the equator, and from side on, in which case the ring all but disappears. Alongside the diagrammatic representation of the planet in each of its phases, Huygens included an artistic sketch to show how it would appear from Earth. This visualization elegantly explained the full range of different observations of the planet and offered a persuasive rationale for its apparent variability in the night sky. Because the plane of the ring is inclined to the ecliptic, Huygens wrote: 'It necessarily follows that the same ring in these different aspects must appear to us sometimes as a more or less large ellipse, other times as a narrower ellipse, and finally sometimes also as a straight line.'

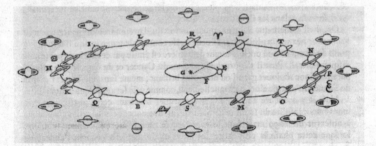

16. Huygens's diagram of Saturn in sixteen phases, with an impression of the view from Earth in each case, showing how a ring could explain the different apparent shapes of the planet.

The real power of Huygens's interpretation was its ability to explain those times when Saturn's 'handles' simply disappeared from view, as they had done in 1642, finally defeating the aged

Galileo's attempt to understand the planet, and again in 1656. Saturn's period of orbit of about twenty-nine years means that the ring will appear exactly side on and be virtually lost to view every fourteen and a half years. This information and other knowledge of the solar system gave Huygens all he needed to construct a calendar detailing when the ring would appear at its brightest (1663–4, 1678–9) and when it would next disappear from view (1671). Huygens was to be among the many astronomers taking up position at their telescopes on those future dates.

Huygens sent copies of *Systema Saturnium* to some sixty colleagues across Europe. Jean Chapelain was among the first to respond, with his customary effusion of praise: 'I received it the day before yesterday in the night and yesterday I read it avidly and attentively, ceasing everything else, not only with satisfaction but still more with admiration . . . I found exquisite invention, judicious sequence and solid doctrine.' The 'bizarrerie of the ring' suddenly made sense to him in Huygens's explication.

This was flattering, but Huygens doubtless set greater store by the response of more expert astronomers, and this was a little slower in coming. Ismaël Boulliau stalled for time: 'I have begun to read your System of Saturn [and] when I have read it all I will tell you honestly what I think of it.' When he had done so, he told Huygens: 'You set out your hypothesis very well . . . nevertheless you still need some experiments to demonstrate absolutely that which you posit.' The French astronomer's chief concern was that the ring – which Huygens had explicitly stated was separated from the body of the planet by a distance a little greater than the width of the ring itself – was simply left hanging, unsupported as it were, in space. Having fought off a 'furious headache', Huygens wrote back confidently to Boulliau in December 1659: 'The greatest proof of the truth of my hypothesis will be when it is found that my predictions about the circular phase agree with observations in the year 1671 and 72. To do which I hope that you and I will have sufficient life.'

Reading the letter, Boulliau may have regretted that he had once spoken on the subject of prognostications, for Huygens next made a desperate effort to offload the sort of chore that from time to time falls to a learned person with connections in high places. This was a request to provide an astrological calculation for an anonymous personage:

> I have a favour to ask of you and a thing that you would
> never guess. There is here a woman of high rank who has
> asked me urgently and seriously to draw up her horoscope,
> thinking that I understand it well. I have had great
> difficulty in making her understand that on the contrary
> I have never got mixed up in this, nor even attached any
> credence to it.

Boulliau duly obliged, and Huygens revealed the subject of the request as Countess Albertine Agnes of Nassau, one of the daughters of Frederik Hendrik and Amalia van Solms.

The existence of predicted dates on which Saturn would appear in certain phases provided a programme of work for many astronomers during the following decades. Huygens himself was able to give more time to Saturn a few years later when he published some further observations in the French *Journal des Sçavans*. And then, through the summer and autumn of 1671, Christiaan in Paris and his brother Constantijn in The Hague separately had the satisfaction of observing the ring gradually disappear from view in conformity with his prediction. But in 1659, many people still remained to be convinced of the existence of a ring at all, and continued to advance alternative interpretations. For those who did acknowledge its existence, there were new matters to be investigated by means of better telescopes, such as the exact dimensions and composition of the singular phenomenon. The ring had made Huygens the most famous astronomer of the age.

8

TIME AND CHANCE

No word came from Prince Leopoldo de' Medici in response to Huygens's tribute offered in *Systema Saturnium*, and no drawing of Galileo's bruited pendulum clock. Had it done so, Huygens would have seen that his own invention was substantially different from, and in fact considerably superior to, anything in the possession of the Florentine court.

Although a prize for a clock that could be used to determine longitude at sea had been first announced in 1567 by King Philip II of Spain, it was the offer by the States General of the Netherlands of a reward of 10,000 florins in 1627 that had drawn Galileo's attention to the problem. Occupied with other matters, however, it was not until 1636 that he turned to consider it properly, writing a design proposal that was duly translated and delivered in November that year. He communicated with the Dutch government through a Calvinist friend in Geneva, Elia Diodati, who had supported him in his disputes with the Vatican. Diodati in turn chose a family friend as his point of contact in Holland – Constantijn Huygens. They corresponded intermittently for the next four years, so it is entirely possible that Constantijn's 'little Archimedes' became aware that a pendulum mechanism might be used to regulate the recording of time before he was even in his teens.

Galileo had devised a mechanism, but died in 1642, leaving

his son Vincenzo to continue his work. Vincenzo died only seven years later, having altered his father's design but never having built a working device. His pendulum may never have been intended to serve for a conventional clock in any case, as Galileo's personal interest had been in timing intervals of importance in astronomy, such as transits of Jupiter's moons, rather than daily timekeeping.

Christiaan Huygens, too, was interested in timekeeping because of his astronomy. He had in addition significant practical experience of constructing machines and a good understanding of mechanical theory. The synthesis of these skills had begun in earnest during his correspondence with Mersenne, sustained from September 1646, when Christiaan was just seventeen years old, until Mersenne's death two years later. Their intense exchange of mathematical problems and solutions combined pure geometry with mechanical examples that gave the geometry a physical dimension. These exercises served primarily as visualizations, but also hinted at a utilitarian connection with the real world. Huygens's analysis of the curve of the hanging chain known as the catenary, for example, proceeded by imagining weights hanging from points along the chain. Prompted by Mersenne, Huygens had moved on to investigate the 'centre of percussion' of different geometrical shapes, which was a problem of obvious relevance to the operation of pendulums.

A basic clock needs two things: a method of displaying the time and a means of changing the display at the correct rate. Medieval clocks used the force from a slowly descending weight or uncoiling spring to drive gears connected to hands on a clock face. A mechanism known as an escapement, often comprising a verge and foliot – a weighted spindle arrangement that repeatedly engages and releases one of the timekeeping gears – was used to regulate the speed.

Galileo had discovered that a pendulum swings back and forth with a constant period related to its overall length but notionally independent of the amplitude of its displacement from its resting position. This property of isochrony immediately suggested the pendulum's applicability to timekeeping. The period of the swing is not completely invariable, however: an excessively high displacement oscillation takes a little longer to return to its starting point. In addition, Galileo's law applies to a mathematically ideal pendulum based on a point mass at the end of a weightless rod or thread suspended from a frictionless pivot operating in the absence of air resistance. In reality, of course, these conditions cannot be met, and any pendulum necessarily loses energy owing to the friction in the pivot mechanism and other factors. Moreover, the rod or thread supporting the pendulum bob oscillates more slowly in summer than in winter because of the slight increase in length produced by thermal expansion.

Huygens overcame some of these difficulties by introducing a number of modifications to conventional designs. The swinging pendulum replaced the oscillating foliot. He employed a metal fork to engage with the pendulum at its centre of percussion so that the movement was as smooth as possible. It was now via this fork, rather than through gears attached to the pivot axle, that the pendulum regulated the timekeeping mechanism. Finally, Huygens affixed a pair of small, curved metal plates either side of his pendulum thread just below the pendulum pivot. These 'wings' (*alae*) or 'cheeks' (*joues*) slightly modified the path of the pendulum bob by effectively shortening the thread at the extremes of its swing, thereby correcting for the increase in period found in large-amplitude pendulum swings, and improving the overall isochrony of the mechanism.

Although his thinking about clocks appears to have begun in pure mathematics, Huygens was well aware of the important goal

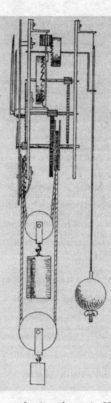

17. Huygens's clock mechanism shown in *Horologium* (1658)
shows the metal fork engaging the pendulum, but not yet
the 'cheeks' to alter the pendulum swing.

of solving the problem of determining longitude at sea. As long ago
as 1530, the surveyor Gemma Frisius had suggested that seagoing
timepieces might solve this perennial problem of navigators, and in
1614 Adriaan Metius, brother of the telescope pioneer Jacob, had
revived the idea. In March 1655 Huygens himself even commended
to the States General a promising invention for determining
longitude, devised by a Polish Cartesian, Johannes Placentinus.
Now, he felt he had his own breakthrough. On 12 January 1657

he bragged to his mentor Frans van Schooten: 'In recent days I have invented a new piece of clockwork that measures time so accurately there is no small hope that it will permit the determination of longitude with certainty if taken to sea.'

Although Huygens's confidence with regard to marine chronometry turned out to be misplaced, and it was to be another century before the longitude problem was finally cracked, his innovations did bring a very great improvement in timekeeping accuracy, such that he can fairly be described as the inventor of the functional pendulum clock. His design was accurate to within a few seconds a day, roughly a hundredfold improvement in accuracy over previous clocks. Perhaps, though, the improvement was not much appreciated in a period when a sandglass with hourly gradations was thought adequate for household timekeeping, and the sound of a ticking clock in the next room was popularly believed to be a harbinger of death.

Huygens assigned the patent that he was awarded for his design to the Hague clock-maker Salomon Coster. Several tidy little clocks that Coster made in 1657, which are now on display in various museums, have short pendulums with spherical brass bobs and all the workings hidden inside ebony cases. In larger clocks, makers sometimes chose a pendulum of about nine and a half Rhenish inches (units of length commonly used in Holland at this time) because it gave a period of oscillation of exactly one second. A thirty-nine-inch pendulum – four times the length – has double this period, two seconds. In an indication of the importance he attached to the accurate measurement of time, Huygens even hoped to redefine distance in terms of the standard foot in relation to this fundamental unit of time, proposing that one third of this length be called the *pes horarius* or 'time foot'. However, most of the clocks that Coster continued to make did not incorporate Huygens's improvements, as great accuracy was not an

18. Huygens's sketch of curved cheeks used to effectively shorten
the length of the pendulum at the extremes of its range,
thereby producing a more isochronous swing and improving
timekeeping accuracy.

important consideration for his customers, and it is likely that
Huygens turned to other makers bound by agreements to secrecy
in order to further his experiments with precision timekeeping.

Huygens wrote up a brief history of time along with a detailed
description of his invention in a treatise titled *Horologium*, which
he presented in 1658 to the governors of Holland, who by this
time included his former mathematics classmate, Johan de Witt.
In it, he made a disparaging reference to the 'so-called science of
longitude' and promoted the merits of a clock-based solution to
the problem. He knew that his innovation would be of immediate
interest to astronomers, though, and he sent copies of *Horologium*

to astronomer friends such as Hevelius in Danzig. But perhaps nothing gave him more pleasure than the praise of a senior figure such as Pascal, who wrote that, since receiving *Horologium*, 'I have been among its leading admirers'.

Horologium and the clocks built according to its specification were to become almost immediately obsolete, however, and it would be thanks to a further discovery made by Huygens himself.

One of the curves that Huygens had discussed in his youthful exchange of mathematical problems with Mersenne had been the cycloid. Known as the 'Helen of geometry' for its beauty and its many surprising properties, as well as for its capacity to provoke disputes among mathematicians, the cycloid was a puzzle because, like the catenary, its form could not be described by algebra. The cycloid is generated by the path of a point on the circumference of a circle as it rolls along a straight line, but the sharp cusp produced in the trace each time the point meets the line makes it immediately apparent that it does not belong to the family of curves that are always smoothly changing, like the ellipse of a planetary orbit or the parabolic path of a cannonball. Yet this curve, so readily generated by the rolling of a simple wheel, surely had some fundamental relation to nature, which might be revealed by the discovery of some correspondingly simple mathematical expression.

Pure geometry had a significant allure for practical investigators as well as mathematicians. Just as a wheel rolls best when it is a perfect circle, forms pure in shape were thought to have intrinsic physical merits. Pure geometries promised conceptual clarity – and perhaps optical clarity, too. When spherical lenses – lenses with one or both surfaces comprising part of a sphere – were found to be inadequate for some purposes owing to spherical aberration, lens-grinders tried the next most perfect curves they knew, such as paraboloids and ellipsoids. Around the time, in 1635, that Huygens's father and Descartes were struggling to make a hyperbolic lens, a Dutch astronomer, Martin van den Hove, known as Hortensius, once a student of Snel and Beeckman, even tried to make a lens in the form of a cycloid.

For his part, Huygens had long ago set aside the cycloid problem. But his interest was renewed in 1658 when Pascal pseudonymously issued a challenge to Europe's mathematicians to characterize the cycloid by expressing attributes such as its centre of gravity and its area in algebraic terms. Christopher Wren came closest to finding a solution, when he succeeded in rectifying the curve (that is, he found a way to describe the area under the curve as a sum of squares of different sizes), but no prize was awarded. For a moment it even seemed that Pascal himself – under his real name – was claiming to have found the answer, until it became clear that he was merely communicating some additional thoughts on the problem.

Huygens was busy at the time trying to improve technical aspects of his clock design, in particular its accuracy. This was largely a matter of tedious trial and error, for example involving setting two clocks running side by side for comparison, with large and small pendulum swings respectively. Eventually, he was able to write that he had 'so finely adjusted two clocks in this way that in three days they never had a difference between them even as much as seconds'.

Huygens found this phase of work troublesome because, although he was the inventor, he was not the maker of these clocks. While he could doubtless improvise for himself simple modifications such as adjusting the curve of the metal 'cheeks', he was dependent on his skilled suppliers, who included not only Coster in The Hague, but makers in a number of other Dutch cities. Keeping control of their work was not always easy. One maker called Josina in Amsterdam owed him some components but would not reply to his letters. 'I don't know how the sow can be so insolent when she well knows that I am still due to pay her 50 guilders,' Huygens complained to his brother Lodewijk. There is also a possibility that Huygens instructed the leading clock-maker in Paris, Isaac Thuret, a more skilled and scientifically minded man than Coster.*

A fine draughtsman, Huygens was able to supply drawings giving enough information for such craftspeople to work from. But because he did not actually make the devices himself, he was always vulnerable to unscrupulous behaviour by his suppliers. Coster, for example, carried on his trade making unimproved clocks, but incorporated features resembling Huygens's 'cheeks' as a kind of modish branding. When another supplier, Simon Douw of Rotterdam, simply copied the clock Coster had built, claiming it as his own, it enmeshed Huygens in a legal dispute between the two makers – 'a most thankless business these thieves have caused me'.

His Parisian correspondents only added to his worries. Rumour came that Roberval had a clock of his own design, but Huygens could find out no more about it because, as Boulliau told him, Roberval had insulted his host at a recent gathering, telling Montmor in his own house that Descartes had 'more wit than he', and as a consequence no longer went along to the meetings. Clock-makers in Paris were just as rapacious as those in Holland, and friends warned Huygens that there too he was in danger of being 'robbed of . . . the glory of the invention of the Pendulum'.

---

* Although there is no written evidence that Huygens met Thuret until 1662, stylistic details of some of the clocks that Huygens developed in the late 1650s are characteristically French, and may have originated from earlier contact (Whitestone 2017).

All this intense competition shows that Huygens was more closely involved in 'trade' than might be expected of a man with his social connections (and than his father may have wished him to be). He fielded frequent enquiries from his scientific friends in Paris wanting to know if he had catalogues and price lists for his clocks. Boulliau wrote under the clear impression that Huygens was selling direct: 'I beg you to tell me the price of your clocks both ringing and without ringing, with weight and with spring.' Huygens responded promptly with prices of five different designs: a thirty-hour spring clock would cost 80 Dutch silver pounds, for example, 120 with ringing.

In his continuing effort at improvement, Huygens had discarded the metal 'cheeks', having realized that they would lead to inaccuracies in any shipboard clock when it was tilted over, as it would be aboard a heeling vessel. Meanwhile, he sought other means of regulating the pendulum swing. The core problem remained: what pendulum path would produce perfectly isochronous oscillation? There had to be an answer, for modifications could be made to the path to produce swings both that were disproportionately slow and that were disproportionately fast. Somewhere in between lay the tautochrone, the curve for which the displaced pendulum would take the same time to return to the centre no matter where it started from. Huygens finally discovered the happy medium by geometrically comparing the fall of an unaltered pendulum with the acceleration of a body in vertical free fall. On 1 December 1659 he was able to pronounce that the tautochronous pendulum path was in fact nothing other than an exact cycloid. It was 'the happiest of all the discoveries that ever fell to me'.

The distraction of Pascal's purely theoretical challenge must have contributed to Huygens's revelation. He was pleased to have found the key to perfectly uniform timekeeping, and still more satisfied to find that pure mathematics lay at its heart. Whereas his 1657 pendulum with its 'cheeks' constraining the path of the bob had been arrived at empirically, this new knowledge was the product of rigorous analysis and possessed a fundamental truth as well as great potential utility.

More was to come, however. There now arose the question of how to make a pendulum bob move along this ideal path. As Huygens went back to trials with various metal plates, he soon found that the shape that would produce a cycloidal pendulum path was itself another cycloid. For a man of Huygens's acute sensibility to the connections between mathematical geometry and the physical world, this was a true epiphany. Huygens had his reward for his loyalty to the classical tradition of geometric analysis. It was this visually based methodology, rather than newer algebraic procedures based on numbers and functions, that kept him mindful that curves with special properties often had manifestations in the physical world, and that enabled him to find synergetic connections between theoretical forms and practical applications. It was surely most satisfying for him to find not only that a mathematical exploration prompted by a matter of practical necessity had delivered up a geometric revelation of great intrinsic elegance, but also that this revelation immediately suggested both further practical applications and new theoretical avenues to explore.

Alongside the mechanical project of drawing and fabricating precision cycloidal cheeks for his clocks, Huygens began to pursue a more general, purely mathematical, investigation of the theory of curved lines and their fundamental, but not always obvious relations to one another.* Having found that one cycloid can be used to generate another cycloid, he went on to characterize the curves that could be likewise evolved from the parabola and the ellipse. From there, he was able to establish a general formula for the relation between any curve and its evolute companion. As in his work on probability theory, Huygens moved with inexorable logic from simple cases to more complex cases, and from there to the general rule in a way that reveals him as a progenitor of modern scientific thinking. He also described the relation between the length of a pendulum and its period of oscillation – as found by Galileo – in terms of a mathematical equation for the first time, although, in typical fashion, it was to be many years before Huygens published the results of these practical and theoretical investigations.

The clocks on Dutch churches are so ubiquitous that it is easy to assume they are as old as the buildings themselves. Mostly, they are not. The church in Scheveningen now known as the Oude Kerk was constructed just behind the beach in the mid-fourteenth century. It had already stood for two centuries when it appeared in an illustration made by Adriaen Coenen for his celebrated *Visboek*, a fabulous bestiary of marine life. His sketch, made from a boat out at sea, shows the church spire with the hummocks of sand dunes rising up all around it. There was certainly a clock in the tower by 1500, because on the morning of the fourth of March that year the Hague magistrate went to the beach and issued a proclamation that fish were to be traded

---

* For example, it was later discovered that, as well as being the tautochrone, the cycloid is also the brachistochrone, the curve of fastest descent, a phenomenon of interest to winter sports enthusiasts.

there only between the hours of six and nine in the morning and two and seven in the afternoon as rung by the clock bell.

Huygens's work with clocks was not restricted to small devices for scientific purposes. In January 1658, he and Salomon Coster embarked upon a refurbishment of the Scheveningen church clock, making a number of alterations to the mechanism in an effort to improve its timekeeping. The two men knew each other well enough by this stage in their collaboration to use familiar forms of address, although their relationship was still clearly one of client and supplier. A short letter from Coster addressed to 'Mijn Heer Christiaen' reports on progress thus: 'The mechanism at Scheveningen is now running, and ran all night, the Bob has a weight of 50 pounds, But I think to hang something less and to alter its spring and chain somewhat. It has as a guess lost a quarter in 14 hours. I am thinking of going there again tomorrow afternoon.'

Huygens described the improvements he was making in a long letter to Jean Chapelain. A short length of silk thread supported the pendulum at the top to allow it to swing freely. The pendulum itself was a twenty-four-foot iron bar and the fifty-pound bob at the bottom was made of lead. The swinging action of the pendulum was converted to the rotary motion of the balance as frictionlessly as possible by means of a sliding brass rod between the two. A clock that Huygens and Coster went on to develop for the imposing cathedral tower at Utrecht, the tallest in the Netherlands, was on an even grander scale, although both church clocks had simpler wheel-trains than the men's smaller brass timepieces. With the addition of several precisely drawn diagrams, Huygens's description of the Scheveningen design was admirable in its clarity, perhaps because he knew he was writing to a poet and not to one of his more mechanically minded confreres. However, if clear communication – to Chapelain, and through him perhaps as a salutary example to the sophists among the Montmor circle – had been his intent, he was not entirely successful, for Chapelain replied to Huygens: 'As for the construction of the clock, I admire more than I understand, either because of the obduracy of my intelligence or because of your too succinct exposition.'

To a physicist, as opposed to a clock-maker, a pendulum is simply a special case of a body in motion. To Huygens, it was a tool to develop his thinking about masses and the factors that cause them to move. From practical consideration of real pendulums and their mathematically ideal analogue, a weightless rod with a bob at the end whose mass is compressed to a single point, he developed a thorough analysis of the centre of oscillation of any object. He generalized his analysis with reference to other practical situations, such as the motion of bodies of various shapes floating in water (perhaps an especially Dutch interest).

Another priority related to pendulums and to any falling object was to establish the value of acceleration due to gravity, which was then conventionally framed by natural philosophers not as a constant of acceleration but as the distance travelled in one second by an object in free fall. Galileo, Mersenne and others had been defeated by this problem, not least because it had been difficult to agree on the precise magnitude of a second. Though hampered at first by an attempt to fit his analysis into a model of Cartesian vortices, Huygens eventually succeeded in measuring the quantity using a special conical pendulum mechanism; it came to fourteen feet (later revised to over fifteen feet, which corresponds well with the modern value of the gravitational constant g equal to 9.8 metres per second per second). The swingball-like device he used for this task also enabled Huygens to relate the outward impulse of the rotating bob, described qualitatively by Descartes, to the downward (and therefore inward) tendency owing to gravity studied by Galileo. This swiftly led him to a new description of both circular motion and fall under gravity in mathematical terms far beyond those of his distinguished predecessors. Implicit in his work was the idea that force gives rise to acceleration, which Isaac Newton would later articulate as the second law of motion. By uniting the practical work of clock construction with a readiness to conduct experiments and make measurements and a belief in the mathematical basis of mechanical phenomena, Huygens had become the first person to produce an accurate measurement of a physical constant.

Despite this breakthrough, a true understanding of the concept of force still lay some way in the future, even though Huygens did invoke the word in the title of the account of his work at this time, *De vi centrifuga* ('On Centrifugal Force'). He was still worried, however, about some of the implications of his findings, and once again put off publishing. At the head of an early draft of the work he put a quotation from one of Horace's epistles to Maecenas: 'I was the original, who set my free footsteps upon the vacant sod; I trod not in the steps of others.' He sensibly omitted the Roman poet's vainglorious next sentence: 'He who depends upon himself, as leader, commands the swarm.'

It is a characteristic of Christiaan Huygens's intellectual life that he worked on many projects at once that might seem diverse to us now, but which to him were all facets of the same urge to learn more about the mechanics of the physical world. This was never more the case than during the first flourish of his maturity in the late 1650s, when he was working in parallel on Saturn, clocks and mathematics, constantly shifting from telescope to drawing-table and drawing-table to workshop. It is clear how much the working out of Saturn's ring and his clock's pendulum owe to his facility with the geometry of curves. Yet at the same time, his mind was also wrestling with a mathematical question of a very different hue: the problem of chance.

Three men are at the gaming table: a soldier and a middle-aged burgher are seated, smoking their pipes, gazing intently at the tabletop; the third braces himself against the table and prepares to throw the dice. The play has reached a pitch of expectation, and two capped peasants have come to peer over the men's shoulders at the action, thickening the air with tension.

David Teniers II produced this painting around 1640. It is known as *The Gamblers*, or more exactly in Dutch as *De Dobbelaars*, the dice players. It was a standard artists' subject in genre painting, and the setting, depicted in shades of brown and softly lit, suggests only the world of routine activity and simple pleasures. The painting is like many others made by the same artist after he took his own gamble and switched from grand biblical panoramas to domestic scenes. There is no movement to ruffle the calm; just the suspense of the throw about to be released.

Games of chance had a particular status in the Calvinistic republic. Whereas chess was deemed virtuous, demanding intelligence, wisdom and stamina, dice was clearly the opposite: trivial, ephemeral and governed by chance alone. Play might be willingly entered into by both sides, but it involved needless risk and was the antithesis of the prudent enterprise encouraged at all levels of Dutch society. The best that could be said for it was that it was a harmless indulgence and a way of keeping youngsters amused. University students were said to be especially fond of dice.

Yet the element of randomness made it something more, too. Gambling was routinely denounced from the pulpit, but the more imaginative clergy understood that casual demonstrations of the operation of laws of chance might serve to impress upon people a notion of divine fate, and so quash dangerous thoughts about the existence of free will. Teniers at this time was the serving *kapelmeester*, or director of music, at Sint-Jacobskerk in Antwerp; he would have understood these nuances.

Then, of course, there was money. Throwing dice quickly becomes dull unless there is betting on the outcome. The pecuniary aspect, and the chance that it brings of undeserved gains or considerable losses, makes play immoral, though only relatively so if the stakes are kept low. The artist is not taking us to one of those dives where more reprehensible games were played, obliging players to down drinks to the value of the number they had thrown; there is no liquor here. Teniers does not want to portray anything sinful, according to one biographer. He is fully aware that 'thoughtful reasoning is useful for sharpening the mind', and that the game 'teaches one to go with fortune or adversity'.

Although these genre paintings often contain the sanctimonious or satirical implication that more is at stake than a wager on a table, there is no moralizing here, no pregnant symbol of foolishness or depravity lurking in the corners of the painting. Yes, an extinguished candle rests on the mantel and a broken pipe lies on the floor, but clay pipes break all the time, don't they?

And the money? Similar, but unequal, wagers lie on the table, along with a piece of chalk to keep the tally. Many coins circulated in the United Provinces – ducats and national dollars of the independent republic, pieces of eight dating back to Spanish days, and numerous smaller divisions of all of these. Each of the players has some gold pieces and some silver. It is hard to be sure, but the player on the left appears to have wagered two ducats, three rijksdaalders and one schelling, the equivalent of 21.8 guilders. The other player has put down two ducats, one rijksdaalder and four schellings, a total of 17.2 guilders. (These sums would have a purchasing power of about €200 today. If you cleaned up here, you might have enough to buy a small Teniers.) At all events, we can see that the play is finely poised. It has reached a crucial stage. Why else the unbidden spectators?

Playing dice soon passed from fashion in smart Dutch society, although this was not the case in France, where church and state were more indulgent and gambling of many kinds flourished along with other modes of conspicuous consumption. On his first visit to Paris, in July 1655, Huygens may not have spent time at the gaming tables, but he was swiftly caught up in the lavish whirl, soon writing to his father to complain that his money was running out with all the clothes he was having to buy. 'By arranging so many visits for us, I pray you not be astonished at the expense we are incurring, it being a sure thing that it's necessary to do it in order to mingle with men of quality.'

The mathematicians Huygens met on this occasion – Roberval, Mylon, Carcavi and others – were chiefly engaged in developing methods of calculating complex geometric curves, a field in which he too was notably adept. He was readily drawn into their discussions, therefore, but these must have soon turned to other mathematical topics, for he returned to The Hague in December determined to make his own contribution to the very different matter of probability and the laws of chance.

Huygens had not sought out Pascal when he was in Paris,

because he had been advised, a little misleadingly, that this greatest of the French mathematicians had lately given himself over entirely to theological study. Pascal was in effect the founder of the mathematical study of probability, which had begun naturally enough at the gaming tables of Parisian salons, but which he soon extended into matters of greater existential significance. The philosophical argument known as Pascal's wager uses betting logic to support the existence of God: why would anybody not take the gamble of faith, Pascal suggests, a gamble where, if you win, you win everything, and there is nothing to lose? Huygens's visit to Paris came just a year after Pascal had taken up the subject in a fruitful correspondence with Fermat over the 'problème des points', or how to divide the stakes fairly in games of chance. This is a clearly a matter of great practical concern to regular gamblers, but it is an appealingly tricky problem for mathematicians, too, because of the uncertain advantage that may accrue to one or another player if unequal amounts have been wagered or an unequal number of turns taken by the respective players during the progress of the game.

Teniers's picture, painted some fifteen years before Christiaan's visit to Paris, hangs in the Rijksmuseum in Amsterdam, and in rooms nearby there are fine engraved glass and silver dice cups from the same period: gambling was clearly not only the vice of low-lifes and addicts. Yet Huygens surely had no need, as the Lombard polymath Girolamo Cardano apparently did a century before, to use his superior awareness of the odds to augment his income. It was the quality of the problem that drew him; that and the wish to prove himself to his illustrious Parisian friends.

In his renowned exploration of play, *Homo Ludens*, Johan Huizinga explains that a game is an entirely voluntary transaction between the actors, more so than most trades in daily life. More important still, it is *inconsequential*: it has no repercussions beyond the gaming table. Nothing rides on the outcome beyond that which is staked – even if this is all one has, as in the case of the reckless gambler. It is this very inconsequentiality which makes it such a good subject for painters. And for mathematicians, too: for a game to be a true game, and not merely a fiasco or the prelude to a fight, it must have an unpredictable outcome, but an unpredictable outcome within certain bounds of expectation. It is these two factors in combination that make dice a rewarding problem for mathematical interrogation.

In that summer of 1655 when Christiaan was in Paris, Pascal and Fermat had not yet published any of their work on probability theory. However, the gossip he heard there was clearly enough to stimulate his interest in this new direction. Back in The Hague, he spent the next few months in playful correspondence with his new Parisian friends, tossing out little problems and chasing them for their answers, keen to know whether his methods were correct. By April 1656 he was able to tell Roberval, Mylon and Carcavi that he had written down his 'foundations of the calculus of games of chance', and that his old tutor Frans van Schooten wished to print it.

The work appeared as an addendum to van Schooten's fifth volume of mathematical exercises, first in Latin, and a little later in Dutch, as *Van Reckeningh in Spelen van Geluck* ('On Calculation in Games of Chance'). Huygens leads the reader through fourteen worked examples of gradually increasing difficulty. He begins by pointing out an important truth about the simplest chances, which is that if a number of outcomes are equally likely, then the *value* of any outcome is the reciprocal of the number of possibilities. Because there is a finite number of possible outcomes of a particular action in a game, such as tossing a coin or throwing a die, the sum of the values of each outcome must equal the total value, namely unity. In other words, if two outcomes are equally likely, such as heads or tails of a coin, then the value of either outcome is simply one half. When a sum of money is at stake, then the value of either outcome is half of that sum. Huygens quickly develops his analysis for any number of equally likely outcomes and for situations where the outcomes have different known chances of occurring.

This is important groundwork for the more complicated situations that he presents next. What if a game is stopped after I have played one more throw than my opponent? How is it fair then to divide the stake? For the simplest possible case, where one player has taken two turns and his opponent only one, the answer depends on who went first, and the stakes should be split either fifty-fifty or in the proportion of three-quarters to one-quarter. Again, Huygens builds up from this simplest case to look at more complex situations, such as where a player falls two turns behind, or where both players have missed turns, or where one player out of three at the table misses a turn. This leads him stepwise to a general calculus capable of handling any number of players missing any number of turns.

Combining the known probability of an event with its monetary value allows for a fair division of the spoils. This concept would be articulated in greater mathematical detail more than sixty years later, in 1718, by Abraham de Moivre as 'expectation'. It is striking that Huygens's imagination is clearly stimulated by scenarios that would arise in real game play – a player might easily leave the table for a turn or two. But the fact that he then moves on to generalize his method for situations far less likely to arise in reality shows the mark of a true mathematician.

Huygens next turns from the problem of fair division of the stakes to games where the exact numbers thrown up by the dice are important. We know that the odds of a given number turning up when we throw a true die are one in six. But how many throws should we *expect* to make before getting a six? He works it out: by the fourth throw the odds tip in your favour; by the sixth throw, you have an almost two-to-one chance of having obtained the desired result. How many throws to get a double six using two dice? Huygens works it out again in longhand arithmetic; although he could have got the answer more readily by employing logarithms, it seems that this relatively new technique was not included in van Schooten's teaching.

After these warm-up exercises, Huygens considers a regular game played with a pair of dice. The scenario is that I win if I throw a total of seven, but my opponent wins if he throws ten, and if neither of these numbers comes up the stake is split equally between the players. For those of us who avoid casinos, the proposition appears distinctly unappetizing: the odds of a seven are one in six; the odds of a ten, one in twelve; it would seem clear who holds the advantage. But the important thing here is the order of play. I go first: if I fail to roll seven, my opponent still has some chance of recouping his stake. My *combined* chance of *either* winning on the first throw of the game and taking both stakes, *or* of losing that throw and relying on my opponent to lose his throw too and then recovering his stake only, leaves my opponent with the overall *expectation* that it is in fact he who will finish to the good (in a ratio of thirteen to eleven).

The final scenario that Huygens examines concerns the more finely balanced game that ensues when players continue by turns until either the first player throws a seven or the second player throws a six. This is a variant of the game of hazard, popular throughout Europe in the seventeenth century, from which we get the expression 'at sixes and sevens' as well as the word 'hazard' in its general sense. Here, although the first player is slightly more likely to throw a seven than the second is to throw a six, the overall expectation after repeated turns works out to be narrowly in favour of the second player by a ratio of thirty to thirty-one.

Huygens ends the *Reckeningh* by considering a few more problems involving dice, *schijven* (discs that are black and white on opposite sides), and cards, in response to challenges issued by Fermat and others. Thanks to van Schooten's enthusiasm for his protégé, the long paper appeared in print promptly, which was not the case for all of Huygens's work. In Dutch, it would reach a local readership; in Latin, it would be universally accessible to scholars. It won praise from the French mathematicians for whom, in a sense, it was written, and from the English mathematician John Wallis, and later also from the German mathematician and philosopher Gottfried Leibniz, and for more than fifty years it remained the only introduction to the theory of probability in existence.

In later life, Huygens returned occasionally to the mathematics of chance events, as the need to understand matters of risk gained importance in many areas of Dutch commercial life with the growth of sea trade and the introduction of marine and other forms of insurance. On one occasion, in 1669, his brother Lodewijk, then employed in the retinue of the Dutch ambassador to Spain, raised the problem of predicting mortality given a person's present age. His enquiry was prompted by the recently published *Natural and Political Observations Made upon the Bills of Mortality*, a work of early epidemiological importance written in 1662 by John Graunt, who had used historical weekly death notices in London parishes as a statistical database to predict the likely progress of the plague if it were to break out in the city, as it did three years later, in 1665. Christiaan advised his brother on a better way to interpolate between infrequent data points. For example, if tables of mortality show the probability of newborns surviving to age six, sixteen, twenty-six and so on, how does one estimate the life chances of a person already aged ten or twenty? Christiaan also suggested some more entertaining variants of such problems: 'A man of 56 years marries a woman of 16 years, how long can they live together without one or the other dying? Or if I had been promised 100 francs at the end of each year they both survived together, for how much would it be right to redeem this obligation?' Such puzzles might seem flippant, or even distasteful, but they begin to demonstrate the applicability of mathematical methods of estimating probability to matters of social consequence, which were to lay the foundations for scientific demography and provide a rational basis for the calculation of life insurance premiums.

In the late 1670s, during his prolonged residency in Paris, Christiaan found a new game in town. Recently introduced from Venice, basset was a card game in which punters bet unlimited stakes against a *taillère* or banker on such ruinous terms that play had to be confined by law to the nobility, and even wealthy players were frequently bankrupted. Christiaan joined with the leading French mathematicians to try to calculate just how great the house advantage was. This concerted effort to establish the certainties underlying games of chance helped to transform a delinquent pastime into a sound commercial prospect. On this new footing, gambling could be taken over and run, professionally if not always legally, by operators who could be confident of making a steady profit.

At the same time as demonstrating that the mathematics of probability might have social consequences both trivial and profound, Huygens was at work on a practical invention that would prove to be of similarly ambivalent application. This was the magic lantern. Nothing Christiaan invented ever delighted his father more. The device appears to have been demonstrated by him at the family home one dark evening in 1659, a time when he was also experimenting with arrangements of lenses for telescopes as well as making pendulum clocks. It must have brought to his father's mind the hazy optical projections produced by Cornelis Drebbel in London more than thirty years before.

Huygens's apparatus displays the functional elements of a twentieth-century slide projector. A curved mirror is placed behind a source of illumination, a candle, in order to bring the maximum amount of light to pass through a condensing lens, which disperses the light onto a translucent glass plate. Two further lenses then project the luminous image on the plate for viewing on a distant screen. A rough diagram sketched by Huygens much later and labelled 'laterna magica' shows a similar but simplified design that makes use of only two lenses.

Drebbel's early-century contraption appears to have been a somewhat superior version of the camera obscura, already in widespread use by artists. Many variations were attempted before Huygens addressed the subject. In 1646, for example, the Jesuit

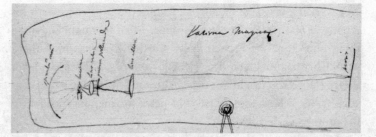

19. Huygens's sketch diagram of a simplified 'magic lantern' shows a curved reflector behind the light source, two lenses with the transparent 'slide' between them and the projection screen on the right.

polymath Athanasius Kircher described a means of projecting an artist's drawing using artificial light, which may have worked by capturing the light reflected from a sheet of polished metal engraved with the design. Huygens's innovation differed crucially from these precursors in bringing together both a source of artificial light and a transparent glass plate bearing the image to be projected.

With his experience of lens-grinding and clear understanding of the optical principles involved, Huygens was able to produce images of superior quality, even using nothing more than candle-power. Primitive projections and shadow-play were popular features of fairs and markets at the time. Trompe l'oeil and theatrical image-making were facets of a broader culture in which the boundary between truth and illusion was blurred for public titillation. Such entertainments were not without a subversive and dangerous aspect, feeding popular scepticism about what was real and what was not, and even raising atheistic doubts in the minds of the faithful.

Perhaps seeing no serious use for it, Christiaan was soon bored by the possibilities of the lantern. But Constantijn's fascination with the device developed into an obsession. In 1662, he suggested that his son make a new instrument for demonstration to Louis XIV. When Christiaan demurred, feeling the project beneath his dignity, it seems that Constantijn hinted that he might show the thing himself at the French court. Unable to refuse his father's wish, Christiaan reluctantly set about the work with the engineer Pierre Petit. 'You would not believe how much trouble I take over such trifles, which are already old news for me,' he moaned to his brother Lodewijk.

In keeping with the melodramatic expectations of the medium, Christiaan even drew a series of jaunty figures of a strolling skeleton for transfer to glass plates, loosely copied from Holbein's famous series of woodcuts *The Dance of Death*. Petit started to refer to the device as the 'lantern of fear'. Huygens's drawings show slight shifts in the skeleton's pose as it doffs its skull and performs other gestures, which suggests that he may have intended the plates to be switched in rapid succession in order to produce an illusion of animation. Huygens may also have produced plates depicting members of his family, as he requested a drawing of Lodewijk at this time, 'so that I can see what you look like without a wig and greasepaint'.

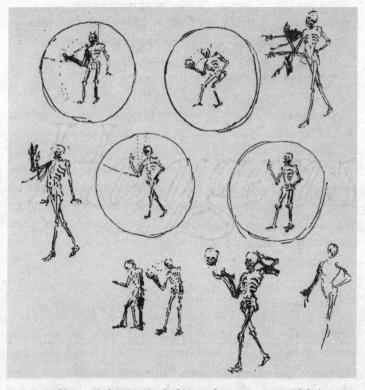

20. Huygens's drawings made for transfer to transparent 'slides' for projection using his magic lantern.

A few weeks later, 'since I cannot think of an excuse for avoiding it', he packed off the requested magic lantern to his father in The Hague. Then he had an idea. He added a note to Lodewijk:

> But when it arrives, if you see an opportunity, you could
> easily render it incapable of working, by removing one of
> the 2 lenses that are close to each other, so that 2 still
> remain, since there are 3 in all. I will plead ignorance
> that anything is missing . . . this is all for the best;
> because I think it does not befit my Father to make such
> puppet-play at the Louvre . . . As for the plates, I do not
> see why they make you so very afraid, for at least he will
> not show them to the King.

Christiaan's exasperation with the magic lantern only grew. It did not appear to him to open up any scientific possibilities, and did little to inform his later work in optics. The basics of the design were in any case easily copied, and many did so. On this occasion, at least, Huygens was more than happy to relinquish his claim to the invention. 'I am prepared to make [my father] Spyglass, microscope and anything he would wish, except the lantern, the creation of which ought to be counted *inter artes deperditas* [among the lost arts].'

After his intense period of activity during 1659 – when, besides the trivial pursuits of dice and lantern shows, he had described the concept of centrifugal force, established an accurate value for the constant of gravitational acceleration, perfected his pendulum clock and prepared detailed texts on both topics, as well as publishing his treatise on Saturn – it seems that a family wedding was not what Christiaan Huygens felt he needed in his life. He bemoaned the several days lost on the 'solemn follies' of his sister Susanna's marriage to her cousin Philips Doublet in April 1660.* His mind was clearly elsewhere as he accompanied his aunt and other members of the family in following the bride up the aisle in the New Church by the Delft canal in The Hague.

---

* His father was rather more excited to see the first of his children married off. He gives a brilliantly colourful description of the ceremony in a letter to his friend Béatrix de Cusance, who had hoped to attend the wedding (OC3 67–72).

9

SCIENTIFIC SOCIETY

Christiaan Huygens returned to Paris in October 1660, travelling
with an ambassadorial mission sent to congratulate Louis XIV on
the successful arrangement of his marriage to the infanta Maria
Theresa, which was designed to bring peace between France and
Spain. If he momentarily doubted his new-found standing as one
of the leading minds of the age, then he must have been reassured
to overhear a man on the Delfshaven ferry speaking in respectful
tones of the clocks of Huygens.

Once in the city, he embarked on a frantic round of social
calls and cultural engagements, happy to be back to civilization.
His brother Constantijn wrote immediately demanding to know
the fashions for the winter season. Christiaan replied in great
detail – suits of 'mouse-grey cloth' with white canions, 'but not
as big as they were before' – but still his brother wanted to know
more. Christiaan took up lodgings at the Hôtel de Venise on the
rue du Bussy in the Saint-Germain district and began to make
adjustments to his style of living appropriate to his status. A
personal carriage was too great an expense, so he planned to buy
a sedan and employ porters when he needed it, once or twice a
day, 'for there is no way of going by foot'.

Many of the friends he made during his first visit to the city
five years earlier had maintained a lively correspondence with
him since then, always eager to learn of his further successes in

observing Saturn and designing clocks. His first call was to his greatest champion, the poet Jean Chapelain. Perhaps his gushing wonderment at Huygens's scientific ideas now found at least a little reciprocal admiration from Huygens, who had recently prepared his father's verse collection, *Koren-bloemen*, for publication. As he made his rounds, Huygens carefully noted when people he called upon were out ('point trouvé') so he would remember to try again. His circle of correspondents expanded, too, as Chapelain and others introduced him to new French colleagues. Even the renowned mathematician Pascal at last made Huygens's acquaintance; he found it 'a surprise and extreme joy', and expressed the hope that they would become friends. But a few were disappointed not to be able to meet him at all. Fermat wrote: 'I learn with joy but not without a little jealousy that my friends in Paris have the honour of having you for some while. I assure you, Monsieur, that if my health were strong enough for journeys, I would with great pleasure join in their good fortune.'

For Huygens, these months provided a vital opportunity to compare theories, observe demonstrations and take part in experiments with true equals, who valued his contribution as he valued theirs. At the home of the physicist Jacques Rohault, for instance, he 'saw experiments made with mercury which confirmed the weight of air, and how that which surrounds us is always elastic'. He also experienced the habitual peril of the reputed scientist when an eager astrologer forced his book on him: it was 'quite mad'. In addition, he made numerous practical enquiries of clock-makers, lens-grinders and instrument manufacturers in the city. He attended music recitals, comedies and the ballet, where he inspected the design of the stage equipment. He glimpsed the king at mass in the chapel of the Louvre. By the end of December, two months after his arrival and after several requests, he managed to see Marie Perriquet, but they did not dine alone.

Huygens also began to attend the meetings of the informal Montmor academy. It had always been a source of excitement at the meetings when Chapelain was able to read out some scientific news from the absent Huygens. This was never more so than in March 1658, when, having carefully won over a dubious Huygens, Chapelain was permitted to reveal the contents of the Dutchman's earlier secret communication offering strong evidence that Saturn was surrounded by a ring rather than orbited by close-lying moons or other strange appendages. By acting as the impresario behind such revelations, Chapelain prepared the ground well, ensuring that Montmor would heartily welcome Huygens to the circle, and assuring Huygens in turn that his confidentiality would be respected in discussions about both his clocks and Saturn. Huygens was more relaxed about the latter than the former, where he had good reason to fear that others were working on similar innovations, and where he hoped to obtain a privilege that would safeguard his commercial rights in France.

The troublesome Roberval raised difficulties in both areas. When he saw the correspondence with Chapelain in which Huygens described Saturn in new detail, he claimed that Huygens was merely repeating his own ideas, which the two had supposedly discussed in 1655. However, because Huygens had taken the precaution of coding his discovery in an anagram with an earlier date, it was clear that the priority was his, and Roberval was forced to back down. In addition, Roberval was said to have devised his own clock, as he told Chapelain, who passed the word on to a worried Huygens. Chapelain tried doggedly to prise further details of the design out of Roberval, but after several months was able to glean only that 'it is still very imperfect'. This threat, too, came to nothing.

But now Huygens was present in person. On 2 November, less than a week after his arrival in Paris, he dined with Chapelain at Montmor's house. He attended most of the Tuesday evening meetings at Montmor's from then until his departure in March. He saw for himself Montmor's paintings, drawings by Dürer, toys, magnets, and 'Little bottles in water which rise and fall without one seeing how'. Huygens wrote to Lodewijk that the meetings typically numbered twenty or thirty 'illustres, including state counsels and other blue ribbons'. But his first impressions of proceedings were not favourable. He must have described one of the first meetings that he attended in less than flattering terms to his brother Constantijn (the letter is lost), because Constantijn replied drily: 'We laughed a great deal about this fine gathering at Monsieur de Montmor, and what happened in this synod of illogicality when you were there, leaving us with a not very honourable opinion of the comprehension of these Gentlemen Academicians who have the patience to listen to the prattle of pedants for hours on end on subjects of nullity.'

Fortunately, perhaps, Montmor was not the only host in town, nor the only follower of Descartes, and Huygens also attended other regular salons, including a number organized by women learned in science and philosophy, as well as more informal gatherings. Even before his arrival, Montmor's academy had explicitly attempted to steer discussion towards things that could be unambiguously known and towards practical advancements. But it was clearly not able to go far enough with its existing members and leadership to suit Huygens, who found it overly rhetorical in its deliberations compared to what he knew of what was then the only other scientific society, Leopoldo de' Medici's Accademia del Cimento in Florence.

Chapelain's announcement of Huygens's further descriptions of Saturn in 1658 illustrates the academy's difficulty in attempting to limit discussion to things that could be definitely known. Chapelain reported back to Huygens that

although everybody did not agree with your interpretation
as something completely certain, the majority nevertheless
deemed it most probable and infinitely praised your
sagacity and judgement in a matter so far removed from
the reach of the senses, rejoicing to see you so perceptive
and so reasoning at such a young age as yours, promising
so much for other mathematical discoveries in the future.

As we have seen, the precise nature of the form or forms
around Saturn remained highly debatable even after publication
of the *Systema Saturnium* in 1659, and Huygens was obliged to
continue defending his interpretation against those of astronomers
in Paris and elsewhere in Europe.

One of the most important new contacts that Huygens made
as a result of his correspondence with French colleagues was
Pierre Petit, the cartographer to Louis XIV and a superintendent
of fortifications. As a military engineer, he possessed the eye for
mechanical detail that many in the Montmor circle plainly lacked,
but he was also interested in questions of mathematics and
astronomy. He was in addition a manufacturer of automata, the
elaborate mechanical toys that became something of a craze in
France at this time. He was thus perfectly equipped to discuss
the most detailed aspects of Huygens's designs for telescopes and
clocks. His letters to Huygens were typical engineer's reports,
long and detailed and precise in their use of language, and their
conversations together must have had some of the same tenor.

But there was another reason for Huygens to want to spend
time with Petit – he had a beautiful daughter, Marianne. Huygens
first visited Petit on 16 January, when Petit showed him his
observatory and lent him his treatise on a proposed River Seine
canal. Thereafter, he was at the house frequently until his departure from Paris, spending days on end working on a portrait of

Marianne. As he drew and painted, they got to know one another a little, and Marianne learned enough about his outlook on life to declare that he was 'heretical' in his views, a pronouncement that gave Huygens pause for thought. When at last the portrait was done, Huygens was dissatisfied with his efforts, certain that he had failed to capture Marianne's beauty.

During these weeks, Huygens was in limbo while he awaited news of arrangements being made by his father for him to travel on to London. He returned the harpsichord he had rented and began a long round of adieux. Chapelain asked him to enquire how John Milton was faring (Milton had been briefly imprisoned for anti-royalist agitation upon the restoration of Charles II). One

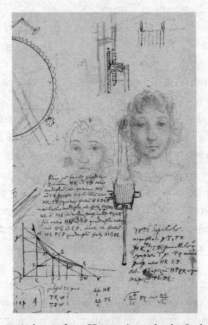

21. A not untypical page from Huygens's notebooks. It shows the face of a young woman, who may be Marianne Petit, amid a chaos of diagrams, notes and calculations.

morning, he awoke to the news that the Louvre was burning and went with friends to watch as the fire destroyed parts of the historic palace. Huygens had grown to feel thoroughly at home in Paris, and now dreaded moving on. It was not only the prospect of awful food and political instability that put him off England. (Cromwell's corpse had just been exhumed to be 'hanged and quartered, and afterwards his head triumphantly fixed to the top of a pole', as his brother Constantijn reported.*) He wrote to Lodewijk: 'I do not suppose that in England, even though I know the language very well, I could amuse myself so well as here where I have made a great number of acquaintances, whose wishing to learn from one another gives me every kind of pleasure.'

After a delay for the weather, Christiaan Huygens departed France by packet-boat and arrived, 'accompanied by porpoises', in Dover on 30 March 1661. He continued onward to London by ferry and carriage, idly noting as he passed through Chatham and Gravesend the number of English frigates lying at anchor.

He was following in the footsteps of his anglophile father and his brothers Constantijn and Lodewijk. But Christiaan had his own reasons to visit London, which was then second only to Paris in natural philosophy. Having previously sent copies of his treatises on Saturn and clocks to selected correspondents, here too he found that his reputation preceded him. On 11 April,† the

---

* Constantijn Huygens was perhaps not aware of the full horror of the story. Cromwell's head was in fact severed from his rotting corpse summarily exhumed in January 1661, shortly after the Restoration, and more than two years after his death, probably from pneumonia related to chronic malarial fever.

† English dates have been converted to New Style to accord with those used in Huygens's travel journal and letters. Evelyn gives 1 April.

English diarist John Evelyn recorded: 'I dined with that great mathematician and virtuoso, Monsieur Zulichem, inventor of the pendule clock, and discoverer of the phenomenon of Saturn's annulus: he was elected into our Society.' In fact, Huygens was formally elected to the Royal Society of London only in 1663; Evelyn's overcompensation here gives some indication of the enthusiasm with which Huygens was received.

Nothing about London could match Paris. After he had been there for a couple of months, he wrote to Lodewijk:

> I . . . don't find the stay in London as delightful as it seems you found it when you always swore you wanted so much to return there. I foresee that we shall have a big argument about that, for I will always maintain that the stink of the smoke is unbearable, and most unhealthy, the city badly built, the streets narrow and poorly paved, with nothing but mean buildings, and finally that piazza and all the common [Covent?] garden is not much and nothing compared to what one sees in Paris. The people there are melancholic, the persons of circumstance civil enough but hardly sociable, the women very shabby and not at all as witty or lively as in France: but perhaps everything was otherwise when you were there, and there is the suggestion that after the re-establishment of the court, some manners will return. I can say anyway that I have had dealings with the most decent people, most of whom have travelled in France and elsewhere, who have treated, entertained and supported me overall most nobly.

In this short time, Huygens visited Windsor and Oxford. In London itself, he called on clock-makers and ascertained the price of telescopes. But much of his time was taken up with visits to and from the leading natural philosophers. He attended the scientific meetings at Gresham College in Bishopsgate, and met many of the founders of what would become the Royal Society, including Sir Robert Moray, a Scottish soldier and diplomat with an interest in engineering projects who had good contacts with many French savants; Alexander Bruce, the Earl of Kincardine, another Scot, who would collaborate with Huygens in developing seagoing clocks; the English astronomer Sir Paul Neile, whose thirty-five-foot telescope was deployed at Gresham College; the royal cryptographer and mathematician John Wallis; and the Irish mathematician William Brouncker, who would become the society's first president.

Huygens must have been especially pleased to make the acquaintance of the astronomer and geometer Christopher Wren (his architectural career was yet to take off at this time). In 1658 Wren had proposed one of the more plausible models to explain the appearance of Saturn, but graciously gave way when he learned of Huygens's conception of a planetary ring, instantly recognizing both its visual appeal and its mathematical credibility. He also met several times with Henry Oldenburg, a German-born theologian whose linguistic skills would equip him to become the 'foreign secretary' of the Royal Society and a vital conduit between natural philosophers in Britain and elsewhere in Europe.

Many of these encounters took place at the foot of a telescope or at the laboratory bench. For Huygens, the evident Baconian dedication to observation and experiment must have made a striking contrast with the airy theorizing of some members of Montmor's group. Neile's thirty-five-foot telescope, Huygens found, 'does not seem as distinct as mine of 22', and he promised to have his lenses brought over from The Hague so that they might be compared. Bruce and Moray repeated the demonstration of Prince Rupert's drops staged at Gresham College the month before Huygens's arrival in London, allowing Huygens to discover more about the curious glass beads in addition to what he had doubtless already gleaned from his father. 'I learned that the glass tears which break are made by dipping in cold water and quickly pulling them out.' He inspected other men's pendulum clocks, too. Following a meeting with Wallis at Gresham College, they went out to observe an experiment to test whether artillery pieces begin to recoil before the projectile flies out – 'which was verified'. As he had found in Paris, his scientific fame came at a price. A clergyman came to visit him, presumably to question him on some question of astronomy or physics – 'a right pedant', Huygens found him.

More stimulating for Huygens was the opportunity to attend a demonstration of the air pump made by Robert Hooke for the Irish chemist and physicist Robert Boyle. Huygens observed the response of animals and objects placed in the vacuum generated by the device, and the following day received a visit from Boyle during which the two men 'discoursed for a long time'. The conversation inspired Huygens to set about constructing his own air pump when he returned to The Hague, which was to prove a crucial action in advancing pneumatic science.

On 3 May, Huygens passed up the opportunity to be present at the coronation of Charles II, preferring instead to observe a similarly rare event – the transit of Mercury – with the telescope-maker John Reeves. Knowing his father would be appalled that he had missed the ceremony, he gathered enough information to be able to describe the anointment by the Archbishop of Canterbury, the diamond-encrusted crown and the crowds of people inside and outside Westminster Abbey crying out in acclamation. 'All this by report, for I, however, was with Reeves to observe Mercury in the Sun, as I did. Being 30 years since

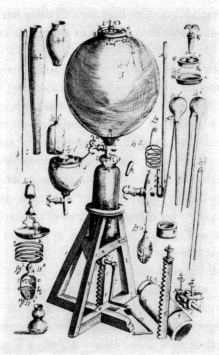

22. The air pump made for Robert Boyle in 1660. Its spectacular results inspired Huygens to make his own version of the apparatus.

M. Gassendi saw the same thing.' His father seems to have excused his son for his omission; he wrote to literary friend: 'He amused himself, I think, while others witnessed the excessive pomp of the coronation. But you will not hear without laughing what he tells me in these terms . . .'

Meanwhile, Huygens's friends in France were eager to have his critical opinion of work that they had only heard about in vague reports. He had hardly been in London a month before Thévenot wrote to him: 'I imagine that you will have been shown many curious experiments in the London academy . . . I would say it is a great advantage to have seen all these fine things with eyes as knowledgeable as yours and to know your assessment of them.'

Huygens had arrived in Paris as an exotic foreigner, a celebrated astronomer and marvellous inventor, and had quickly become a fully engaged and accepted participant in its learned gatherings. In London, he became something in addition: a scientific intelligencer and a usefully neutral agent who could be relied upon to say as he found. His record of achievement and great proficiency across disciplines – mathematics, astronomy, mechanics – meant that he was equipped better than anybody to form an accurate picture for his French colleagues of the state of scientific development in England, and likewise to acquaint the English with the latest thinking in France.

Chapelain believed that Montmor's academy, running in an informal way since 1657, had inspired the English to follow with their own society at Gresham College. Now, he felt that the scientific progress the English were making in many fields could be used to galvanize renewed efforts in Paris. He told Huygens: 'It seems our Academy is warming up by the emulation that they [the London savants] give it, and the wish is to turn to experiments in preference to all other activity where there is the mind to do it.' A similar effect was observed at the Accademia del Cimento in Florence. Such virtuous circles made it possible to compare observations and experimental results more effectively between scientific centres in different countries. They also made it easier to compare experimental methods and the design of instruments, which was perhaps of even greater significance for the progress of science. These comparisons depended not only upon a sustained international exchange of letters, but also increasingly upon the physical presence of expert practitioners, whose witnessing of, and sometimes active engagement in, the design and conduct of experiments gave them a practical knowledge not always communicable even in the most detailed letter.

All the time he was in England, Huygens was mindful of how the science he saw practised there might be emulated in France. He may not have held London in high regard at first, but the men who assembled at Gresham College clearly had much to teach the Montmor circle. Years later, when he was living in Paris, he told a visitor of his belief that the members of the Royal Society were 'an assembly of the Choisest Witts in Christendome'. He reported back to Chapelain on his visit to London with the clear objective of trying to steer his French colleagues towards a more rigorous experimental approach, but his initiative received an obstinate rebuff.

Huygens had hoped to return directly to France after his stay in England, accompanying his father and brother Lodewijk on business for the House of Orange, but he was obliged to return home to represent the family at another no doubt tedious wedding of some cousins. He busied himself at home in The Hague with a typically eclectic set of investigations, observing Saturn, grinding new lenses with Constantijn, making calculations of the pendulum, modifying Boyle's air pump and thinking about musical theory.

He gave Lodewijk a letter to take to Marianne Petit, expressing the hope that she remembered the man who came and drew her, and assuring her that he 'goes over in his memory everything that is beautiful in your face, and charming in your conversation'. But it seems that the sketches Huygens made of her in the hope that a painter might use them to make her portrait had caused her some embarrassment. After delivering the letter, Lodewijk, duly struck by her beauty, reported back to his brother that she wanted them destroyed. Christiaan wrote back: 'I told you that Mademoiselle P. was more beautiful than in the drawing I showed you, and you have seen too well to judge otherwise. However I will not do what she wishes me to do with these portraits, although in every other thing I would wish to render complete obedience.'

On 1 August 1661 Christiaan Huygens received a visit from Henry Oldenburg, who was passing through The Hague on his way back to London from his home city of Bremen. Oldenburg was able to inspect Huygens's telescope, and to review his plans to build an air pump like the one he had seen Boyle demonstrate at Gresham College. The two men also spent some time discussing the reproductive habits of the crayfish, because Oldenburg was carrying with him a new natural history of the animal by Sir Kenelm Digby, who claimed to have generated the young by distilling the putrid remains of the adults. Huygens found the work 'admirable but I admit that I have trouble believing it'.

Oldenburg and Huygens were kindred spirits. Adept in several languages, curious about all manner of natural phenomena and accustomed to the etiquette of court life, each was eventually able to rise to a senior position in the early scientific societies of Europe in a country other than that of his birth. Both believed that progress would come faster if there was dialogue between investigators everywhere. Though Oldenburg made no important observations and performed no experiments himself, he was to prove one of the most significant figures in the development and spread of scientific ideas during the second half of the seventeenth century.

Heinrich Oldenburg, as he was once known, had set out from Bremen for London first in 1653 as a civic envoy commissioned to negotiate the release of a Bremen trading ship carrying a cargo of French brandy, which had been seized by the English during the First Anglo-Dutch War. He settled in England, working as a tutor, as he had done in the Netherlands and elsewhere on the continent, until Robert Boyle employed him as an assistant. He travelled again to France and Germany in his new capacity and became well acquainted with the leading natural philosophers in both countries, including members of the Montmor circle, whom he doubted would ever be able 'to produce any great matter in point of Tubes [telescopes] or chymistry or any mechaniques'.

Oldenburg was elected to the future Royal Society in December 1660 and listed – with equal status, not as a foreign member – among the forty founding names. As one of two secretaries of the society, he had many duties to perform, salaried initially by the wealthy Boyle. He attended the weekly meetings and recorded the minutes. He served as a translator, especially of English work into Latin, so that it might be more readily understood abroad, but also of scientific news that he received from his German contacts into English. He also launched the *Philosophical Transactions*, which quickly became the official journal of the society.

But his most important task was to oversee the scientific communications of more than thirty regular correspondents in the British Isles and on the continent. In addition to Christiaan Huygens and his father, these included the Danzig astronomer Hevelius, Boulliau and Petit in Paris, the telescope-maker Cassini and the physician Malpighi in Italy, and the microscope pioneers Swammerdam and Leeuwenhoek as well as Spinoza in the Netherlands. It was arduous work, involving the hand copying of their often long and detailed letters, as well as writing to colleagues summarizing the recent work of others in whom they might be interested. These progress reports typically comprised a short paragraph on each person's work and would cover a wide range of topics. A typical letter to Huygens, for example, included intelligence on new observations of Saturn from Italy, on Hooke's work on pendulum clocks, on Barrow's forthcoming book on optics and on Wren's progress with lenses.

Oldenburg soon adapted his role into a more subtle one that made greater use of his diplomacy. He became a vital intermediary, able to effect introductions that would prove significant for the progress of science. For example, in later years, he introduced Huygens to Leibniz and, with the help of the Huygenses, brought Leeuwenhoek's work with the microscope to the attention of the Royal Society. He helped experimenters obtain the materials they needed. From his broad knowledge of progress in many fields, he was often able to advise his correspondents when another already had the priority for a particular discovery, or when an experiment had already been performed, so that they might avoid unnecessary duplication. His secretarial duties involved much chasing of tardy and hesitant writers, and his judicious persistence saw to it that much work which might never have been finished or published was in fact added to the body of scientific knowledge. He developed a useful ability to handle both the arrogance of overconfident claimants to this or that new discovery and the offence of those who felt themselves cheated, and duly found himself having to moderate numerous, often international, disputes between rival scientists. (Huygens was to experience his share of these disputes.) Here, his skill often lay in advising agitated correspondents on ways to respond temperately, thereby defusing tensions between them.

Oldenburg was put on his mettle especially during times of international conflict when letters were often rerouted or intercepted. An ignorant official might easily construe the scientific and mathematical content of Oldenburg's typical correspondence as being of military value or in some form of code. During the Second Anglo-Dutch War, therefore, Oldenburg asked Huygens to address his letters to 'Monsieur Grubendol' (a decidedly pregnable anagram of his name), and gave him a new address from which items could be forwarded without the risk that Oldenburg might be accused of spying for the enemy. Huygens, like most of Oldenburg's scientific correspondents, did what he could to ignore the exigencies of war, and Oldenburg repeatedly expressed peaceful sentiments in his letters to Huygens: 'I wish always an end to the war and the plague, with an unequalled passion, so as to re-establish study and good relations.'

Perhaps, though, Oldenburg's elaborate precautions merely encouraged the suspicion that he might be a spy. When the Dutch launched the audacious Raid on the Medway in June 1667, torching a number of English warships and towing away others including the flagship, HMS *Royal Charles*, the government feared that an intelligence failure was to blame, and suspects were indiscriminately rounded up. Oldenburg was imprisoned in the Tower of London, accused of 'dangerous desseins and practices' because of his copious foreign correspondence. Samuel Pepys, a fellow of the Royal Society as well as a senior official on the Navy Board, wrote in his diary that Oldenburg had been held for 'writing news to a Virtuoso in France with whom he constantly corresponds in philosophical matters'. Although Oldenburg had other correspondents in France, if the concern was the leak of intelligence to the Dutch, then this can only refer to Christiaan Huygens, who was living in Paris at the time. After two months, Oldenburg was released without charge or apology, and promptly resumed his former duties.

Huygens's professional association with Oldenburg, augmented in its later stages by a personal friendship struck up by his father while staying in London in 1671, lasted until the German's sudden death from ague in 1677. With that, a link was broken between scientists everywhere, and especially between scientists in Britain and their continental counterparts. Oldenburg left behind him the visionary prospect of an international network in which it was understood that the best hope for the advance of science lay in speedy, concise, honest and civil communications between like-minded participants irrespective of their nationality. No other personality emerged with the combination of scientific, linguistic and diplomatic abilities to continue what Oldenburg had started, before rising nationalism and a succession of European wars made such a thing seem hopelessly idealistic. The connection was arguably not reforged until the foundation of international scientific associations in the twentieth century.*

Huygens had returned home from London in May 1661 with Boyle's description of the air pump. He was keen 'to do some yet new experiments in the vacuum, and to have the pleasure of trying some of those which are in his book'. The air pump was an exciting invention for both practical and theoretical reasons. The nature of the vacuum, and whether it was truly an utter void, was a topic of lively speculation not only by scientists, but also by philosophers, theologians and lawyers. Control of the vacuum promised to lead to improved technology for pumping out mines and polder land. It was 'seventeenth-century "Big Science"'.

Huygens followed Boyle's instructions closely, paying particular attention to the airtight seals where different parts of the apparatus were joined. However, he found it hard to get parts made with the necessary precision by his errant instrument-maker, and early

---

* Of organizations pertinent to Huygens's interests, the International Astronomical Union was established in 1919, and the International Union of Pure and Applied Physics in 1922. Their European equivalents were not set up until after the Second World War.

results were not encouraging. 'The pump does not work,' he wrote to Constantijn in October in a short letter that also included notification of the family wedding that had prevented him from returning to Paris. 'The tube was so uneven in width that little or no air could be worked from the bottle.' After another month, however, he dared to make some modifications to Boyle's design, and found that the bladder used to show the goodness of the vacuum in the bottle would remain inflated overnight – 'which Mr. Boyle was not able to effect'.

He was then able to repeat many of the experiments he had seen Boyle perform in London, for example demonstrating the rapid boiling of water in the vacuum, and showing, by placing an alarm watch inside the evacuated chamber, that the transmission of sound was greatly impaired. He bought some mice and birds, and then in triumph sent the corpse of one bird, 'which died in the same way as described by Monsieur Boyle', to Lodewijk in Spain. Huygens had greatly admired Boyle's rigorous experimental approach, which stood in marked contrast to some of the scenes he had witnessed in France and to the English Hobbes, who had long held a fascination with the vacuum but who remained ill-equipped as a scientific investigator.* He was soon emboldened to bring his own further work with the air pump to Boyle's attention. Boyle was pleased to find that Huygens's objections to his

---

* As Huygens wrote to Moray in September 1661: 'M. Hobbes is about as good a geometer as Jos. Scaliger. [the sixteenth-century French scholar who insisted that π was equal to the square root of ten]' (OC22 71–3). In response, Moray sent Huygens a copy of Hobbes's treatise on the nature of air. It was no better. Huygens replied: 'In the dialogue of Monsieur Hobbes [in which he attempted to apply the methods of geometric proofs to the vacuum] I find nothing solid, only pure visions. It's a mental error or because he likes to contradict what he doesn't accept as real reasons in the effects of the vacuum given in Monsieur Boyle's book . . . besides it is a long time since Monsieur Hobbes has lost all credit with me in the field of geometry' (OC3 383–5).

work were 'soe few, as well as soe judicious', and made a positive response to Huygens's suggestions.

At the Royal Society, however, members were unable to replicate the improvements Huygens had made, despite his letters describing what he had done, and even the visit of an emissary sent by Oldenburg to view the apparatus directly. It was not until Robert Hooke was made 'curator of experiments' at the society, and Huygens himself came over from The Hague in 1663, that the London air pump could be made to work well enough to repeat Huygens's new experiments.

Many scientists lack the 'green fingers' needed for building apparatus and carrying out successful experiments that Hooke and Huygens clearly both possessed, and Huygens remained the only investigator who succeeded in building an air pump independently of Boyle's original. But even he resented the time eaten up by the temperamental device, and sometimes pretended it was broken so that he did not have to set up another demonstration. His achievement was nevertheless an important next step in validating Boyle's experimental results, spreading them on the continent, and establishing a robust theory of air and the vacuum. Huygens himself had initially disagreed with Boyle's suggestion that the pressure exerted by a given mass of air is inversely proportional to its volume (the relationship now known as Boyle's law), but he was obliged to change his thinking when confronted by his own experimental results.

What happens, though, if the volume of air is infinitely expanded and the pressure falls to zero? This was the broader context of experiments with the air pump ever since Torricelli had invented the barometer by creating a vacuum in a tube of mercury in 1643. What actually occupied the void thus created? Descartes and his followers, including Huygens, believed some form of 'subtle matter' must be present, but others, including

Pascal and Roberval, were prepared to believe that the space was truly empty. The potential emptiness of the void created by the vacuum pump raised important theological questions, because of the belief that any space not filled by God's spirit might be occupied by the Devil. The philosopher Hobbes could conceive of a vacuum in a physical sense. (He imagined two bodies pulled apart to leave a space between them. Did they then still touch? Of course not; therefore the vacuum exists.) But he strongly resisted the idea in a metaphysical sense, believing that the void left behind created a space for political dissension.

Boyle and Huygens were not much interested in such sophistry. However, another of Huygens's experiments with his own air pump introduced a new confusion. During the winter of 1661–62, Huygens placed a Torricellian tube of water (a filled glass tube inverted and placed in a larger open container of water) inside the chamber of his vacuum apparatus. As the air was evacuated from the chamber, he noticed that bubbles began to form in the tube, and as they rose to the top the water was gradually pushed out. This observation could be explained by Boyle's theory, which predicted that the level of the water in the tube would gradually fall to the level in the larger container. Alternatively, it could more simply be due to the release of air naturally dissolved in the water. Huygens therefore repeated the experiment using water that had been purged of air by prolonged suction. This time, the tube of water remained full as the air in the chamber around it was pumped out. A tube of mercury, on the other hand, did behave in accordance with Boyle's theory. The 'anomalous suspension', as Huygens termed it, of the water column appeared to suggest that it might contain within it some additional, unknown 'subtle fluid' responsible for the unexpected behaviour.

Experiments with the air pump were always challenging, and Boyle had run across various 'hydrostatical paradoxes' before as he refined his experimental setup. But when he tried to replicate Huygens's experiment, he was unable to get the same result. The apparent fact of the new 'discovery', and his respect for Huygens, the only other person who had been able to construct a functioning air pump, led Boyle to fear that Huygens now had the superior apparatus and technique. This latest development seemed to put Boyle's law in jeopardy and his scientific credibility on trial. Boyle, who, along with Oldenburg, had done so much to crystallize the rules for dealing with disputed claims in science, now found himself with a personal opportunity to see how well the rules worked in practice.

There was no falling out on this occasion, but the matter was resolved to both men's satisfaction only when Huygens visited the Royal Society again in 1663. With Huygens personally present, and even then with Hooke's involvement too, Boyle was finally able to repeat the Dutchman's work with success. The fact that both men fully understood at this relatively early date the importance of being able to replicate experimental results in order to establish scientific facts must surely have helped to overcome any tensions between them. Both could now agree that water behaved differently from mercury in the Torricellian tube, even if they could not explain why. (Later, Newton was able to explain 'anomalous suspension' as an effect of capillary action.)

This outbreak of scientific concord may well have been the most pleasing aspect of the visit for Huygens. But there was an official accolade awaiting him too, as, on 22 June 1663 (OS), just one month after the admission of its first fellows, Huygens himself was elected to the Royal Society.*

---

* Huygens was admitted at the same time as Samuel de Sorbière – the first foreign fellows of the Royal Society, although their foreignness was never thought worthy of note until the society tightened its election rules in 1682 (Hunter 119).

New Music

Music always sounded through the Huygens family home. On the Fridays when he was in residence, Constantijn hosted musical evenings, with neighbours and guests joining him in playing and singing while he himself stood ready to take up whatever instrument the ensemble required. When they were old enough, the children, too, were encouraged to participate in the amateur *collegium musicum*, and sometimes, it was noted, they performed even better than their father. As he grew up, Christiaan did not exclude music from the range of his scientific explorations, and the curious innovation he was to make in this field surely has its genesis in these domestic entertainments.

His father had shown an extraordinary aptitude for music from the beginning. At the age of two, he could repeat his mother's singing the Ten Commandments to him in French. His formal musical education began at four, with a tutor who helped Constantijn and his brother Maurits identify the notes of the scale by assigning their values to the gilded buttons that ran up the sleeves of their winter coats. Constantijn proved to have the better ear and voice, and at six years old he began to learn the viol, or viola da gamba, which was soon followed by lessons on the lute, the harpsichord and the organ.

On a visit to Amsterdam as a young boy, Constantijn attended the soirée of a family friend where the music-making was led by

Jan Sweelinck, the organist at the Old Church in the city and the most influential Dutch composer of the seventeenth century.

> It happened that in the presence of a good company of men and women (who were surprised that I, while others had been making all kinds of mistakes, had never made an error the whole afternoon), I lifted my eyes from the line in the score and in my confusion and shame had no possibility of finding my place again. This little misfortune – indeed, I remember it exactly – dealt my childish enthusiasm such a heavy blow that I burst out in bitter tears and could not be persuaded to take up my viol a second time.

This mishap did not hold the young Huygens back for long. When, in his twenties, he travelled to England – the major centre for viol consorts – on his first diplomatic assignments, he fell in easily with the musical circles there and met other composers, such as Nicholas Lanier, the Master of the King's Musick. In Venice, he heard Claudio Monteverdi conducting his work in St Mark's, 'the most perfect music I ever heard in my life'.

Music served many purposes for Huygens. It opened the door to friendship wherever he went, and the unforced appreciation with which his efforts were rewarded was perhaps all the more welcome for coming without the further obligations that tended to follow upon the heels of praise for his work at court. Often, it gave him access, both to intellectual and creative figures and to people with real influence who would be in a position to help him realize his master's diplomatic goals. Constantijn had learned this useful lesson early on when he suddenly found himself playing for a delighted King James I. Huygens – a young man from a young republic – was clearly thrilled to have experienced a royal audience, but, equally clearly, he was not overwhelmed, either by the occasion or by the monarch himself, as he continued to play on unbidden in his presence.

Throughout Huygens's long life, music-making was also a means of escape from daily cares. Often, as he told an English musician friend, he would use it 'to fiddle myself out of bad humour'. When troubles mounted, music became more important, not less. At the end of the 'disaster year' of 1672 when France betrayed the Dutch Republic by joining England in war against it, Huygens wrote to this same correspondent that he found it vital to 'sweeten the bitter displeasure of these times with some harmonious practise, as I can tell you I doe, and never will give over while I breathe'.

The recipient of these confidences was Utricia Swann, née Ogle, the most enduring of Huygens's many female companions in music. For it was undoubtedly the case that Constantijn Huygens saw musical gatherings as a congenial way of meeting educated and talented women. The pattern was set in his bachelor years in the Muiden Circle of poets, where the beautiful Tesselschade would sing and play, and became more important for him after the death of his wife in 1637. These social occasions offered abundant opportunities for flirtation while sharing a keyboard or accompanying a song.

From Tesselschade onwards, many of Huygens's women friends served him in the role of muse. But Utricia was clearly something rather more than this. A likely portrait of her by a follower of van Dyck shows a woman looking imperiously at us down her long nose and sidelong across her bare shoulder. With lips pursed, ringlets falling across her forehead and a rope of pearls around her neck, she projects the air of a diva. They first met in 1642 when Huygens was overseeing a visit to the Dutch Republic by Henrietta Maria, the wife of King Charles I, and mother of Mary, the child-bride of the future William II of Orange, and Utricia was one of her retinue. She soon charmed him by her appearance and her voice, prompting an outpouring of verse and new keyboard arrangements of songs he had heard her sing. When she sang in his garden at Hofwijck, he likened her sound to a nightingale, although another time she cried off an assignation, saying her tone had 'a certaine tang like that of a paire of skates upon your ice'. The musical friendship intensified even after her marriage to William Swann, who was, like her father, an English officer in the Dutch army. Not three weeks after her wedding, Huygens wrote teasingly to her – crossing out an absent-minded 'Mademoiselle' and putting 'Madame' in its place, and referring ironically to 'a season when I imagine that you would have "much adoe about nothing"' – simply to milk her praise for his 'musical productions'. When Huygens learned one time that they would be unable to meet for an extended period to make music together, he boldly proposed an alternative arrangement. 'Since all our aural communications are ruined,' he wrote, 'I am minded to reconnect us in some manner by the sense of smell.' Enclosed with the letter were a couple of pot-pourri sachets together with an instruction to place them in her bed so that her sheets and his might smell the same. If he overstepped the mark with such tokens, it was soon forgiven. Their contact intensified in the 1650s, Huygens's letters often larded with musical offerings to tempt her to visit again.

His poetic rivals Hooft and Vondel might on occasion tease Huygens for his knotted, unmusical verse, but Huygens could take quiet comfort that he was a genuine composer as well as a versifier. Vondel called him 'Holland's Orpheus'. He wrote more than 800 musical compositions by his own count, although none survives in manuscript form and only a few were copied. The most important of these surviving works is *Pathodia Sacra et Profana*, a setting of twenty psalms in Latin together with a similar number of songs in Italian and French for solo voice with a basso continuo accompaniment that would be played on a lute or theorbo, with the option of additional players on other instruments. It was written for Utricia's voice and is dedicated to her. Huygens arranged to have the work published anonymously in Paris in 1647, perhaps because he had a low opinion of the Dutch music scene – he tried and failed to get the stadholder's court to sponsor more musical activity – but more likely because he felt it would be improper for a servant of the court to be seen seeking attention in his own right. When he played, too, he was at pains to stress his amateur status, not out of modesty but for fear of

23. Title page of Constantijn Huygens's collection of songs and psalms, *Pathodia Sacra et Profana*, 1647. 'Occupati' is Huygens's pseudonym and alludes to his busy work as a diplomatic secretary.

compromising his professional position as secretary to the stadholder. His compositions were, as he wrote to an English courtier in 1648, his 'after-dinner diversions and, as you might say, my taking breath after the day's work'.

The *Pathodia* was an unusual melange of European styles. Its Latin psalms are set in a traditional polyphonic manner with two independent melodic lines, while the Italian songs follow the simpler pattern of madrigals. The French songs are written in a more modern – that is, baroque – style with a freer singing line. Like Sweelinck before him, Huygens avoided setting Dutch to music, preferring languages more usually associated with the singing voice at the time. Some of the song texts are Huygens's own and, as might be expected of a poet, the collection is more notable for its exploration of the possibilities of word-setting than for its compositional novelty.

Nevertheless, Huygens was interested in musical innovation. As a child, he learned music according to the new seven-tone scale rather than the older scale of six notes. In adulthood, he thought fit to offer a list – in verse couplets, of course – of criteria that a musical work should satisfy. The first two items on the list were that a piece should be 'new' and 'pleasant to sing'. He also emphasized the new in purely instrumental music. He boasted that he had 'created music that nobody had ever heard, but that was born in me without the slightest exertion'. In fact, Huygens's musical achievement may have more to do with the enthusiastic way in which he distributed his work, thereby transmitting ideas he had heard in one place to be heard perhaps for the first time in another. The *Pathodia* thus may have played a role in facilitating the spread of basso continuo from Italy, where Huygens was presumably introduced to it, to France. His interest in originality reveals his modern sensibility, but perhaps this should be weighed against his mien as a diplomat, seeking harmony wherever possible, which may have caused him to play safe in his compositions. As in his thinking about scientific problems, Constantijn Huygens remained in his musical ideas awkwardly poised between old ways and the new.

Music in the Dutch Republic was also a moral issue. The Calvinist liturgy was generally confined to psalms sung – usually badly – by a single cantor. Organs in the churches were often owned by the city. Their music did not feature in the body of the religious service, but might sometimes be played before or afterwards. However, Constantijn Huygens, who was certainly as devout a Calvinist as the next man, loved music, and believed that hearing it and making it was 'not unbecoming for persons of good standing and good birth'. He felt there was no reason why music should not be heard in a church of any denomination. In a pamphlet entitled *Gebruyck of Ongebruyck van 't Orgel* ('Use or Misuse of the Organ'), printed in 1641 – in which he took the opportunity to reaffirm that he had been 'born, nurtured and trained against the public divisiveness of the tendency to Rome' – he nevertheless argued for the readoption of organ music as part of Protestant services. Ever the diplomat, Huygens had first sent the potentially controversial text to leading figures whose opinion on musical matters he valued, including Amalia van Solms, Hooft and Descartes. The French philosopher struggled with the Dutch text, but recognized that Huygens was hoping to see the organ used very much as it was in Catholic services, and replied warmly that it might be the instrument 'to rejoin Geneva with Rome'. This did not happen, of course, but during the next few decades, congregational singing with organ accompaniment did slowly come into vogue in Protestant churches.

Constantijn Huygens was fortunate to live long enough to see this transformation under way. He never tired of music-making or musical experimentation. The theorbo remained, as it long had been, his favourite instrument for playing at home, although when he was away on campaign he took with him the more practical lute. Lying on the table in his early portrait by de Keyser is a hybrid instrument combining the portability of a lute with the greater dynamic range of a theorbo. When he was well into his seventies, he tried out the guitar, but found it a 'miserable instrument'. Two months before his death, he complained to a friend that gout was hampering his musical efforts, but he could still play the theorbo, 'at least so, as the saying goes, a drunken farmer would not notice the difference'.

In these last years of his life, with most of his artistic friends long dead, Constantijn Huygens was comforted by the presence once again in the family homes of his favourite son, Christiaan.

Constantijn saw to it that the musical upbringing he had received was replicated for his children. Young Constantijn and Christiaan received instruction together. This education started at an early age when their father played the notes of the scale to each of them – only Christiaan was able to sing them back accurately and wished to know more. Less attention was paid to their younger siblings, even though the youngest, Susanna, may in fact have been the most naturally talented musician among them. Although the Huygens family was undoubtedly more committed than most, music was a standard feature in the education of any young gentleman or lady at this time. Not only was music one of the four subjects in the traditional quadrivium of the liberal arts, but harmonics, or the theory of music, was also considered one of the 'classical sciences', along with astronomy, optics, static mechanics and mathematics. Music thus also played a full and entirely expected part in the range of interests of men such as

Galileo, Mersenne and Bacon. In the Low Countries, both Simon Stevin and Isaac Beeckman sought to contribute to the development of music theory. Prompted by Beeckman, Descartes, too, produced a treatise on music, even though by all accounts he could hardly tell a fifth from an octave.

Christiaan clearly enjoyed music, and musical parties were often held in his Paris apartment – he wrote of one such concert, where there were women 'who sing very well, and play the harpsichord even better'. But he was rather less of a composer than his father. Although he and his brother occasionally wrote when the other was in Paris about the latest songs to be heard there, the practical production of music was not as important to him as understanding its fundamental mathematical structure. Like other scientific figures, he thought this was a prerequisite for further innovation in music. All that survives of any music he wrote is sixteen bars of the courante from a dance suite, along with odd snatches of notated melody scattered through the margins of his scientific notes. He had learned to sing and play the viola da gamba and the lute. However, he most often played the harpsichord, an instrument more usually taken by women in amateur musical circles, and it was at the keyboard that his interests took a more theoretical turn.

The human voice is able to sing by producing sound waves over a continuous range of frequencies. Many musical instruments, however, are designed to emit tones at fixed frequencies, according to the length of individual vibrating strings in the case of keyboards, or resonating columns of air in organs and wind instruments. Players of such instruments are thus governed by rules of temperament, which determine the relative magnitude of the intervals between successive notes in a scale.

Temperament has necessarily always been a concern of musicians, however they choose to acknowledge it. The number and spacing of notes in a musical scale (the graduated series of sounds into which an octave is divided, an octave being the musical term for a factor of two difference in sound frequency or pitch) is clearly a cultural decision. But underlying this is the sense of a natural or divine system of proportion which might guide instrument-makers' choice of where to position individual notes (and which might, or might not, bear some relation to other proportional systems, such as the 'harmony of the spheres' supposedly exhibited by planetary orbits in the solar system). For example, the musical interval of a fifth is produced by the vibration of strings with lengths in the proportion 3:2. A system of tuning based on this ratio alone is attributed to Pythagoras, who is generally credited with being the first to notice that pleasant musical harmonies tend to be produced by vibrations of strings with lengths in simple ratios.

During the Renaissance and early modern period, the lost sounds of ancient music became a topic of avid speculation. Indeed, one of Christiaan Huygens's first forays into musical theory, made in 1655, was a response to a book produced by a Danish music theorist, Marcus Meibomius, who had attempted to re-enact performances from ancient Greece. Huygens believed that the ancients must have used the octave and the fourth (based on string lengths in the ratio 4:3) and the fifth (3:2), but not the smaller interval of the third (5:4), which was thought to be less harmonious than the larger intervals.

The discovery during the seventeenth century that musical pitch (a qualitative index of how high or low a musical note is) is proportional to the frequency of vibration of the air produced by musical instruments – and therefore also inversely proportional to the more easily measured length of a string or pipe – set music theory on a solid quantitative footing. It was now possible to confirm that the musical intervals of Pythagoras were indeed based on exact ratios of the lowest numbers. Using this knowledge, some more mathematically inclined musicians now proposed to divide the octave in new ways that paid little heed to musical tradition. Stevin, for example, favoured a twelve-tone scale in which the pitch of successive notes increases or decreases logarithmically by exactly the same ratio (the twelfth root of two). This twelve-tone 'equal temperament' eventually became the most widely used tuning system in Western music, and is still the norm today. Others explored the division of the octave into even more tones. Such innovations held the promise of music that would sound entirely new. In 1627, in his utopian novel *New Atlantis*, Francis Bacon imagined, among many other scientific inventions, 'sound-houses' wherein 'We have harmonies which you have not, of quarter-sounds and lesser slides of sounds'.

In October 1660 Christiaan Huygens had passed through Antwerp on his way to Paris. There he heard the spectacular new carillon installed by the Hemony brothers, bell founders who made carillons for many churches throughout the Low Countries, which may have stimulated his own decision to explore the division of the octave. Holding fast to Pythagorean fundamentals, he sought to express all the intervals needed by practising musicians in algebraic terms of the octave and the fifth only. After much calculation, beginning by splitting the octave into 100,000 equal subunits, Huygens found that the best fit could be effected by shifting the ratio of a fifth a little, from 100,000 / 66,666 to 100,000 / 66,874 (a shift of two cents in modern musical terminology, which is imperceptible to the human ear).* He reported on his work to his Scottish friend Robert Moray: 'I have been busy for a few days working on music, and the division of the monochord [i.e. octave],

---

* There are 1,200 cents in an octave, and thus 100 cents between each note in Stevin's twelve-tone scale.

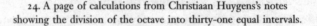

24. A page of calculations from Christiaan Huygens's notes showing the division of the octave into thirty-one equal intervals.

to which I have happily applied algebra. I have also found that logarithms are of great use, which has caused me to consider these marvellous numbers and to admire the industry and the patience of those who have given them to us.'

Huygens's hope was not to discover yet another mathematically ideal scale, like Stevin's scheme, which was widely rejected by the subsequent generation of music theorists including Beeckman, Descartes and Mersenne, but to create something of practical use to musicians, and especially to keyboard players. He was convinced that pure ratios, 3:2, 4:3 and so on, gave the most pleasant-sounding intervals between notes. The difficulty was that not all these ratios could be produced precisely on a conventionally tuned keyboard.

Instrument-makers had to introduce various cheats in order to produce a usable keyboard. One such compromise system, called mean-tone temperament, attempted to privilege thirds by tuning them to lie equidistant between the octave and the fifth. However, this required the fifth to be slightly flattened, which some musicians found objectionable. By contrast, in the newer (but still unpopular) system of tuning to 'equal temperament' based on Stevin's twelve-tone scale, *all* the intervals were somewhat impure, with fifths only slightly off but thirds more substantially so. This system, however, offered important advantages for practical music-making such as greater ease of modulation and transposition between keys.

Huygens did not favour twelve-tone equal temperament because its thirds were so far off true, but neither was he willing to accept the pragmatic alternative of 'well-tempering', which aimed to distribute the impurities in such a way that they affected the important intervals the least, because this approach was so makeshift. Instead, he wished to see a system that would achieve musicality while at the same time being underpinned by mathematical truth.

Proceeding further with his algebraic analysis of temperament during the summer of 1661 – his notes are strewn with little sketches and calculations of ratios to eleven significant figures – Huygens eventually found that he could divide the octave into thirty-one equal segments. In this scheme, each semitone is made up of two or three of these segments. The full octave then comprises the standard seven diatonic (white-key) and five chromatic (black-key) semitones as follows: $(7 \times 3) + (5 \times 2) = 31$. This system produced a close match with the aurally pleasing but previously unmathematical mean-tone tuning, and introduced the possibility of unlimited transposition between keys. It gave the fifth a ratio 31:18, the major third was 31:10 and the minor third 31:8. Other ratios had the potential to produce smaller intervals such as the 'quarter-sounds' imagined by Bacon.

This was not an entirely novel departure. Huygens was aware

from his French mentor, Mersenne, of instruments with many more than the usual twelve keys per octave that had actually been built in Spain and Italy as long ago as the mid sixteenth century, but these reportedly sounded dreadful. The discovery since that date of logarithms made it possible to calculate the proper length of the strings more precisely (even if they could never be fitted to eleven significant figures of accuracy!) and promised a somewhat sweeter result. However, the prospect of sitting down to an instrument with thirty-one keys to the octave, and having to pick carefully among them simply to play an ordinary melody, was hardly likely to appeal to any real musician, and Huygens did not yet consider constructing such a keyboard, which he felt would be unplayable 'without being confused by the multiplicity of white and black keys'.

But in 1669 Huygens returned to the question once more with renewed determination to find a practical solution. His answer was to fasten a conventional twelve-keys-to-the-octave keyboard by means of pins to an underlying keyboard comprising thirty-one pivots, which were connected in turn to the thirty-one strings within a given octave. With contact made between the keys on top and the pivots that came into alignment with them from below, the player would be able to strike the correct twelve strings for a given key as on any normal keyboard. Yet it would also be possible to reposition the top board elsewhere along the row of thirty-one pivots underneath in order to play tunes in other keys. In this way, the player would be able, with a certain effort, to transpose music more freely from one key to another.

Musical history, however, chose a different path. During the eighteenth century, twelve-tone equal temperament came to prevail for the simple reason that composers valued the ability to modulate between keys more highly than tonal purity.

Although the exact details of his design are not known, it seems that Huygens did have a specially adapted harpsichord built for him in Paris. On 10 July he wrote triumphantly to Lodewijk 'that my harpsichord invention has succeeded very well, and I would not be without it for anything'. Sadly, the instrument has not survived, and Huygens's only treatise on music, *Le cycle harmonique*, which was not published until 1691, reveals little additional information, confirming only that he experimented with a variety of instruments:

> I have at other times made adjustments to such movable keyboards on harpsichords as there were in Paris, and even to those which had their ordinary keyboard, where it was necessary that what I put on top matched the heights of the white and the black keys so that the keys could slide without impediment. And this invention was admired and imitated by the great masters who found in it utility and pleasure.

One particular effect of Huygens's retuned keyboard might have provoked a small revolution in music if the new instrument had been taken up more widely. His adjustment of certain musical intervals made some chords that had been conventionally regarded as dissonant sound more pleasing to the ear. Among these was the chord known as the tritone (because it was made up of two notes three whole tones apart; it is now termed an augmented fourth or a diminished fifth). The tritone had been infamous since the Middle Ages when it purportedly acquired the nickname of the *diabolus in musica*, or the 'devil in music', and was prohibited from use by church musicians. In Huygens's tuning system, however, the chord suddenly lost its terror. Huygens himself judged it 'harmonious upon attentive examination' and employed it without compunction when he played his own harpsichord.

Christiaan Huygens's experience with his novel keyboard illustrates some of the similarities and differences between father and son in their musical philosophy. Christiaan was more prepared to listen with an unprejudiced ear and more interested in the principles underlying musical effects. He clearly believed that innovation in music, as in other fields, represented an idea of progress. For example, in *Cosmotheoros*, his posthumously published meditation on extraterrestrial intelligence, he speculated that beings on other planets would hardly have a theory of music less advanced than our own, 'neither can we presume that they want the use of half-notes and quarter-notes, seeing the invention of half-notes is so obvious, and the use of them so agreeable to nature'. Constantijn, meanwhile, was more concerned with musical output, in terms of both composition and performance, and was content to work within existing conventions and genres. As with their contrasting approaches to optical phenomena, the father was more captivated by the spectacle while the son sought fundamental understanding, reflecting the emergence during their two generations of a more rigorous idea of scientific method.

But Christiaan was not obsessed with the principles of music at the expense of its pleasures, many of which, he was sure, would remain inexplicable. He was adamant that the purpose of music was to please the senses and not to make gratuitous display of its artfulness. He favoured mean-tone temperament in the end not only because of its purer intervals, but also because it offered an appealing variation of character to sustain the listener's interest when a melody was played in different keys. In this he was in complete harmony with his father, who in 1680 saw fit to remind Christiaan of 'my couplet which gives the 6 things I require in all composition'. He ended by advising Christiaan to be true to what his own senses tell him: 'I would be led ear-wise rather than nose-wise.'

Although the instrument Huygens had made for him in Paris no longer exists, an organ based on his original concept was built after the Second World War by a Dutch physicist, Adriaan Fokker. This instrument is now installed in the Muziekgebouw concert hall in Amsterdam, where it is regularly used for recitals, both of music from Huygens's period and before, when these musical experiments were originally undertaken, and of contemporary works written specifically to showcase its microtonal capability. In place of Huygens's detachable top keyboard, the Fokker organ has multiple keyboards arrayed in a colour-coded grid so that the player can alter pitch by quartertones simply by moving the hands vertically up or down from one row of keys to the next, while any one row of keys is played just like a standard twelve-note keyboard with its familiar white and black keys.

What does it sound like? In short, not as outlandish as you might expect. The one 'universal constant' of music is the octave – the fact that a tone at twice (or half) the frequency of another has a simple mathematical proportionality that we hear as related – although even this phenomenon is rather odd and without parallel in the other human senses. The only other fundamental consideration is how small a difference in the pitch or frequency of sound the human ear is able to detect. This varies according to the volume, pitch and quality of the sound (its timbre) as well as with the listener's age and training, but is generally agreed to be about 0.5 per cent of the octave, equivalent to around two hertz in the middle of our auditory range. Over time, Western music has developed in such a way that, of the vast permutations theoretically available, only relatively few chords and sequences of notes have been deemed pleasant and permissible. A few other sequences and chords might be used sparingly, to project a sinister mood, perhaps, or to signal avant garde intentions, while many other possibilities have been simply ignored because of the fixed tuning of many musical instruments. Only in the twentieth century did composers begin to explore music based on quartertones and even smaller intervals, although most of their work has not entered the concert repertoire.

On the other hand, we are now more familiar than ever with the music of other cultures, from gamelan to blues, where different tonal laws apply. In addition, electric and electronic instruments are easily able to present sounds drawn from the continuous aural spectrum. This means that much of the music we hear today is microtonal, never more so than when it comes with the visual distraction of action on a screen. A celebrated example is heard in Stanley Kubrick's film, *2001: A Space Odyssey*, which makes use of several works of the Hungarian composer György Ligeti. The 'micropolyphonic' tones of his 1966 *Lux Aeterna* accompany the appearance of the black monolith that signifies the presence of a higher alien intelligence. That their music should sound like this is just what Huygens would have expected.

II

THE PARISIAN

Paris in the autumn of 1663 was a city in ferment. Following the death of his chief minister Cardinal Mazarin in March 1661, the twenty-two-year-old Louis XIV assumed his right to govern France in person. Handsome, physically powerful and acutely image-conscious, he set about numerous reforms, centralizing government power and consolidating the position of Paris as the European capital of learning, arts and fashion. The royal palace of the Louvre, which Huygens had seen burn when he was last in the city, was rising again with new wings and endless grand

25. Louis XIV. This engraving shows him in 1661 when
he took over the personal government of France following the death
of Mazarin, although he had been King since 1643.

facades. Directly across the Seine were the frames of new college buildings of the Sorbonne.

When Huygens arrived in October, he promptly resumed his scientific activities with the telescope and in mechanics and clocks. He went to Montmor's, where his friends had also been defeated in their own experiments with an air pump. As he had found in London, descriptions and diagrams were not enough to transmit knowledge: it required eyewitness and personal demonstration of how to set up and operate the complicated device. By providing hands-on instruction for his colleagues, first in London and then in Paris, Huygens himself became the key vector for the international transfer of knowledge concerning this important new technology.

But Montmor's circle was running out of puff. Spurred by what they heard of the success of the Royal Society, its members had vowed to talk less and experiment more, but they still failed to make much scientific progress, and their windy soirées were satirized by Molière in his new comedy, *Les femmes savantes*. Envious of their brothers across the English Channel, whom they knew enjoyed the sponsorship of Charles II (even if they had an exaggerated impression of the largesse that had been scattered), they put together their own bid for royal patronage. However, their independent programme of work was distrusted by the increasingly powerful and power-conscious government, and they failed to win the king's support. When fire devastated Montmor's palatial home in 1664, it was all over. In June, Huygens told Moray: 'In Paris there is nothing new in the field of Sciences, except that the Academy of Monsieur Montmor is finished for ever, but it seems that something else could be born out of its debris.' That something else would be the government's own academy, the first truly national academy of sciences.

It was the dream of Jean-Baptiste Colbert, who had been appointed Louis's finance minister in 1661, to magnify the power of the government by establishing a series of national academies, which would reflect greater glory on the king. Humourless and irascible, Colbert was dubbed 'Le Nord' – the north wind – by the prominent social commentator the Marquise de Sévigné. He was known for his habit of avoiding spontaneous answers when questioned, always preferring to call upon the data that he had to hand. He believed all knowledge was of practical value to politics, and his administrative power grew directly out of his own curiosity and the seventeenth-century conviction that all might be known. Colbert had long been the owner of the finest library in Paris at his apartments on rue Vivienne. The ground floor of the library contained books on science and natural history as well as classical and biblical texts, while the floor above held government accounts, reports on industry and national statistics. His concept of scientific management *avant la lettre* encompassed land use, industries, population and everything from church doctrine to garden design. He was aware, too, of the potential for the data gathered to his office to be used for purposes other than the obvious one for which it seemed to have been obtained. For example, a comparative survey of cloth factories in the Netherlands and France ostensibly aimed at improving efficiency also served to provide intelligence on the numbers of Protestant workers.

Colbert eventually set up national academies of literature, music, dance and architecture, and controlled others, including the Académie Française established by Cardinal Richelieu in 1635. Unsanctioned academies were banned. His objective was to add lustre to Louis's reign, and for this he sought the greatest minds that he could attract to Paris, even if they came from abroad. Christiaan Huygens's return to Paris was thus well timed. He came well prepared, too, thanks to his diplomat father, who, seeking the restitution of the principality of Orange in Provence to his masters, had presented one of his son's pendulum clocks to Louis XIV. Christiaan was put on an annual pension of 1,200 livres, along with leading literary figures including Corneille, Molière and Racine. With the others who gained this award, Huygens became the first scientist to be directly employed by a nation state rather than a princely benefactor.*

Periodically, leading savants had petitioned Colbert to ask the king to set up an academy of sciences. But official fears of Cartesianism, and squabbles over which disciplines should be admitted and who would hold the rights to intellectual property, held back progress until Colbert was able to persuade Louis that such a body might be beneficial. His vision was of an organization that would be more concerned with the public good than the old salons, and of direct service to the civic authorities, leading to measurable improvement of the French state. As well as having an internationalist view of the arts and sciences, Colbert saw Huygens's Dutchness as a specific advantage. In addition to the high regard in which he was already held by many French mathematicians and natural philosophers, and the valuable network of international connections to which he promised access, Huygens would be likely to share the practical bent that his national project demanded.

Moreover, Huygens had powerful allies to recommend him. Chief among these was his influential old friend Jean Chapelain, who was Colbert's Virgil in the worlds of literature and science. Shortly before Huygens arrived in Paris that autumn, Chapelain wrote to him: 'I have never done anything with more joy than to propose you to Monsieur Colbert as one of the most worthy objects of the esteem and beneficence of the King when he did me the honour of consulting me on the eminent Persons of letters to whom his Majesty has resolved to give firm status.' Knowing that Huygens

---

* The Royal Society agreed to employ Hooke as its salaried curator of experiments in November 1662, but his formal appointment was delayed until 1665 (Inwood 30).

was not quite polished in the manners of the French court, he suggested they meet as soon as possible to discuss the form of thanks that Colbert and the king would require. The mathematician Pierre de Carcavi, Colbert's appointed royal librarian, was another senior friend to recommend Huygens, feeling confident that, aside from the attractions of the scientific appointment, he would be unable to resist the 'mondanités, jolies femmes', and other charms of Paris. If Marianne Petit were not prepared to see Christiaan again, it seemed there would be other young women to meet.

To members of the lapsed salons of Paris, the new academy remained an uncertain proposition. Indeed, the most auspicious thing about it so far as most of them were concerned was Huygens's early enlistment to it, notwithstanding his nationality. However, only a handful of those who had gathered at Montmor's house made the transition to Colbert's organization. Others were doubtless put off by the commitment, based on the best practice of societies abroad, particularly the Royal Society, to observation and experiment and to the exposing of '*tromperies*' – erroneous or even deceitful thinking.

Huygens spent the early months of 1664 waiting for the necessary decisions to be made. He went to plays and ballets and balls. Meanwhile, his family took advantage of his idleness. His new brother-in-law wrote for advice on the latest fashions and asked him to handle an order for a gold watch with a gold-studded case. Susanna placed a request for a pair of chandeliers. He shipped a new desk built by a Parisian furniture-maker for Lodewijk. He was forced to make repeated visits to a perruquier who was supposed to be making wigs for him and his brother Constantijn. Such trivialities clearly irritated him, and every now and again he indicated his disinclination to carry on with these favours. But there were amusing distractions, too. One letter Christiaan wrote using an all-glass pen imported from Venice, 'a new invention and very convenient because the nib of these pens doesn't wear out, the ink doesn't spoil them, and they write in all directions, which is even better for drawing'. He enclosed two of the pens so that Lodewijk might try to make his own, 'if you haven't entirely forgotten this noble art [glass-working]'.

Christiaan made his own modest requests in return: 'I beg you to send me by the first ordinary seeds of cabbage of all sorts, Savoy cabbage, cauliflower, red cabbage and if they are there others too. Cousin Zuerius (to whom I beg you to offer my greetings) will show you them. It is for one of my good [female] friends, which is why I would not have you forget.' The cabbage seeds probably arrived in Paris intact, but a special consignment of three clocks sent from The Hague did not. Taking delivery – after an argument about carriage fees and customs duty – Huygens opened the box to find two of the clocks smashed, with their wheel shafts snapped and broken glass everywhere. Fortunately, the third clock, sandwiched between the other two, had survived.

The difficult journeys of clocks were much on Huygens's mind. One of his instruments was just then aboard an English ship undergoing comparative tests against an English clock in order to judge their suitability for use in calculating longitude at sea. This was exactly the kind of useful science that Colbert wanted to see done by the French Academy of Sciences. But it was a matter of acute interest also to Europe's greatest maritime powers, the English and the Dutch. Could science remain neutral with so much at stake?

Huygens felt that he needed to return home to The Hague to give this work his full attention while Colbert finalized the terms of his employment in Paris. He was not to return to France until 1666.

The determination of longitude at sea requires a form of universal clock that allows the position of celestial bodies seen from aboard ship to be compared with their position at a datum meridian such as a home port. The time difference between, say, noon, when the sun is directly overhead, in the two places is a measure of the fraction of the Earth's diurnal rotation that the ship has sailed. However, the vessel requires a highly accurate timepiece if the datum time that it registers is to be trusted on long voyages.

As long ago as 1657, Huygens had stated his conviction that his pendulum clock might provide the key to determining longitude at sea, thereby greatly improving the accuracy and safety of ship navigation. Sir Robert Moray saw Huygens's early pendulum clocks on a visit to The Hague in 1658, and he and Alexander Bruce then discussed with him how the design might be adapted to make it suitable for use at sea during Huygens's visit to London in 1661. Both Bruce and Huygens set about designing new clocks modified to undergo the rigours of passage. But it was excessively hard work. A vessel carrying two of Bruce's clocks on a short test voyage from Holland to England encountered heavy seas. As the ship lurched, one of the clocks was dashed to the cabin sole from the beam where it had been hanging, and 'the other was so strongly jostled that it stopped immediately'. Huygens, meanwhile, rigged up a simulator at home. This involved fastening a clock to a board suspended by five-foot ropes at each corner, which were connected to weights so that the whole assembly could be roughly moved around to mimic the motion at sea. His clock continued to run even when 'being suspended and shaken'.

Finally, in the summer of 1663, Bruce's and Huygens's perfected test instruments were labelled A and B and put aboard an English naval vessel, HMS *Reserve*, for trials on a voyage to

Lisbon and Tangier under the captaincy of one Robert Holmes. Huygens in Paris received detailed reports from Moray when the vessel returned to England. The short voyage was a qualified success. 'It is certain that the clock made in The Hague (which was marked A) is much better than the other made here marked B,' Moray wrote. This augured well for Huygens's hopes of obtaining an English privilege for his design. But more exhaustive trials were necessary before the quality of his innovation could be fully gauged, not only to test the instruments over a longer duration and in more challenging sea conditions, but also in the hope that the readings might be taken with greater care than had turned out to be the case on the Lisbon voyage.

In November the two clocks, encased in a steel sphere and a copper cylinder to protect them from corrosion, were carefully installed aboard a different ship, HMS *Jersey*, again under Holmes, bound for west Africa. Huygens provided detailed instructions for how the instruments should be managed at sea. Shortly after the *Jersey* had sailed, Huygens wrote anxiously to Moray: 'I hope that for the new voyage to Guinea and Jamaica you will have given orders that everything that happens to the clocks and everything that is done is noted exactly.'

Robert Holmes was perhaps not the best man to undertake such an exacting mission, although he would certainly put the clocks through their paces. Described by one contemporary as 'Swaggering, roystering and corrupt', Holmes is one of the most colourful and controversial figures in British naval history. On land, Pepys found him 'a cunning fellow, and one that can put on two several faces'. At sea, he became known for the independence of his actions, which were marked by bravery but also by rapaciousness and aggression. Even by the standards of the Restoration navy, he often appeared as unscrupulous and out of control. In 1661 he had led a small fleet of ships on a failed expedition to find 'a mountain of gold' in west Africa, during which he forced the surrender of a Dutch fort on an island in the River Gambia, prompting a sharp diplomatic rebuke from The Hague. Similar actions later provoked the Dutch twice into naval action against England, and made Holmes, in the words of Andrew Marvell, 'the cursed beginner of the two Dutch wars'. Although Holmes did have some interest in science, and could see the potential of an accurate marine clock, it was nevertheless hardly ideal that Huygens's delicate instrument was placed in the hands of such a rogue.

The *Jersey* had hardly left port before a new worry engulfed Huygens. Bruce wrote to him indicating that he wished to claim the whole of the longitude clock invention for himself, 'as if', Huygens complained to their mutual friend Moray, 'the pendulum Clocks were no longer my invention ever since I presented them in public. And it seems that the part he wishes to give me, he gives me as alms, and not as that which belongs to me by right'.

Both Bruce and Huygens were keenly aware of the rewards that lay in store for the inventor who could crack the longitude problem. Doubtless in the interests of making practical progress as they worked together on their devices, Huygens had hastily agreed with Bruce that any profits they might make should be evenly split between them. He now had reason to regret this naive arrangement, which was no true reflection of his contribution. He wrote to Bruce, his hurt plain to see:

> However, since you still remain in some doubt about this,
> you do well to wish it clarified before anything else, in
> order to avoid disputes in the future, and so using the
> freedom and openness to which your example invites me
> and which is to be practised between good friends, I will
> say firstly of what happened between us at The Hague,

when we were adjusting our two clocks, that I was very
surprised when I heard you propose that we should go
halves in the above-mentioned advantages, which, however,
I did not wish to contradict in order to avoid argument
with you, not that I believed the division was equitable.
Furthermore, those to whom I spoke afterwards, who were
only 2 or 3 persons, told me I had been unwise to make
this agreement. As to the argument against me, that the
inventors of the ancient clock and pendulum have as much
right as I do in the invention of the Longitudes, it is true
that we can say that these two previous discoveries are the
foundation of my invention of pendulum clocks, just as it
is said that canvas and colours are the basis of the art of
painting, instead of which my clocks are so much the
foundation of the invention of Longitudes that they are
the invention itself, save only carrying them at sea.

Foreseeing greater difficulties ahead, Huygens thought it wise
to safeguard his priority in the marine clock in the Dutch courts,
and wrote to the Grand Pensionary – and his old mathematics
partner – Johan de Witt. De Witt acknowledged Huygens's prior
claim, with the proviso that both clocks would also have to be
tested aboard Dutch ships. In a rare resort to the comfort of
patriotism, Huygens wrote bitterly to Moray: 'At least my friends
in Holland properly understand this.'

Moray meanwhile moved skilfully to cool tempers between
the two claimants, urging Huygens to clarify the extent of his
initial innovation, distinct from improvements made for marine
use introduced later. In a sign of its growing role in such disputes,
Huygens agreed to Moray's suggestion that the council of the
Royal Society be allowed to adjudicate the matter in his absence.
As it was not possible for a foreigner to hold an English patent,

Huygens was forced to give way to Bruce on this issue, with the assurance that his material interest would be protected by the society. Huygens had no compunction about submitting the Dutch patent application to the States General in both of their names, making clear that he was the inventor of the pendulum clocks, and that both he and Bruce had made the additions so that they could withstand motion at sea.

The feeling that he could easily be denied his due spurred Huygens on at home in The Hague. Throughout the summer and autumn of 1664, while these letters flew back and forth, he sought further improvements in his timepieces by following a systematic procedure of commissioning minor modifications to existing designs from his clock-makers and then setting them up in comparative tests for evaluation. 'It is why I continue ceaselessly to experiment with the accuracy of my new timepieces,' he confided to Moray.

At the beginning of December 1664, Captain Holmes reached Plymouth, bringing a trail of trouble in his wake. It was another six weeks before he surfaced in London to deliver news of his exploits in person, but the early word picked up by Moray was that the clocks had performed well, and indeed may have saved his small fleet from disaster.

On the last leg of their return from the West Indies, the *Jersey* and its companion vessels had set sail for home from São Tomé in the Gulf of Guinea in August, planning to travel via Cape Verde and the Canary Islands. The ships made good passage at first. However, during a ten-day spell of flat calm weather, water began to run short. The ships' captains called a conference and, comparing their logbooks and course calculations, reached agreement that the fleet should turn west in the hope of finding stronger winds that would carry them to Barbados where they could take on fresh water. But Holmes overruled the decision, declaring his view that the other ships had miscalculated their positions by as much as several hundred miles, and they were not as far west as they imagined: 'For observing my pendulas I had on board, which I constantly attended, either they could not be true or . . . we must be must more to the Eastward . . . I was resolved (presuming upon my pendula's) to try 48 houres in steering for Fogo [the highest island in the Cape Verde group, where there was water] as it bore by my own Reckoning . . .'

Moray relayed this dramatic story to Huygens, ending with the words: 'And because they had complete confidence in the clocks, he decided it was necessary to continue the route, & the next morning this Island appeared as he had judged that it would.' Within weeks, Moray's narrative appeared in the French *Journal des Sçavans* and in the Royal Society's *Philosophical Transactions*. Picking his way carefully through his colleagues' competing claims, Moray announced: 'These Watches having been first invented by the Excellent Mounsieur Christian Hugens of Zulichem, and fitted to go at Sea, by the Right Honourable, the Earl of Kincardin [Bruce], both Fellows of the Royal Society, are now brought by a New addition to a wonderfull perfection.'

Huygens naturally welcomed the news from England. But his scientific scepticism was aroused. Was it not a little too good to be true? (Chapelain had doubts, too, although his were informed more by Anglo-Dutch tensions in general and his wish to keep Huygens in France.) Huygens replied to Moray saying that he had not expected the clocks to perform quite so well, and asked him to find out what other members of the Royal Society made of the story, and 'whether the Captain seems to be a man of sincerity of whose faith we can be assured. For in the end I am astonished that these clocks would have been sufficiently accurate for it to be possible by means of them to come across such a little Island."

There was every reason to doubt Holmes, not only because of the poor quality of the data returned from the earlier Lisbon voyage, but also because of his dubious conduct on this occasion. The ostensible purpose of the Guinea expedition had been to support English merchant vessels engaged in trade in gold and slaves for the Royal African Company. Holmes went about the task by attacking Dutch forts on the African coast and Dutch ships at sea. Following the capture of one vessel, the *Jersey* suddenly found itself carrying Dutch prisoners as well as Huygens's Dutch clock.

By the time Holmes was giving an account of himself in London, a significant Dutch fleet had already moved to retake the surrendered forts, and England had declared war in response. The war was an immediate impediment to Huygens, who was seeking to send specimen clocks with new features to London and who now found it necessary to arrange for their transport via neutral ports such as Dunkirk and Calais. Judged to have exceeded his remit, Holmes was briefly detained in the Tower of London. It was here that William Brouncker, the president of the

---

* Huygens had reason for doubts on technical grounds, too. Experiments that he conducted in The Hague at this time revealed that two clocks mounted a few feet apart would begin to swing their pendulums in time with one another. This 'sympathy', as he termed it (OC5 246–9), Huygens first attributed to an effect of the air between the clocks, but he later realized that it was due to vibrations passed through the shared support (OC5 255–6). Unfortunately, he had gone public with his first thought, and so felt obliged to publish one of science's first important retractions. 'There is no shame in retracting what one has misunderstood,' he reassured his father (OC5 301–2). The phenomenon remains a puzzle to scientists today. See for example *Scientific Reports*, vol. 6, article number 23580 (2016).

Royal Society, was obliged to go in an effort to extract a more detailed first-hand account of the clocks' performance in the continuing absence of any formal report from Holmes. After further interviews with Holmes, and independent confirmation from an officer of one of the other ships on the expedition, the excellent performance of the clocks was surprisingly confirmed.

Huygens moved swiftly to capitalize on his success throughout Europe. In England, he continued to seek patent protection for clocks with new features and to arrange sea trials for them, despite the war. Chapelain, who was negotiating a privilege for him in Spain and licensing rights to Parisian clock-makers, wrote with one of his periodic reminders that it never hurt to flatter Louis XIV. Huygens replied with his usual discomfiture about such niceties: 'I will write to thank Monsieur Colbert and the King although I hardly know the style of such dispatches.' He finally formulated the necessary effusion of gratitude, adding a significant personal note in reference to his own exceptional status:

> But I will say only that when distributing his benefits Your
> Majesty makes no distinction between his subjects and
> foreigners, considering only the virtuous inclination of the
> person, and their favourable intentions towards the public
> good, [and so] the just recompense that he must expect is
> that the Foreigners have as much veneration and love for
> him as the French themselves, and that they too wish for
> the long duration of his life and reign.

Paradoxically, it was the Dutch who seemed least enthusiastic about his innovation. Huygens went to Amsterdam to pitch his clocks to seamen and navigators, including the cartographer of the Dutch East Indies Company, Joan Blaeu, the son of Willem, who had created the famous atlas. He reported dejectedly to his father that they 'cannot deny the utility. However, I have noticed how our people are slow and reluctant to adopt something new even though the utility is manifest.'

Huygens would be periodically involved in sea trials of new clocks under various auspices for the rest of his life. Rumours first relayed to him by Moray in 1665 that Robert Hooke had devised an alternative approach using a balance spring to regulate the timekeeping in place of the pendulum, which was so obviously problematic aboard a tossing ship, proved accurate. Huygens experimented with this kind of mechanism, too, as well as continuing his work on pendulum-based designs. A Huygens pendulum clock was tested on Colbert's behalf aboard a French ship sent to join an unsuccessful mission to defend Candie (Heraklion on

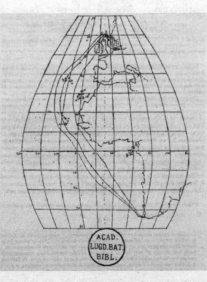

26. Estimated routes of the *Alcmaer* returning from the Cape of Good Hope. Longitudes calculated using Huygens's clocks indicated an impossible route taken through Cape Verde, the Azores and Ireland. The corrected route (westernmost line) allows for the slower pendulum swing near the equator where gravitational attraction is reduced.

Crete) against Turkish invasion in 1668. 'The shaking of the vessel during all the cannonades and the clangour on the vessel *Sainte Catherine*, jolted by its powder-fire, are a good test for the clock and the constancy of the pendulum,' Huygens commented drily when he read the captain's report.

After the French and English declared war on the Dutch Republic in 1672, Dutch trials were his only option. A trial of a seconds-pendulum clock aboard a Dutch East Indies Company ship, *Alcmaer*, in 1686 did not solve the technological problem, but it did yield a result of interest to science. No readings were obtained from the *Alcmaer*'s outward journey owing to the death on board of the man who had been entrusted with recording the data. But on the return voyage from the Cape of Good Hope, the longitude estimations showed, when reconciled with charts, that the vessel appeared to have sailed straight through Cape Verde, the Azores and the island of Ireland. This systematic offset error, greater near the equator than further north, indicated that the weight of the pendulum was effectively less in those latitudes. It proved, therefore, that there was no such thing as a universal seconds-pendulum, and gave a basis for calculating the corrections in length that needed to be made for a pendulum to oscillate with a constant period at each latitude. By the time the *Alcmaer* returned home to Texel, Huygens had received a copy of Newton's newly published *Principia Mathematica*. Huygens believed that the equatorial variation could be attributed to the centrifugal force of the Earth's rotation alone, whereas Newton's theory also took into account reduced gravitational force owing to the oblate shape of the Earth. Because the *Alcmaer* figures correlated well with his own interpretation, Huygens felt confident in using them to refute Newton's suggestion that gravity was universal rather than something applicable only to celestial bodies.

An acceptable marine chronometer – regulated by springs that were independent of gravity – was not made until well into the eighteenth century, by John Harrison.

By the summer of 1665 Huygens was itching to return to France, where there was peace, the company of intellectual colleagues, and the prospect of sustained patronage. Smallpox had swept through the Petit household; Petit's wife died, but Marianne – after three purges – was recovering. Carcavi passed on Colbert's word that it was the king's wish that Huygens should settle for good in Paris. He gave 'plenty of reasons and fine promises,' Huygens wrote to Lodewijk,

> without however yet coming in particular to know what he
> will make my Pension. I have written to my Father to
> know what he wishes for to me it seems that there is no
> cause to refuse, provided that I am given a good mainten-
> ance and that I live with all freedom without being liable
> for anything, as I have been promised.

At length, terms were agreed. Huygens was to receive a stipend of 6,000 livres and an apartment close to Colbert's and Carcavi's in the royal library. He pushed for further information about plans for the academy and his own laboratory, and even enquired as to whether he could be exempted from customs inspections, which often delayed the movement of scientific equipment and papers. At last, though, he – and his father – were satisfied, and on 8 April Huygens wrote to Carcavi and Colbert that he would be departing for Paris in no more than ten days' time.

In the Palace of Versailles hangs a canvas more than three metres high and nearly six metres wide painted by Henri Testelin. It carries the title *Colbert Presents to Louis XIV the Members of the Royal Academy of Sciences Created in 1667*, and shows a score or so of the scientists, courtiers and noblemen closely associated with the foundation of that body. They have gathered to be presented in turn to the king, and are seen framed by giant terrestrial and celestial globes and surrounded by the instruments and subjects of their endeavours. The author Charles Perrault is there, with his brother Claude's Paris Observatory under construction seen through curtains drawn aside in the background. So are the physicist Edme Mariotte, the mathematician Pierre de Carcavi, the astronomer Giovanni Cassini and many others. To one side, servants are putting up a large map of the planned Canal du Midi. In the gloom below a suspended armillary sphere lurks the skeleton of a lion preserved after a dissection. There is uncertainty over the identity of some of the figures, who look notably similar in their wigs and robes. Christiaan Huygens may be the one in the fitted lavender coat, standing behind the table overflowing with books and drawings that also supports a small clock; or he may be immersed in the crowd, engaged in casual conversation with Cassini.

It was never like this, of course. Cassini did not arrive in Paris until 1669. In fact, Testelin painted the work several years after the event it depicts, which in any case did not happen. Louis XIV paid only one reluctant visit to his Academy of Sciences, in December 1681, when he saw water being frozen and wine being distilled, and admired the illustrations of animals and astronomical instruments in its publications. He departed with the words: 'gentlemen, it is not necessary for me to exhort the Academy to work; it is working well enough on its own'.

Christiaan Huygens, however, was presented to Louis XIV only a few weeks after his return to Paris in April 1666. The king

said some greatly pleasing things, and I to him nothing of
worth, as I think, because besides the emotion that my
audience could have caused me, I still had a little bit of
fever which the movement of the carriage had given me.
The best was that with the King there was only Colbert
who afterwards took me dine with him, where I found also
Madame Colbert and Mademoiselle his daughter. I was
made most welcome. He asked me if on returning to Paris
I would like to pass by Versailles, which I readily agreed
to, and the sight of that beautiful place and the fresh air
one breathes served to cure me.

He wrote this letter to his brother-in-law Philips Doublet as
he wrote most family letters, in French. But towards the end he
broke into Dutch to express the hope that other members of the
family would one day visit this country awash with 'pots of jam
where syrup runs along the streets'.

Huygens was soon comfortably installed in an apartment above
the royal library in Colbert's house on rue Vivienne directly across
from the royal palace. The rooms next door were occupied by
Colbert's librarian Carcavi. He must have felt safe, temporarily
at least, when he heard news from home of the plague and English
naval raids on the Dutch coast, and from England, where the
Great Fire had destroyed much of London and threatened
Gresham College. In Paris, he immersed himself once more in
the endless round of entertainments – the balls and comedies,
banquets and concerts, with occasionally a piquant contrast, such
as when he attended the anatomical dissection of the body of a
woman who had been hanged for letting her child perish.
Suddenly, too, there was time for ceremony of the kind that
Huygens had found tedious in Holland and England. Meanwhile,
very close to where he lived, building work continued on the

palace of the Louvre. The centre of Paris must have been a teeming construction site for much of the time that Huygens lived there, and an energizing source of stimuli of all sorts. It was a city instilled with a sense of purpose and progress, a place dedicated to making the new, and doing so in style, and Huygens wanted to be a part of it.

Huygens had always been on Colbert's shortlist for the first scientific academicians, his name included based on Chapelain's consistent recommendation and perhaps also his father's subtle lobbying on his behalf. Now that the terms of his appointment and administrative details, such as the location where the academy would meet, had been agreed, Huygens would remain in Paris, with the exception of short breaks for illness, for the next sixteen years. Perhaps there were Frenchmen who had opposed his appointment, contributing to the delay, but now Huygens's arrival in Paris symbolized the arrival of the academy itself. It was an astonishing feat, given his alien nationality. For Huygens personally, this was the crowning moment of his career in terms of recognition by his peers and public esteem. He was thirty-seven years old.

To understand why Huygens was not just acceptable but close to being the perfect candidate for the figurehead of the Academy of Sciences, it is essential to appreciate Colbert's plan for the aggrandizement of France and of its king. Colbert's belief was that all facts could be known, marshalled and harnessed to increase the economic and social potential of the state. His library, with its great books and national statistics, was just the beginning of his attempt to bring all learning under his direct control. To run and maintain such a continuous 'big data' project would require an intelligent hub where the information gathered could be analysed and moulded into a programme of action. It would demand rigorous organization to be sure that all the important data were being systematically captured and utilized. In its comprehensive scope and range and its logical structure, it was a dream of totalitarian control that was Cartesian in all but name.

The inventions and theories of science were to be gathered just as ruthlessly as figures for crop yields or church attendance. Although Colbert was spurred on by the examples of Florence, Rome and especially London – he was, as Voltaire later wrote, 'jealous of that new glory' that was Louis's rival Charles II's Royal Society – his Academy of Sciences was far closer in conception to Bacon's model of Salomon's House in *New Atlantis* than its London analogue, which had no great ambition to control the use of scientific knowledge. Colbert envisioned an institution without statutes whose members would never own the rights to the intellectual property they developed. It would be a clearing-house for ideas, but all ideas would ultimately accrue to the state.

To eliminate wastage, Colbert excluded from the academy dilettantes, social climbers and those who were seen as mere artisans. Jesuits and Cartesians (ironically) were also excluded because of the perception that they were not primarily interested in scientific truth. Thanks to the revelation of his own discoveries, Huygens by this time was clearly no dogmatic Cartesian, and easily escaped this stricture. Colbert also instinctively understood that national greatness is not created by nationals alone. He attracted foreign scholars to a number of the academies he set up, including the Danish astronomer Ole Rømer to the Academy of Sciences in addition to Cassini and Huygens. It is a sign of his unique vision, as well as of changing times, that no further foreign academicians were appointed after Colbert's death. The fact that, by the end of the seventeenth century, French was the chief international language of science stands as vindication of Colbert's cosmopolitan outlook.

Since the early membership of the academy already included able mathematicians and astronomers, it is likely that Colbert regarded Huygens primarily as a mechanical scientist and inventor. Apart from his remarkable observations and discoveries in a range of scientific fields, he had a proven record of devising novel machines and pursuing them through construction and testing to the point where they could demonstrate their utility. In short, he embodied the best of Baconian and Cartesian traits. The long wrangle over his salary and benefits in France is perhaps explained by the fact that he would have stood to lose more than most in earnings from privileges and patents on his own inventions, which he would be obliged to forfeit by joining the French academy. With his terms agreed, however, Huygens submitted happily enough to Colbert's demanding regime. The annual budget for the Academy of Sciences was 88,000 livres, of which 51,000 livres was for scientific projects. (At the time, a labourer's weekly wage was less than one livre.)

For Huygens's more cosmopolitan colleagues in other countries, his new status as a scientist in the employ of the Colbertian French state represented a potential opportunity rather than a loss. Only weeks after Huygens arrived in Paris, Oldenburg wrote idealistically from London to exhort him to get on with the formation of the French academy, which might then join an international fraternity of science:

> I wish to be persuaded that you will do your best to
> request and obtain a fellowship of skilled persons in
> France, in order to employ their minds and a part of their
> time and faculties in the careful investigation of Nature,
> and to advance Mechanics more and more; in which things
> you will not fail to contribute a very considerable portion.
> I hope that, with time, all nations, now so ill-mannered,

will embrace one another as dear companions, and will conjoin their forces, both intellectual and material, to drive out ignorance, and to install true and useful Philosophy.

The Académie royale des sciences gathered for its first proper assembly on 22 December 1666 in the royal library. The members who had met less formally earlier in the year were Auzout, Roberval, Carcavi and three other Frenchmen, plus Huygens. They were joined by nine more founding members. Regular meetings were held twice a week – mathematics (including mechanics and astronomy) on Wednesdays, physics (including natural history) on Saturdays. Although Carcavi chaired the inaugural meeting and introduced the fledgling academicians to Colbert's programme, it was Huygens who emerged as the natural driver of the agenda. He was in effect, though not named as such, the scientific director of the academy, a role he was able to fulfil owing to his practical mien, and because he was proficient in so many fields and relatively indifferent to rival ideologies of the kind that had marred the meetings of the Montmor circle and other groups.* Though some of his closest colleagues were never admitted to the academy and others drifted away, Huygens continued to participate in the meetings during all the years he lived in Paris, knowing that he would never have been able to find its like in the Netherlands.

Huygens began work by making a list of research topics for Colbert's consideration, something he had rehearsed several times since his visit to Paris in 1663. He summarized the task facing

---

* Not that there were not disputes. One especially futile squabble concerned the Paris Observatory that Colbert had ordered to be built. The architect Perrault wanted a frieze of the zodiac carved into its walls, but the newly arrived astronomer Cassini judged the notion vulgar and unscientific (George 394).

the academy in a covering letter to Colbert as follows: 'to work on natural history somewhat following the scheme of Verulamius,' which meant to institute a Baconian programme of experiments to discover the causes of all – 'gravity, heat, cold, the attraction of the magnet, light, colours, the constituents of air, water, fire and all other bodies, what is the purpose of animals' respiration, how metals, stones and plants grow, and all those things of which nothing or very little is still known . . .' Colbert scribbled '*bon*' in the margin against each paragraph.

Huygens's list is worth reviewing today as a reflection of the range of scientific questions considered important during the second half of the seventeenth century. It naturally plays to his own strengths and interests, but it also reveals Colbert's statist agenda, and the growing belief that progress in science might lead to material benefits for humanity.

1. To identify the meridian line and the level of the pole from Paris, which are the foundations of all other astronomical observations.

2. To reinstate the fixed stars, in which lies the foundation of astronomy.

3. To measure the diameters of the sun and the moon at their various distances, which will serve to reveal new hypotheses of their motion and better than those available today.

4. To observe the extent of refraction in the atmosphere, which must be known in order to correct observations of the altitudes of the sun and the stars.

5. To observe the inequality of the days, and to establish their equalization, which is so necessary for calculating the motion of the moon and for eclipses.

6. To perfect spyglasses and microscopes.

7. To observe refraction in all kinds of transparent bodies.

   7.1. To observe whether light is communicated from afar in an instant.

8. To observe the diameters of the planets, in order to determine their sizes relative to one another and to the sun.

9. To observe the spots on the planets and discover from their motion the rotation of their axis.

10. To observe the motion of the satellites of Jupiter and make tables of them.

11. With the help of these tables, to observe here and in other places of the world, such as in Madagascar, the occultation of some of these said satellites behind or in front of Jupiter, in order to discover thereby the true longitude of the said places, and to correct the maps.

    11.1. To observe the magnetic declination [of the Earth, i.e. the angle between magnetic north and true north] and the variation that affects it.

12. To send pendulum clocks by sea with the necessary instructions and a person who can take care of them, in order to apply the invention of Longitudes, which has already succeeded so well in the experiments that have been made.

13. To measure the times and rates of descent of heavy bodies in air.

14. To measure the size of the Earth. To inform the means of making geographical maps with greater accuracy than up until now.

15. To establish for ever the universal measure of dimensions by means of pendulums, and in consequence also those of weights.

16. To find exactly the ratio of the heaviness of metals, and of all kinds of solid bodies and liquids.

17. To find the heaviness of air, by means of the vacuum pump, which serves for an infinity of other fine experiments.

18. To observe the strength and speed of the wind.

19. *Item* the speed and strength of flowing water and their relationship with the slope.

    19.1. To advise the best and simplest means of raising water.

20. To investigate the power of gunpowder.

21. *Item* that of fulminant gold [a spontaneously explosive chemical compound].

22. *Item* that of water evaporated by fire.

23. To investigate the force of percussion or the transmission of motion when bodies meet, knowledge of which is very useful in mechanics.

24. To investigate the force which bodies have to move away from the centre due to circular motion.

25. To investigate the relationship between tones and the size and shape of sounding bodies.

26. *Item* the relationship between the sound of strings and their length, thickness, weight and tension.

27. To determine what is the best tuning for organs, harpsichords, carillons of bells &c.

[unnumbered later addition]. To observe and define various degrees of heat and cold and their effect by means of thermometers. To have tubes of mercury under continuous experimentation to investigate their varying heights and their relationship with the present and future constitution of the air.

Colouring this list is a sense of the relative nature of physical variables such as length, time and mass, and the need to establish absolute standards against which important dimensions, such as the speed of light and the sizes of planets, could be reliably measured.

His official duties for the academy left Huygens with little time for his own work. In addition to the endless whirl of court activity, the twice-weekly scientific debates and his continuing correspondence with Oldenburg and others abroad keen to know what the French were up to, he found that his growing success bred new problems as he dealt with the commercial challenges of rival clock-makers and letters from unknown admirers with their own wild theories about the planets.

Astronomy was where his heart led him. Early on the morning of 2 July 1666, he stood yawning with Carcavi and others in Colbert's garden to observe a solar eclipse – he had hardly slept the night before for fear of not waking in time. He admired Cassini's measurement of the rotational period of Mars, not least because it came close to his own unpublished measurement from several years before, and looked forward to his joining the academicians in the future. But he held back from making new observations of his own for want of a telescope with improved lenses with which he might achieve 'something exceptional'.

Sample lenses produced by new works set up in the Faubourg Saint-Antoine to make mirror glass (including, later, that for Versailles) were murky and full of blemishes, and he turned for help to his brother Constantijn, and to Baruch Spinoza, then pursuing his trade as a lens-grinder only a few yards away from the Huygenses' Hofwijck house. But Constantijn disappointed him too. Critical of his brother's workmanship, he attempted to goad him on by reminding him that Spinoza was able to achieve a higher polish on his lenses. Nevertheless, he was pleased to have his brother's help, even if it was at a distance. 'I rather admire that the mood for telescopy has returned to you, without your having any companion in the work, because I find that it helps a lot, and if I had had one here, I think I also would have done something, even though we are still lacking the thick glass to make good objectives.'

Christiaan's hope was that Constantijn might be able to grind the components for a new arrangement of simple lenses that would achieve the same goal as the elusive hyperbolic lens by eliminating spherical aberration. (Chromatic aberration, which they were unable to exclude, Christiaan now came to think must be a fundamental property of refraction.) After numerous exchanges and trials – with Constantijn unafraid to modify Christiaan's design when he felt he knew better – Christiaan finally assembled what he wanted, a 'composite lens emulating a hyperbolic lens', based on a biconcave lens placed next to a planoconvex lens. As he sometimes did when he was especially satisfied, he added to his notes in Greek letters the word 'EUREKA'.

After a few months of this work, Christiaan realized he had been somewhat insensitive towards his brother, who, at the age of forty-one, was preparing to marry Susanna Rijckaert. He wrote ruefully:

So then marriage there be, I wish you all the success and
happiness that your heart desires. You have had enough
time to consider for and against and you had to bring to
an end an affair that had lasted so many years [Constantijn
had had a liaison with another woman, which produced a
child, to whom he provided support]. It is a great thing to
have a wife who is at your will, and that can weigh against
many other considerations.

With Spinoza serving – whether he knew it or not – as a kind
of sparring partner, the brothers also sought to make improvements in microscopes. Here, very small lenses were the key, but
these had the disadvantage of requiring very precise positioning,
which meant 'that one cannot see the top and underside of a hair
at the same time'. Relying on Kepler's *Dioptrice* of 1611, which
he found was still the best primer in optics, Christiaan was able
to confirm by his own experiments Spinoza's counter-intuitive
finding that smaller objective lenses could show objects with
greater clarity; 'without doubt the reason can be given, although
neither Sieur Spinoza nor I know it yet'.

Huygens also worked on problems in mechanics, prompted
by members of the Royal Society who had proposed new experiments to investigate the laws of motion. In January 1669 he sent
Oldenburg the 'beginning of a treatise on motion owing to collision', in which he considered the action of colliding bodies of
various masses and velocities. The work was read at the Royal
Society along with another on the same topic by Wren, and the
two men's ideas were found to be in pleasing agreement. Oldenburg
also pressed Huygens to publish his own long-awaited treatise on
optics, 'lest you be pre-empted by some other, who, to my knowledge, is now hard at work on it, and is a most clever man'. He
was alluding to the lectures on optics of Isaac Barrow, Newton's

predecessor as the Lucasian professor of mathematics at Cambridge. Barrow's collection appeared in 1669, instantly becoming the most advanced treatise available on the subject. Huygens could only respond with an apologetic note: 'I wish I could apply myself with a little more assiduity, but the range and number of jobs I have is a great obstacle.'

And what was Huygens working on for his masters in Paris? The answers were many: longitude clock tests, cartography, surveying and waterworks, in particular a scheme to divert water from the River Loire to Versailles. He became deeply involved in another inventor's novel means of suspension to reduce the jolting suffered by a carriage being drawn along a bumpy road. He devised surveying levels, barometers, anemometers, water-flow meters and a kind of depth sounder for ships. Like many others, he was involved in the development of thermometers. Anticipating the work of the Swede Anders Celsius in the following century, Huygens saw the merit of 'an agreed universal measure of cold and heat,' and recommended 'as a starting point the degree of cold at which ice begins to freeze, or the degree of heat of boiling water'. He experimented with a method of duplicating text and drawings by means of stencil plates that could be inscribed directly 'right-reading' rather than reversed as in printing. He developed an engine powered by gunpowder, but it proved too dangerous to use. And there were yet more waterworks: finding ways to make one man's fountain spout higher than another's on Colbert's estate, and using water to drive an ornamental clock, which Perrault thought might look pretty in his grotto.

All this was entertaining, maybe, but it was hardly the stuff of which scientific reputations were made. If Huygens seemed agreeable to this work, then this was not only because it was what his employer required of him, but also because he enjoyed an unusual technical challenge. And if it took him away from more important tasks, then that only reveals his great strength and his great weakness: the ability and willingness to tackle a dazzling array of scientific problems all at once.

The truth was that, despite the considerable merits of many of its individual talents, the academy failed to live up to the ambitions set for it. Although the direction had been clearly pointed out from the beginning, little work of real merit was produced in its early years, and membership soon stagnated. Only sixty-two academicians in all were admitted in the years up to 1699. The fly in the ointment was Colbert himself, who found that the ideals of science were at odds with his vision of a totalitarian state. His exclusion of Cartesians and other perceived dogmatists shut out some of the best minds from the academy's deliberations. He also found it hard to accept his academicians' determination to share their results with colleagues and similar institutions elsewhere in Europe. Oddly, given his openness towards employing foreign men, he had badly misread the primary commitment of the scientist, which is to the science rather than to the nation.

This was not the 'toute liberté' that Huygens once told his family he expected to enjoy if he moved to Paris. Might he have felt more at home in the gentlemanly and less authoritarian scientific atmosphere of the Royal Society? Perhaps. But then Huygens loved Paris, too, for its cultural vibrancy as well as for the sense of purpose that seemed to spring out of Cartesian idealism.

Huygens faced more specific difficulties, too. The most serious of these came from the always abrasive figure of Gilles de Roberval, who had once challenged Huygens's ideas about Saturn and claimed to have designed his own pendulum clock, before falling out with the entire Montmor circle. In 1668 Huygens conducted a series of experiments using a revolving table in order to learn more about centrifugal force. A tank of water was placed on the table with small blocks of wood and lead immersed in it. By analysing the motion of these blocks, Huygens thought he had discovered the cause of gravity. The project had begun simply, with the French government's desire to know the force required to lift a wheel of a given size over an obstacle in the road, and so unintentionally shows how, in the hands of one such as Huygens, a banal problem of Colbertian applied science might occasionally lead on to things of greater consequence. Huygens's results led him to reject Descartes's conjecture that gravity is caused by a single universal vortex, and he envisaged instead a system of multiple vortices acting through 'subtle matter' in space. The anti-Cartesian Roberval forcefully rejected this hypothesis, launching a sustained series of attacks, which seems to have led to Huygens falling seriously ill during the cold early months of 1670.

He had complained of periods of illness before, usually colds accompanied by severe headaches that struck during the winter months. His own first description dates from 1652, when he was twenty-three years old, and refers to a 'capitis dolor'. This can be translated simply as 'headache' or, literally, as 'sadness of the head', which could mean anything from the fashionable melancholia of the seventeenth century to modern clinical depression. Whatever it was was typically exacerbated, if not actually brought on, by the attacks of fellow scientists, which Huygens tended to take personally. On this occasion, he was so drained that he had to be carried up to his room by a servant, and believed himself to be at the point of death. In The Hague, his brother Constantijn consulted the Huygens family doctor, Diederik Lieberghen, who was more sanguine. Knowing Christiaan as he did, he confidently diagnosed 'Hypochondriac melancholia pure and simple'.

In his doleful state, Huygens made preparations for putting

his scientific papers into safe keeping. The repository he desig-
nated was the Royal Society in London and not the French
Academy of Sciences. Francis Vernon, an attaché at the English
embassy who was keeping Henry Oldenburg informed of the
activities of French scientists, was summoned to Huygens's room.
He found him in bed, looking pale and weak, and 'yet there was
something worse which the eye could not perceive nor sense
discover, which was a great dejection in his vital spirits'. Huygens
told Vernon he believed he was about to die, to which Vernon
remonstrated: 'God would not sett up soe great a light meerly to
extinguish it.' Huygens brushed aside the compliment, and
explained which documents he wished to bequeath to the Royal
Society. Finally, according to Vernon's possibly not unbiased
account, Huygens said he thought the French academy was bound
to dissolve 'because it was mixt with tinctures of Envy', and
depended upon 'the Humour of a Prince & the favour of a
minister'.

Despite Lieberghen's breezy diagnosis, his brothers were sufficiently worried to dispatch Lodewijk to Paris to supervise
Christiaan's treatment. As winter turned to spring, they discussed
between them the likely effect of milk, then thought to be a cause
of melancholia.* The doctor had recommended skimmed milk
only, but was French milk different from Dutch; and what about
breast milk? Christiaan's continuing black mood led him to express
religious doubts, and he began, to his brothers' great distress, to
debate with himself the existence or not of the immortal soul.

Huygens recuperated only slowly. Ironically, it was the melancholy planet that helped to restore him to health as, by the end
of May 1670, he found himself observing the phase of Saturn's

---

* 'Milk, and all that comes of milk, as butter and cheese, curds, etc., increase
melancholy', according to Robert Burton's *Anatomy of Melancholy*. Bad news
for many Dutch people (Burton, vol. 1 219).

ring once more with fellow academy astronomers. In fact, though, there was every reason for gloom. France and the Dutch Republic had been loosely allied against England during the Second Anglo-Dutch War, but following the peace, Louis XIV saw his chance to fill the power vacuum that had opened up in the Low Countries since the defeat of Spain. By late 1669 he had quietly made an arrangement with Charles II to invade the Netherlands, which was sealed with a secret treaty in June 1670. It now seemed certain that the two countries where Huygens's closest scientific colleagues were to be found would shortly attack his homeland. At first, Huygens hoped to remain in Paris, 'where I am employed in nothing that has anything to do with the war'. But he changed his mind, and that September he withdrew to The Hague.

SCIENCE DURING WARTIME

In The Hague, Christiaan Huygens had his portrait done by Caspar Netscher, a leading society painter and an artist he personally admired. It is a sumptuous painting, but there is nevertheless something cold and impersonal about it. Huygens is wearing a bronze-and-blue padded silk robe in a fashionable Japanese style and resting his arm on a velvet cushion positioned on a socle with a carved relief. He is extravagantly laced at his collar and his sleeves, which are tied with ribbon knots. The golden curls of his wig tumble heavily around his face.

Ignore, if you can, all this disguise. What do you see? His wide, dark eyes look straight out of the canvas with disarming directness. There is a little colour on his lips. If anything, he looks younger than his forty-two years, that girlish look from his childhood still present, despite the faint line of a moustache on his lip. You would not think he was ill.

You certainly would not know he was a scientist. There is no telescope or clock resting casually next to him to remind the viewer of his achievements, no chart of the heavens on the wall to hint at the scope of his thoughts. Even books and papers are banished from the scene. It could have been otherwise: Netscher painted the astronomer Nicolaas Hartsoeker in 1682 with a globe, telescope and geometry instruments crammed onto the plinth next to him, and he painted himself holding his palette and brushes.

Clearly, Netscher was following Huygens's wish to be shown above all as a man of the world. As portrait painter to the Huygens family, he was perhaps also acting on the father's instructions; the picture has a strong formal resemblance to the portrait he produced of Christiaan's sister Susanna a couple of years earlier.

Is Huygens trying to suggest that he is not a scientist at all, or telling us that he is so much more than a scientist? He may have felt, in 1671, that his great discoveries and inventions lay in the past. Perhaps he no longer had any need for the tools of the trade. He was as secure in his position as he had ever been; he felt needed by Colbert and the king. He was an officer of the state, all but, a senior functionary of the French court just arrived from Paris.

In the late spring Huygens passed what must have been a pleasant reacquaintance with his native land, sailing up and down the IJssel and Lower Rhine rivers near Arnhem, engaged in a survey of the vital waterway with Johannes Hudde, who had been his fellow student with Frans van Schooten. The project was on commission from the States General, in which Johan de Witt, another graduate of Schooten's classes, had for many years held the position of States of Holland Grand Pensionary. In the past, the three men had occasionally worked together on mathematical problems of national relevance, such as the calculation of state annuities based on life expectancy. The hydrological survey involved measuring the river depths, channel widths and the stream gradient in order to establish whether works were necessary to avoid excessive silting. Hudde and Huygens made a few recommendations of where to dredge and where to cut new channels, but decided that major interventions were not needed.

The rivers also possessed strategic significance as a forward defensive line against attack from the south. However, the next year, during what the Dutch came to know as the *rampjaar*, the year of disaster, Louis XIV and an army more than 100,000 strong was to cross them without difficulty, taking advantage of the weakened political and military situation in the republic.

By then, Huygens had already been back in Paris for several months. He returned to find an enlarged apartment waiting for him, and the unpleasant surprise that in his absence Carcavi's son had taken out his carriage without permission and wrecked it in an accident. This strained relations with his neighbour for a time, but there were new friends to see him, among them the remarkable Perrault brothers, the architect Claude, Charles the renowned writer of fairy tales, and Pierre, who would shortly be the first to describe the hydrological cycle.

There was also a brilliant young polymath eager to make his acquaintance. In November 1670 Oldenburg had written to Huygens to make him aware of 'a certain Doctor Leibnitzius of Mainz, who is advisor to the Elector there, but is also involved in philosophy, principally in speculations on the nature and properties of motion. He claims to have found the very principles of the laws of motion, which others, says he, have stated only simply, without proofs a priori.' The twenty-six-year-old Gottfried Wilhelm Leibniz had come to Paris with the Elector's backing to put forward an elaborate plan for peace, but, overtaken by events, he stayed on and began to study mathematics with Huygens. Leibniz would eventually surpass his tutor by developing differential and integral calculus.

Huygens busied himself with the grand waterworks that were under way at Versailles, proposing modifications to the windmills erected to pump water to its ornamental lakes and fountains, and introducing design details familiar from the

Netherlands, such as small holes in the bottom of the water buckets, which help the wheel restart after a lull in the wind. The major embellishment to the Academy of Sciences in his absence was Perrault's observatory on the Left Bank of the Seine, which was nearing completion. It looked, said Huygens, 'most handsome and magnificent'.

Cassini was appointed as the first director of the Paris Observatory, and given a salary of 9,000 livres, which peeved Huygens, on his 6,000. However, Cassini adopted French citizenship, became thoroughly naturalized in France and ensconced himself deeply at the observatory. He died in Paris in 1712 after more than forty years at the helm, the first of a dynasty of Cassinis who would lead the institution right up until the French Revolution. He was industrious in cartography as well as astronomy, and published numerous astronomical tables and almanacs, which were popular at court. Cassini was always a more systematic watcher of the skies than Huygens, and a herald of the growing professionalization of scientific work. He was rewarded in 1671 and 1672 with the discovery of two new satellites of Saturn in addition to Titan, which Huygens had found more than fifteen years earlier. (He found two more in 1684.) 'He . . . does not miss a clear night to contemplate the Sky,' Huygens complained to Lodewijk, 'to which I would not wish to subject myself, being content with my old discoveries, which are worth more than all those made since.'

Cassini's discovery was assisted by the disappearance of the planet's thin ring that Huygens had expected to occur as it fell edge-on to the sun's light as seen from Earth. On 5 November 1671 Huygens wrote to his brother Constantijn: 'I observed yesterday in the evening Saturn which I had not seen for 10 or 12 days, and I found its arms so greatly diminished that it was all I could do to perceive them, so much so that I said goodbye

27. An engraving of Giovanni Domenico Cassini.
The Paris Observatory is under construction through the window,
and a long telescope has been set up in the grounds.

to them this time. And there is my prediction verified . . .' He
was doubtful that Cassini had identified a true satellite, however,
thinking it might be simply a passing comet, but Cassini was
quickly shown to be correct.

In 1669, Huygens had written 'EUREKA' in his notes, thinking
he had finally found a practical alternative to the supposedly ideal,
but apparently unmakeable, hyperbolic lens. Now, three years on,
he went back and deleted the word in a series of angry looping
slashes. The reason was that Huygens had just learned, from the
ever-reliable Oldenburg, that an Englishman by the name of Isaac

Newton had designed a reflecting telescope and devised a convincing new theory of colours, which together instantly rendered obsolete much of his work on optical aberration.

Newton's theory of colours based on his famous experiments with a prism explained the coloured fringes that were seen when looking through many lenses, and which Huygens had been unable to explain or get rid of in his own. These unwanted artefacts were now revealed as an intrinsic property of light itself refracted through the thickness of the glass. Oldenburg sent Huygens a brief description of the reflecting telescope, along with a promise to enclose fuller details and a diagram in his next letter. The instrument worked by transmitting the image from the objective to the eyepiece by means of two mirror surfaces, one concave and one planar, and thus completely avoided the separation of the light into colours that would have occurred with glass lenses. When the Royal Society's telescope-maker had become aware of the invention, Oldenburg added, he had ceased his own lens-grinding on the spot.

Huygens greatly admired Newton's innovation when he saw the detailed design, and passed on the exciting news to his friends in the French academy, calling the instrument 'beautiful & ingenious'. As well as obviating the need for a thick lens ground to the shape of a complex curve, he pointed out, the polished mirror required its metal to be perfect only on the surface and not through its entire depth like glass. Liaising with Constantijn in The Hague, he immediately set about trying to replicate Newton's achievement. But the brothers struggled to a find a metal that could be formed into shape and polished to the necessary high degree. Then, as the political crisis in the Dutch Republic deepened, Constantijn was called away to pressing duties for William III of Orange, in which capacity he had largely succeeded to his father's historic function from before the Stadholderless Period.

Encouraged by Oldenburg, Christiaan Huygens offered some suggestions for improvements that Newton might consider making to his design. Newton at this stage was very much the junior partner in the exchange, while Huygens was acknowledged as the most able optical physicist in Europe, and he was pleased to hear that Huygens 'who hath done so much in Dioptricks hath been pleased to undertake the improvement of Telescopes by Reflexions also'. A positive reaction to constructive criticism was somewhat unusual coming from Newton, and the fact that he responded warmly may be taken as a sign that Huygens's scientific opinion was one of few he felt he could truly respect.

Huygens found greater difficulty with Newton's theoretical work. He wrote back cautiously at first to Oldenburg when he read the Englishman's paper on the generation of colours from white light, calling the work 'ingenious' and 'very likely', but demanding experimental confirmation, not of the separation and recombination of white light by prisms, which he accepted, but in regard to how chromatic aberration arose in lenses, where he justifiably felt himself to be an expert. This time Newton responded abruptly to Huygens, and accused him of seeking an explanation that was entirely superfluous to the working of the theory.

Newton was doubtless exasperated by the fact that Huygens was far from his only critic. Hooke, too, had misgivings about his colour theory, and Newton had huffily offered to resign from the Royal Society in the face of his attacks. His disappointment with Huygens, though, was perhaps all the more intense because the Hollander was assuredly his learned equal. The nub of the problem was that Huygens was still enough of a Cartesian to wish to see a mechanical explanation for colour. He wanted an answer to the question: what motion causes colour? This question did not interest the more empirical Newton in the least. Furthermore,

Huygens's own ideas about colour were highly unsatisfactory – he believed that a hypothesis able to explain 'the most saturated colours', which he held were blue and yellow, would be sufficient to explain them all. As the spat developed, Oldenburg forwarded Newton's response to this notion directly to Huygens, with a cover note to warn him: 'I can assure you that Monsieur Newton is a person of great candour . . .' Newton advised Huygens how to separate and recombine the colours of the rainbow as he himself had done, 'a tedious and difficult task', before repeating: 'But to examine how Colors may be explain'd *hypothetically*, is besides my purpose.'

After more than a year – and thanks in large part no doubt to Oldenburg's careful handling of the affair – both men finally agreed to let the matter rest, holding fast to their original positions. Somewhat gracelessly, given how wrong he was, Huygens wrote of Newton: 'seeing that he holds his opinion so hotly it removes the wish to argue'. Wisely, as it turned out, Huygens never discussed colours in any of his works on light. However, as a Cartesian, he could not remain satisfied, as Newton was, by the evidence of experiment alone, and the question of the original cause of colours continued to exercise him for the rest of his life.

All this time, Huygens remained sharply aware of the deteriorating relations between France and the Dutch Republic. Louis XIV, his patron, declared war on 6 April 1672, with England joining in on the French side soon after. The French army marched into the country the following month. The republic that had grown and prospered through trade, and built up sufficient military power to eject the Spanish and to contain the separate threats from England and France, proved unable to counter their forces combined. Financial collapse and poor efforts by the authorities to defend the country led to riots in many cities. On 21 June Johan de Witt was wounded in a knife attack in The Hague – in Paris the rumour was that he had been killed. Huygens wrote to Lodewijk wishing to know at first-hand 'everything that is happening in our miserable country'. He was worried for his family as well as for the safety of the Dutch forces and The Hague and other cities. He revealed his split loyalties and ambivalence about the war in a pointed further remark to Lodewijk: 'I see that you are not very well informed about the dead and injured on the French side.'

Huygens struggled to separate fact from rumour in Paris, the task made more difficult still when the brothers did not always receive each other's weekly letters. He feared his communications were being intercepted, although he was careful to write 'nothing which could do me ill; so that I regret only the loss of a letter'. Even at this time, his news typically included snippets about his latest experiments in Paris, and he continued to ship scientific equipment when necessary, although some of these items too were seized by doubtless mystified border officials.

Huygens feared his homeland would be brought to ruin by the conditions imposed by its enemies. As control was ceded of many provinces, and Catholic worship reintroduced elsewhere, de Witt resigned his office and other civic officials were purged. The Orangists, who had always opposed Johan de Witt, now promised the reforms necessary to bring peace. William III was appointed as Stadholder of the largest provinces, the first to hold that office since his father William II had died in 1650. But Christiaan could at least welcome Constantijn's appointment as William's secretary, succeeding his father in this role to the preceding two stadholders. 'If the State is saved from this bad situation, the brother is certainly in a good position,' he observed.

The new stadholder quickly mustered support and was able to retake the territory lost to France within a couple of years. But worse was to come for Holland before that. On 20 August 1672, an angry mob set upon Johan de Witt as he visited his brother Cornelis in prison in The Hague. Both men were beaten, stabbed and shot to death. Their bodies were dragged to the gibbet nearby on the Vijverberg, where they were hanged by the ankles and disembowelled, their clothes ripped off, and their body parts auctioned to the frenzied crowd. According to one eyewitness account, pieces of the entrails were cooked and eaten by some of the bystanders. 'The story of Monsieur the Pensionary and his Brother is horrible,' Huygens wrote to Lodewijk when he heard the news.

> I had known it since Friday, but not with the details that
> you relate. When one sees things like that, one would
> surely say that the Epicurians were not wrong to say
> Versari in Republicâ non est Sapientis.* There was much
> imprudence on the part of the Pensionary in going out
> into the open to the angry people, but I strongly remain
> of the opinion that he had not committed any crimes that
> would merit death.

De Witt's republican regime had endured for nearly twenty years, the logical outcome as he saw it of the struggle of the Eighty Years War. During that time, his government had stood for a high degree of sovereignty of the individual provinces, for toleration, freedom of thought and a Cartesian separation of theology and philosophy, ideals with which Christiaan – despite

---

* Huygens paraphrases Cicero: 'It is not the business of a wise man to take part in politics.'

his family's allegiance to the House of Orange and his own to the very different regime in France – could say he was broadly in accord.

Huygens escaped from reality not by surveying the heavens but by peering into the microcosm. He had received an early copy of Robert Hooke's groundbreaking illustrated book on optics and microscopy, *Micrographia*, in April 1665. Some of the optical theory contained in the work gave him pause for thought, but he was astounded by the images he saw there. 'Good figures. Flea and louse as big as a cat. Writes much about Refraction, Coloures &c. but in English,' he summarized to Johannes Hudde, for whom he would later translate parts of the work into Dutch.

Huygens's own interest in the microscope sprang from his personal experience of lens-making, and no doubt also from his awareness of his father's poetic raptures at the new worlds revealed by Drebbel's instrument when he had visited him in London. Christiaan favoured a design modified from a telescope like Drebbel's, based on two lenses, rather than the more recent innovation based on a single glass bead for the lens, which he had tried and found hard to focus.

It was the father, too, who first became acquainted with Anthoni Leeuwenhoek of Delft, not far from The Hague, and encouraged him in the work he was doing with the microscope. Apprenticed as a cloth merchant, but lately employed as a sheriff's clerk in the city, Leeuwenhoek's duties left him ample time to pursue his unusual obsession. Largely self-taught, he may have acquired the technique of making bead lenses from Hudde or from seeing *Micrographia*. To construct a microscope, Leeuwenhoek mounted a good specimen from among the many beads he made in a hole drilled through a simple oblong plate of brass about the size of a folded letter of the time. The object to be viewed would then be fastened to a threaded pin which could be gradually brought up close to the lens. Leeuwenhoek made hundreds of such microscopes for his own use. The best lens that survives today is no bigger than a mustard seed and has a focal length of less than one millimetre, but gives an impressive magnification of 270 times.

Scepticism about bead lenses from reputed scientists such as Christiaan Huygens and Hooke was understandable. The very first microscopes, such as those made by Drebbel and the lens-grinders of Middelburg, likely employed optics developed from spectacle lenses. At a time when optical ray diagrams were rudimentary at best, it was not at all obvious that in fact smaller lenses might produce much greater magnification. Both Huygens and Hooke had tried making very small lenses, but exceptionally tiny, clear beads – far simpler than any ground lens – were the innovation of Leeuwenhoek in the 1670s.

Constantijn Huygens had supposed that the microscope would be of interest mainly to artists, craftsmen and traders – a cloth merchant already accustomed to using a thread-counter might want one to inspect the fabric he was buying, for example. Leeuwenhoek, though, was the first to use one to make a systematic study of nature. He observed red corpuscles in his own blood, and discovered *diertgens* – little animals – swarming in a drop of water from a lake. 'The motion of most of these *diertgens* in the water was so swift, and with such varied movements, up as well as down and around, that it was truly astonishing to see,' he wrote.

We do not know what Mevrouw Leeuwenhoek thought when one day in 1677 her husband dashed from the conjugal bed and, 'before six beats of the pulse had intervened', placed his own semen under the microscope for examination. He announced the discovery of human spermatozoa with some circumspection, although it made him famous and gave him a fruitful topic for much further research. He revelled in his success, and later added 'van' to his name. His election to the Royal Society occasioned him 'some small vanity', Constantijn reported to Christiaan, as he wondered whether he would now no longer need to bow and scrape before a doctor of medicine. Based on his discovery, Leeuwenhoek rejected the concept of the mammalian egg, believing instead that 'the seed of plants and animals contain animals which in their seed contain still others in infinitum, and that no new creatures are made in the world but those which are made already merely enlarge and grow, which would be a wonderful thing'.

A little later, Johannes Swammerdam, a physician in Amsterdam, began to make bead lenses too – forty an hour, he boasted, though not all of optical quality. He made further pioneering microscopical observations before turning to the priesthood. His work included an attempt to introduce a classification of insects based on their varied and intricate anatomies, which showed, he said, that they should not be regarded as inferior, but were as miraculous as any of God's creatures.

For a brief moment, it seemed that microscopy was an exclusively Dutch art. Though they were more powerful than the two-lens alternative, bead lenses were prized as much for their clarity of image as for high magnification. Their main use was to reveal the surface of things in greater glory, with perhaps little conception that, if you kept zooming in, there would be further new levels of detail to be seen. Instead, the unfamiliar scenes that people were able to glimpse fed a growing hunger for visual richness and novelty. Visible complexity had grown to be a matter of curiosity in its own right, offering something new to admire beyond simple form and colour. Patterned shells and minerals were coveted for collectors' cabinets, and tulips were bred with ever more outlandish variegated flowers. Even before the instruments were invented, Dutch paintings seemed to require the eye to operate simultaneously as telescope and microscope, bringing the distant close and rendering the minuscule large. We see the entire city in Vermeer's *View of Delft*, but we draw close to inspect the glittering drops on the roofs after the rain. In the numerous paintings that show a terrestrial globe – de Keyser's portrait of Constantijn Huygens, for instance, or Vermeer's *The Geographer* – we see the Earth as if from a distance across space even as we crane forward to inspect the detail of its coastal outlines.

Now, however, Leeuwenhoek and Swammerdam had gone much further. They had revealed things that were utterly new, and which demanded to be understood and described accurately for what they were, just as the mysterious form around Saturn had once demanded to be decoded as a planetary ring. Neither man had any pretensions as a philosopher, yet the primacy that their work afforded to what was simply being seen for the first time, with no theory a priori ready to receive it, represented a major challenge to the Cartesians' lofty lack of interest in the observable world.[*]

Christiaan first became aware of Leeuwenhoek's work in June 1673, presumably through his father, when he was sent some of his observations and a description of his microscope. He was sceptical of both the instrument, which, he told Oldenburg, 'seemed to convert everything into little balls', and of Leeuwenhoek's observations, which he thought might be 'deceptions of his sight'. When he tried to replicate some of the observations made in aqueous media in Paris, he was unable to

---

[*] In 1691 Leibniz wrote to a by then entirely sympathetic Huygens: 'I like better a Leeuwenhoek who tells me what he sees than a Cartesian who tells me what he thinks' (OC10 49–52).

see what Leeuwenhoek claimed to have seen, and asked Oldenburg what credence was being given to his observations in England. Eventually, Leeuwenhoek told Huygens's father that he was using fine glass capillary tubes to draw up the liquid to be viewed under the lens, and sent him some tubes to try for himself. Even using these aids, Christiaan still struggled to get results, but after persistent effort he was rewarded, and in July 1676, when he returned to The Hague, he began to take a more serious interest in this form of microscope.

He made a French translation of Leeuwenhoek's account of his major achievements for communication to the French Academy of Sciences. (An English translation appeared in *Philosophical Transactions*.) The paper described in vivid tones how Leeuwenhoek had seen in rainwater 'little animals which appeared to me more than ten thousand times smaller than those of which M. Swammerdam has described the form, which he calls flea or water lice', including organisms with bodies comprising clusters of transparent balls, some with little horns 'like horses' ears', and tails three times longer than the body with another ball at the end. Other results concerned well water taken from his courtyard, sea water and Delft canal water (which was used to make beer), as well as peppered water and waters infused with ginger, cloves and nutmeg. All these revealed animal life teeming in such profusion that 'there could be several thousand in a drop of water'; only water melted from three-year-old snow revealed no living creatures. Leeuwenhoek wrote delightedly to Christiaan when he heard that news of his discoveries had been well received in France, and hoped he would repeat the favour for some 'trifling observations' he had made since.

Christiaan Huygens probably learned of Leeuwenhoek's discovery of human spermatozoa from him directly when he was back in Holland. This work in particular seems to have kindled a wish to take up experimental microscopy for himself. In addition to his exchanges with Leeuwenhoek, he also visited Swammerdam. A young Rotterdam lens-maker, Nicolaas Hartsoeker, who had seen Leeuwenhoek's public demonstrations of the microscope, schooled Christiaan and his brother Constantijn in a new way of making the bead lenses, and devised a better means of controlling the illumination of the object to be viewed. Christiaan may have improved or altered Hartsoeker's setup sufficiently to satisfy his conscience in claiming it as his own when he took one of these microscopes with him back to Paris in 1678. Constantijn even had one of the highly portable devices with him at the bloody battle of Saint-Denis in August 1678, the last major engagement of the Franco-Dutch War. Huygens made his own observations of the spermatozoa of various animals, and experimented by heat-treating various waters and then counting the animalcules that appeared under the microscope. He found that the microorganisms reappeared both in water that had been frozen and in water that had been boiled, but they multiplied much more slowly in the latter.

As he had hoped he might, Hartsoeker accompanied Huygens back to Paris, where he was able to take advantage of the senior man's introductions to French scholars. Their work together, along with the Dane Rømer, culminated in the development of a microscope with a rotating stage capable of holding six different lenses. With renewed patriotic consciousness after the war (and briefly forgetful of where his funding came from), Huygens reminded his French colleagues that spermatozoa were a Dutch discovery, and that for this new invention 'none of the honour is to be attributed to the French nation, for it played no part in it'.

After just a year in Paris, Hartsoeker was disappointed to find the novelty of the instrument already wearing off. In September 1679 he reported to Huygens: 'I doubt not that the French interest in microscopy has already evaporated . . . people will have to persist with it for longer ere they become much wiser.' But Huygens, too, soon set aside the microscope in favour of other activities. Hartsoeker departed for Holland; he would return to settle in Paris in 1684, but this time it would be to build a series of ever larger telescopes for the observatory. Microscopy fell from fashion, limited by the difficulties of seeing and of knowing what one was seeing, and scholars' interest turned back to the venerable science of astronomy.

'Cursed and banned from the people of Israel' in Amsterdam, Baruch Spinoza was earning his living by grinding lenses in the small village of Voorburg outside The Hague when he wrote one of the greatest of all works of modern philosophy. *Ethics* was published in 1677, a few months after Spinoza's death at the age of forty-four.

He had known it would cause controversy. Using mathematical rigour, he sought to disprove the existence of a sentient god. For good measure, he declared that good and evil did not exist in any absolute sense, but were to be regarded objectively as forces leading to individual human betterment or impairment. He reasoned that emotional investment in past traumas and in future hopes or fears is irrational because they are only products of our artificial conception of time. His greatest influence had been Descartes, but now he argued contrary to the Frenchman that the body and mind are not mutually independent like a machine and its controller. Instead, they are metaphysically identical, so inextricably bound up with one another that there are forms of knowledge that can be said to be embodied – held within the body, not only by the mind. This is credible when one considers sporting prowess or the physical aspect of creativity displayed by a painter or a sculptor, but in Spinoza's conception it extended to more quotidian areas of expertise, such as writing or sewing or washing down the stoop. These ideas have taken their place in the philosophical canon. But what was the background that encouraged Spinoza to nurture these so humane, and at the time so revolutionary, thoughts?

'[A] good-looking young man, with an unmistakably Mediterranean appearance', according to the monk who left us the best contemporary description of him, Spinoza was small in stature, with black eyes, black hair, and a 'beautiful face'. Sephardic Jews, his ancestors had fled from Spain to Portugal and then, when the Inquisition pursued them, to the Dutch Republic, where Baruch was born in 1632 and spent the entirety of his quiet life. By his virtuous conduct and work, he became the man whom Bertrand Russell called 'the noblest and most lovable of the great philosophers'.

In Amsterdam, he gained the education – absorbing ideas from literature, dissenting Protestant thought and even the theatre – that ultimately caused the elders of his synagogue to exile him for the 'abominable heresies which he practised and taught'. It was probably after this that he learned to grind lenses. This practical skill stood him in good stead when he moved away, initially to Rijnsburg outside Leiden and then to Voorburg, where he sustained himself by making lenses for the new optical instruments – the telescopes and microscopes of gentleman scientists, camera obscuras for artists – as well as lenses for spectacles. Lens-grinding, though it might seem like an antiquated handicraft, was also a wholly modern activity, and Spinoza became very good at it. His lenses were sought out by leading astronomers, including Christiaan and Constantijn Huygens. Even though they ground their own lenses, they prized Spinoza's above those of other makers. The Huygens brothers regarded Spinoza both as an occasional associate and as a competitor. They swapped calculations and books, and compared methods of bringing lenses to a high polish, but did not share everything, especially not the 'Huygens eyepiece', an arrangement of two lenses which Christiaan had found would overcome chromatic aberration.

Spinoza was greatly interested in the new sciences, especially optics, and was aware of the work of the Huygenses early on through his correspondence with Henry Oldenburg, who had visited Spinoza in Rijnsburg. When Spinoza moved to Voorburg in 1663, his little house in the town was only yards from the Huygenses' Hofwijck estate. There, in the spring of 1665, Christiaan was able to show Spinoza the shadow cast by Saturn's ring onto the planet's surface. We may picture them. The telescope is set up in the garden, perhaps on the gravel forecourt across the moat bridge from the house, where it would not be too shaded by trees. The night is moonless and clear, for to see Saturn at all it requires optimal conditions, and the air is not yet warm. The two men, both in their early thirties, huddle before the telescope in their heavy clothes, shapeless in the dark. Equals in some respects, both were free spirits without the ties of family life, both comfortable in their chosen pursuits. But there may have been some awkwardness. Christiaan might have been patronizing, even a little bumptious, eager to show off *his* planet. Spinoza, necessarily, would have been deferential, hesitant to follow his host's lead, as well as captivated and enthralled not only by what he was able to see, but also by the marvellous apparatus by which he could view it.

This unlikely companionship between a banished Jewish tradesman and an aristocrat with links to the highest levels of Dutch society was strictly scientific. The practical empiricist in Christiaan found Spinoza's tendency to abstract thinking hard to take, and their relationship remained cordial but distant, with Huygens referring to 'the Jew of Voorburg' and 'the Israelite' in correspondence with his brother rather than using any familiar name.

Spinoza had found an occupation that would maintain him, and one that made good use of his scientific acumen. Perhaps he had found something more, too. 'Grinding and polishing lenses, in Spinoza's day, was a quiet, intense, and solitary occupation, demanding discipline and patience – in a word, an occupation perfectly suited to Spinoza's temperament,' according to his biographer Steven Nadler. What connection might there be then between this work and Spinoza's philosophical output?

Spinoza believed that lenses were best ground by hand: 'a spherical surface is more safely and better polished freehand, rather than by any machine,' he told Oldenburg. It was tedious and hazardous work for those not fully attuned to its rhythms and demands. Understandably, efforts were made, not least by the Huygens brothers themselves, to mechanize the process in order to lessen the drudgery and reduce the risk of making a slip. But Spinoza felt it was better to use his sense of touch to adjust the pressure and friction of the glass against the dish with minute precision, and resisted the use of machinery.

In *Ethics*, Spinoza identifies three ways in which we gain knowledge: by random sense experience or imagination; by the use of reason, which may include formal instruction; and by intuitively grasping the essence of a thing, by which Spinoza means gaining an appreciation of its place within God's overall creation. It is bold enough to say that we can learn simply by grasping the essence of a thing, but it would surely be more bold for an 'armchair' philosopher to make this claim, than one like Spinoza, whose own hands were routinely engaged in close physical labour that transformed the appearance of matter.

Using methods of proof taken from Euclidean geometry, Spinoza asserts that God cannot stand outside nature, and so must be in nature and in all of nature. To the extent that God exists, God *is* nature, therefore. He observes further that our bodies are subject to the laws of nature – a fact readily appreciated from working at a grinding table or undertaking any exhausting physical task. When we remember this, Spinoza argues, we become free from evil passions and the fear of God because we understand the futility of resisting these laws.

Departing from Descartes, Spinoza believes that mind and matter are not inherently separate, but that the two are intimately connected. In particular, he thinks that expert knowledge belongs with the realm of ideas and thoughts, rather than residing in the mind, and is therefore accessible to the body as much as to the mind. 'Thus,' according to Aristides Baltas, a philosopher of science (and recent Greek minister of culture), 'expert action manifests the merger of mind and body and displays how this merging works: a body-mind, that is, a person *as* body-mind, knows on his or her own, by his or her inseparable body and mind, what the body should do and what the mind should do and how to act with both as inseparable.' Watching craftspeople at work, it is easy to appreciate this from the tactile feedback they obtain as they form an object. It is as if the eyes can see through the shaping fingers.

Voorburg seems an unlikely setting for revolution of any kind, and yet only yards away from where Christiaan Huygens was beginning to understand the relativity of motion, Spinoza was claiming that time itself was an unreality. This is surely another realization stimulated by the sensations of his work. In *The*

Craftsman, the sociologist Richard Sennett describes how craft workers lose their self-awareness and in a sense merge with the object they are making: 'We have become the thing on which we are working.' The successful accomplishment of a task, arrived at through intent concentration and the application of expertise, is 'invariably accompanied by a feeling of being at one both with oneself and with the world at large', according to Baltas. Working in this way, body and mind together realize in the object or experience created a manifestation of the creator's 'whole and undivided nature', and show that he or she has 'taken in the world as it really is and hence that he or she has been in full harmony with it'.

Aside from the crafted object, the effect of this process on the maker is to bring a profound sense of satisfaction that infuses both body and mind – a 'feeling that,' as Baltas puts it, 'he or she has fully lived the moment of success as a *present* moment, the feeling, precisely, of having experienced eternity'. It materializes the sense of a 'job well done', where body and mind have worked together without, as it were, thinking about it. Perhaps this sensation seeping through Spinoza's body in his little workshop in Voorburg informed another doctrine of his, that of 'the eternity of the mind', or the idea that mind exists outside time.

Spinoza's daily work of grinding lenses may not have directly inspired his specific ideas about the human mind and its place in nature, but it is clear from his own writing that there was little to separate the development of his philosophy from this practical activity. In a series of letters written during the first half of 1666, just a few months after Huygens had shown him the ring of Saturn, the philosopher recapitulated his conception of God. In the last of the three letters, he summarized his position – 'there is nothing outside God that is', he wrote – before moving seamlessly on to discuss aspects of refraction, complete with an optical diagram.

Despite the social distance between them, Huygens and Spinoza were more attentive to each other's ideas and methods than they were sometimes prepared to let on. Though he found its abstract language challenging, Huygens was undoubtedly influenced by Spinoza's philosophy as well as by his optics, as is apparent in the nature of the speculations he made late in his career about life on other planets. And, while Huygens was always eager to learn how Spinoza obtained such transcendent results with his apparently primitive technique of grinding lenses, the dedicated handcraftsman Spinoza was actually desperate to know more about Huygens's machine for doing the same job.

13

FEUDS AND TRIALS

In August 1671, ever ready to remind his son of his place in God's clockwork, Constantijn Huygens penned two short verses in ambiguous commemoration of Christiaan's achievements in chronometry. Each drew a parallel between inventor and invention, between the 'bold discoverer' and his 'weak mechanism' that had been unable to conquer the oceans, fearing that the tables might be turned and Christiaan's ingenious success might be his physical undoing. The final stanza of 'On My Son's Timepiece' finds a moral lesson even in the simple pendulum:

> Start it going, and let it die down,
> Finally to nought: only those who go likeways
> By doing constant good all their days
> And enduring to the End, shall inherit the Crown.

If the poems were prophecy, they were well timed, as the publication of Christiaan's treatise on the pendulum clock, *Horologium Oscillatorium*, in 1673, and his next major invention of a balance spring as an alternative means to regulate the running of clocks and watches in 1675, were to plunge him into the most testing professional disputes of his career.

Horologium Oscillatorium stands as 'one of the masterpieces of seventeenth-century scientific literature', perhaps second only to

Newton's *Principia Mathematica* of 1687. Its publication immediately sealed Huygens's reputation as 'the recognised leader of European science'.

Much of the content had been ready for years. As long ago as September 1660, Huygens admitted it had been 'complete for a long time'. But while Chapelain and other colleagues in Paris exhorted him to publish, he prevaricated, making excuses about his imminently expected departure to Holland. In fact, Huygens was hesitating for his usual reason of wishing to be more comprehensive. In the end, therefore, it was only the first and shortest of the work's five parts that was devoted to the design and operation of pendulum clocks. This section included his mathematical proof that the pendulum weight must trace the path of a cycloid in order for it to swing with an equal period, and that this path is the evolute of another cycloid, thereby providing a theoretical foundation for the empirically established fact that cycloidal 'cheeks' added to alter the pendulum path improved the timekeeping accuracy. Above the proof Huygens wrote, quoting Ovid: *Magna nec ingenijs investigata priorum* ('This great thing has not been investigated by prior geniuses'), thus frankly identifying himself as a genius.[*]

The remaining four parts of *Horologium Oscillatorium* covered the related physics and mathematics in a more general way, dealing respectively with the uniform motion of bodies and the way they fell under gravity (of which the pendulum can be considered a special case); the geometrical properties of curves and their evolutes; and the theory of the centre of oscillation for lines, planes and solid bodies. The final part comprised a discussion of the motion of the conical pendulum and the announcement of

---

[*] The word was only just beginning to acquire its now usual meaning as applied to scientists. One of the earliest examples cited of this usage in the *Oxford English Dictionary* is from the diarist John Evelyn, writing in 1662: 'Hugens . . . so worthily celebrated for his . . . universal Mathematical Genius.'

his hypothesis of the existence of 'Centrifugal Force – it is thus that I wish to call it', which was stated in theorems only, with the promise of proofs to follow at a later date.

Huygens was so convinced that pendulum action was central to the physical world that he reiterated his proposal that it should be the basis of a fundamental unit of measure that would relate dimension in space with time. His 'universal perpetual measure of length' would be the length of the pendulum required to oscillate a given number of times during 'a complete revolution of the sky'. Such a measure might be easily promulgated around the globe by the comparison of clocks in different locations, starting naturally with his own as the reference. If such a measure had existed in antiquity, Huygens noted, it would have avoided disputes between mismatched Roman, Greek and Hebrew foot measures.

Huygens necessarily dedicated *Horologium Oscillatorium* to his patron, Louis XIV. Chapelain checked over the text to ensure that it was sufficiently fulsome. Observing that he hardly needed to demonstrate the utility of his invention to the king, who already had his clocks in 'the private areas of your palace', Huygens praised 'that great and extraordinary liberality with which you protect the sciences and those who excel in them; liberality which, despite the great cost of wars, far surpassing ordinary expenditure, is in no measure diminished, and on which your land of France places no limit'.

He distributed copies to Colbert and twenty-five French academicians, as well as to important colleagues in England, Holland and elsewhere in Europe. One was sent posthumously to Johan de Witt because he had seen the manuscript before publication. Newton admired the work, finding it 'full of very subtile and usefull speculations very worthy of the author'. He was eager to know more about centrifugal force, especially as applied to astronomy, about which Huygens had said nothing. Newton sought to draw Huygens out with a promise to send him a more elegant method of treating evolute curves (based on the method of fluxions, or calculus, which Newton had developed some years before but was keeping secret). But Huygens, perhaps recalling Newton's sharp response to his criticism of the separation of light into colours the year before, did not reply.

Other members of the Royal Society were less emollient. John Wallis felt that Huygens had not given sufficient credit to the work of English scientists, and fired off a bigoted diatribe: 'I find, since his being Frenchified, his humour is strangely altered from what it was wont to be.' The message was softened somewhat as relayed to Huygens by Oldenburg, and the internationalist in Huygens responded stoutly that in his experience the English were no less covetous of praise than the French or the Dutch.

But it was Robert Hooke who found most to irk him in *Horologium Oscillatorium*. In 1666 he had devised a conical pendulum, using a bob that moved in a horizontal circle rather than in a vertical arc, and whose string therefore described a cone. He was alarmed, therefore, to see Huygens apparently claiming this invention for himself. On 12 June 1673 Oldenburg wrote cautiously to Huygens to point out that Hooke had had the idea before him. Huygens wrote back, pleading ignorance, and asking Oldenburg to assure Hooke that he had not been copied. Pressed further, Huygens admitted that he *had* heard Hooke mention the concept when he visited England in 1663, but the idea was by then already in his head.

Hooke and Huygens were of very different social standing – Hooke being the son of a humble curate who never received much formal education. While this disparity undoubtedly added to the tension between them, they were in another important respect very similar. The men were two of the finest experimentalists of the age, each being mechanically minded and skilled in designing, making and operating the most demanding of precision instruments. They were both skilled observers, fine draughtsmen and bold theorists, although Huygens certainly had the edge in mathematics. And each was keen – sometimes excessively keen – to safeguard his intellectual property. This impulse was perhaps nowhere stronger during the seventeenth century than in the matter of clocks, where there was a powerful commercial incentive forcing the pace of innovation.

The predicament that Hooke and Huygens found themselves in was a familiar one at a time when the conventions of scientific priority were yet to be firmly established. Like many of their peers, both men were frequently guilty of holding back findings when they thought they might have commercial value or were too premature to publish. In the course of their angry correspondence, however, they pushed the crux of their dispute back a decade, shifting the onus of proof away from printed documents with a definite chronology and onto hazy memories of verbal assertions made at an uncertain date.

What in these circumstances constitutes evidence of a discovery made? What amounts to 'publication'? A journal article, a delivered paper, a dated anagram, a private letter, an experimental demonstration witnessed? In this case, there was no satisfactory answer. However, thanks once again to Oldenburg's adroit mediation, a truce was agreed the following year. Hooke asked Oldenburg to send Huygens the text of his recently delivered lecture, 'Attempt to Prove the Motion of the Earth', as a peace offering. Huygens replied that he found it 'very fine and of great consequence', and they could at least agree that they were firm allies in the wider war against the 'ridiculous subterfuge' presented by anti-Copernican philosophers.

The peace between Huygens and Hooke was to prove short-lived. In January 1675 Huygens announced a new invention. This was a mechanism in which an iron spiral spring attached to the axis of the balance wheel of the clock was used to provide the same time-regulating function as a pendulum. (Spiral springs were already widely used to power the drive mechanism in clocks.) Lacking the pendulum's critical dependence on orientation in relation to gravity, the balance spring was potentially a highly practical innovation in chronometry, not least for use at sea. Huygens's device was fabricated by Isaac Thuret, his preferred clock-maker in Paris. The work was published in the *Journal des Sçavans* the following month, and Huygens applied for a privilege for the design in France. At the same time, he communicated news of the invention to Oldenburg and the Royal Society, initially in the form of an anagram, in the hope of protecting his design in England as well. He suggested that the patent might be put in Oldenburg's name since Huygens, being a foreigner, was ineligible to hold it himself. Any income might then be divided between Huygens and the Royal Society. He rated the new mechanism very highly. To his brother Lodewijk, whose wife had just given birth to a son, he wrote: 'If your boy is beautiful [then] my girl, my new invention, is also beautiful in her way and will live long, like the children of the good Epaminondas,* with her sister the elder pendulum, and her brother the ring of Saturn.'

Three weeks later, Oldenburg read the members of the society a new communication from Huygens, which revealed the solution to his anagram. Hooke reacted immediately by claiming to have devised a similar mechanism years before, but others at the meeting doubted this. Positioning itself as an independent arbiter,

---

* The Theban general Epaminondas died on the battlefield in 362 BCE. According to legend, a friend, commenting that he died childless, received the riposte that he was leaving behind him his daughters, his victories.

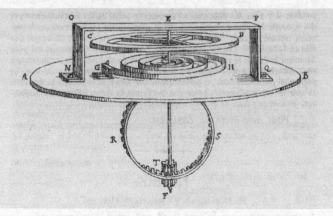

28. Huygens's diagram of a spiral balance spring was published
in the *Journal des Sçavans* in 1675.

the Royal Society took Huygens's side, but offered Hooke the
opportunity to prove himself. In April, Oldenburg explained the
developing situation to Huygens: 'Hooke has also requested a
privilege for his own watch, which he claims depends on the same
principle, and which he is said to have had for several years. We
shall see by their effects which will be the best.'

That summer, Hooke raced with London's leading clock-
maker, Thomas Tompion, to bring his own concept to fruition.
Tompion built the instrument to Hooke's specification with two
springs fitted to the inside of the balance wheel. In an effort to
underwrite Hooke's claim to priority, the clock was retrospectively
inscribed 'R. Hooke invenit an 1658. T. Tompion fecit 1675'. At
the same time, Hooke began to fear that Oldenburg was using
his official position at the Royal Society for profit, and that he
and others were in alliance with Huygens against him.

Huygens, meanwhile, set about having specimen clocks made
for favoured persons, including William of Orange and William
Brouncker, the president of the Royal Society. (He entrusted the

safe carriage of Brouncker's clock to a famous Italian commedia dell'arte actor, Dominique Biancolelli, Louis XIV's favourite Harlequin, who then happened to be touring Paris and London.) He wrote to Oldenburg:

> M. Hooke's proceedings seem neither good nor honest if
> one wishes to make oneself the author of everything new-
> found, and particularly as regards that invention he has
> bad grace when he says that he has had it for a long time,
> having produced nothing when I sent you the anagram
> that you know, which I told you contains a new time-
> keeping invention. Those who are capable of producing
> fine inventions themselves do not behave like this.

In reply, Oldenburg was sympathetic, though mindful too of Hooke's key role as the Royal Society's curator of experiments, which he was anxious not to jeopardize. 'As for M. Hooke, he is a man of extraordinary mood; I would wish, however, that you write a few words to Mylord Brouncker to explain to him that you knew nothing of M. Hooke's invention before you sent the anagram.' This Huygens duly did, only to learn that Brouncker was already won over to his cause, and convinced that Hooke had not taken up the work until he heard of Huygens's breakthrough.

All Hooke's fears now seemed to be confirmed. He launched a furious attack on Huygens and accused Oldenburg of acting as his spy. (Oldenburg had, of course, been briefly held on suspicion of spying at the time of the Second Anglo-Dutch War.) He also compiled his own detailed chronology of his career in inventing timepieces from 1660 onwards, belittling Huygens's achievements at every turn, in particular the seagoing clocks, and hinting that he had always been in possession of improvements, which he had discussed on each occasion with members of the Royal Society, while withholding crucial details of the innovations.

Huygens also faced legal and competitive challenges from rival inventors in France. The most serious of these came from his own maker, Thuret. Even before he got Oldenburg's letter warning him of Hooke's claim of priority, Huygens learned from friends that Thuret had copied the design he had realized for him, making some slight improvements, and was claiming the invention for himself. Huygens dealt with the matter roughly, stealing back the changes Thuret had made and forcing Thuret to retract his claim in bald terms. Thuret issued a statement whose grudging tone suggests that it may have been imposed on him as a condition to continue trading: 'I claim nothing of the glory of that invention, which belongs entirely to you.' Made aware of the threat posed by Thuret, Huygens, too, prepared a diary of the rapid sequence of events surrounding his invention, for presentation as evidence to Colbert and the Academy of Sciences. The day-by-day affidavit records that Huygens came up with the idea on 20 January, a Sunday, and went the next day to inform Pierre Perrault, the Paris receiver of taxes, that he had made a significant invention, but without revealing the crucial details. Huygens then says that Thuret made the clock for him on Tuesday while he waited (!), but, by the next day, he had also made his own copy of the new design – 'I could not imagine why, unless that he was fond of it.' According to Huygens, Thuret even urged him not to rush to apply for the privilege until the mechanism was proven to work properly, presumably in order to buy himself time to make his own working clock and apply for the privilege himself.

This unusual record is revealing for what it shows of the sheer pace of work, the level of improvisation and the close collaboration between inventor and artisan maker that was often required in the field of timekeeping. Huygens, the peerless scientist, comes across as naive and somewhat clueless in this new milieu of cutthroat competition, a gentleman accustomed to the ways of the court and international diplomacy suddenly adrift in dealings with men for whom earning a living was the most pressing need. The fact that he felt betrayed by Thuret at precisely the same moment that he was locked in conflict with Hooke may help to explain why he appears to treat Hooke, too, as little more than a tradesman.

Huygens's special pleading had the desired result. Colbert awarded him the privilege for the balance-spring clock less than four weeks after he had conceived of it. The document did not mention Thuret's name. Huygens also obtained the patent for the Dutch Republic. In England, the unauthorized dissemination of his new design via Thuret's copy rendered the anagram Huygens had sent to Oldenburg obsolete almost from the moment it arrived, which is why he quickly followed it up with a more straightforward announcement.

Despite support from Oldenburg and others at the Royal Society, he did not gain an English patent, however. When his prototype finally arrived from Paris, it was missing its minute and second hands and did not perform well when demonstrated. Despite having his own powerful backers, Hooke was not awarded a patent either. Interest in the balance-spring mechanism quickly waned when it was found that it did not perform at sea as well as had been hoped. Both Hooke and Huygens emerged from the episode deeply scarred, without the rewards they had hoped for and still at odds over their priority. If it is impossible in the end to disentangle who heard what from whom at what moment, and to declare a definite 'winner' of the 'race', then that must be considered as a fair price to pay for the emerging open(ish) culture of scientific communication.

Hooke continued to blame Oldenburg in particular for his

woes. He suspected Oldenburg as a foreigner (although the German had taken English citizenship, married an English woman and not been out of the country for fifteen years) of colluding with the other foreigner, Huygens. Oldenburg for his part had always tried to ignore differences of nationality when promoting dialogue between scientists across Europe. For example, he had sensibly ignored a warning in 1669 from Francis Vernon, the English embassy official in Paris, that Huygens 'hath a great Zeale for the honour of the French academie even to Jealousie & I believe . . . that hee is a little suspitious of the workings of the English nation'.

As he had done when he came under attack by Roberval several years before, Huygens relapsed into illness and took refuge in The Hague. He suspended his communications with Oldenburg as well, although he was able to rely on his father's continued correspondence. Oldenburg died in September 1677, and for a long time thereafter Christiaan was to have very little contact with English scientists.

On 24 November 1680 Huygens rode to the Château de Chantilly some twenty-five miles north of Paris, drawn there by word that one of the water features was emanating a strange and musical tone. People who visited were puzzled by the effect, and Huygens was not satisfied by the wild attempts at explanation he had heard. On the chateau's vast parterre, partitioned by little canals and dotted with reflecting pools, he found the source of the mysterious sound, a small waterfall designed to let in water from the River Oise that ran close by the estate. 'Echo of the water splashing on the ground from the steps of the grand stairs of the terrace, made a tone like a distant trumpet, which comes from the successive echoes of the steps,' he noted. He measured the tread of the steps, 'which are 17 inches deep'.

Huygens went back to his rooms and promptly constructed a tube seventeen inches long, open at one end, and confirmed that 'blowing into it gives the same tone as the echo from the stairs'. Nine days later, he returned to Chantilly to make some further measurements, only to discover that 'the stairs were full of snow which covered the steps up to half their height. The sound could no longer be heard at all.' Nevertheless, he reasoned correctly that the musical tone must arise from the reflection of sound waves from the waterfall off the steps in such a way that the pitch related to the depth of the treads was amplified at the expense of other tones.

The invisible means of conveyance of the sensations of sound and light were not well understood in the seventeenth century. But that they both travelled analogously to one another, and in some respects like a wave, was a widespread hypothesis, based on the obvious fact that both seemed to radiate from their sources like the ripples made when a stone is dropped in water. The musical Huygens, aware of Mersenne's measurements of the speed of sound and of Boyle's demonstration that sound cannot travel through a vacuum, correctly intuited the wavelike behaviour of sound travelling through the frosty air that day at Chantilly, and no doubt saw, too, how this might be pictured in terms of particles in the air connected by vortices in accordance with Cartesian doctrine. Light, which can traverse a vacuum, was more of a puzzle.

In 1668 a Danish ship returning to Copenhagen from Iceland unloaded pieces of a curious transparent mineral, and a specimen was installed in the Wunderkammer of King Frederick III. The mineral, a highly pure form of calcite, came to be known as Iceland crystal or Iceland spar. It may be the same as the 'sunstone' mentioned in Norse sagas by which the Vikings were able to navigate on long voyages even when the weather was overcast. Its property of birefringence, or double refraction, means that its appearance alters when it is turned in relation to the direction of the strongly polarized light of the low sun, even when that light is diffused by clouds. This would have given Viking navigators a means of knowing their direction. Most strikingly, Iceland spar disobeys the general rule of refraction that an incident ray of light perpendicular to a surface is not deflected as it passes through it; in this case, the ray splits and travels on through the crystal in two new directions. The Danish geometer Rasmus Bartholin analysed the crystal's strange optical behaviour in different orientations and measured its anomalous angles of refraction, which he hypothesized (wrongly) were due to pores in the crystal.

Huygens acquired a brick-like chunk of the mineral for himself in the summer of 1676, when he was back in The Hague recovering from the stresses of life in Paris. His father described it to Oldenburg as being

> white and transparent as water, in the form of a lozenge ◊
> but a little scratched as they all arrive from the country of
> Iceland, where it originates. What is remarkable apart
> from its great pellucidity in three inches' thickness is that
> its refraction is always double and, what is more surprising,
> it can be cleaved in any direction, lengthways and cross-
> ways . . .

The material certainly made for a good exhibit in a cabinet of curiosity. But these obsolescent displays did nothing to answer the growing need to properly understand the basis of strange natural phenomena. The irregular refraction of Iceland spar challenged Snel's law and the whole theory of dioptrics, which had exercised the greatest minds for most of the century. Christiaan therefore set about investigating its properties in more detail.

After his dispute with Newton over colours, Huygens felt compelled to produce a thoroughgoing mechanical explanation of light. Rømer's discovery that light travels at a finite speed, deduced from observations of the orbital speed of Jupiter's moons and published in June 1676, proved timely for Huygens, and led to a productive correspondence with the Danish astronomer during the winter of 1676–77. Rømer's theory confirmed an earlier hunch of Huygens's that 'light extends circularly and not in an instant'. If light travels at a finite, albeit very great, speed, might it not also travel at different speeds through different kinds of transparent matter? After all, Descartes had proposed that light, like sound, would be shown to travel faster through mediums denser than air.

Having established the angles of refraction of Iceland spar, Huygens was ready to pursue his enquiry by means of analytical geometry with no further need for experiments. He was convinced that the puzzle could be resolved using mathematical methods alone to discover the regularity that surely lurked within the curious substance. His earlier work on light had depended on linear rays, but here he found it helpful to think of light as a spreading wave. If light radiates spherically from its source in a uniform medium such as air, he argued, might it not radiate on a *spheroidal* front in a medium with different material properties along its different axes? It might: Huygens's elegant geometric diagrams turned out to explain the behaviour of the crystal in all regards, implying that light travelled through it at different speeds in different directions. On 6 August 1677 he wrote in Greek and Latin in his notes: 'EUREKA. Cause of marvellous refraction in Iceland Crystal.'

Huygens may not have recognized that what he had before him was a new theory wholly consistent with everything then known about the nature of light apart from colour. The wave-based model offered a mechanistic explanation for travel in a straight line, circular or spherical radiation from the source, finite speed, and reflection and refraction, including the curious refraction of Iceland spar. But his fidelity to Cartesian theory, based on the notion of a particle-filled space able to transmit the light impulses only by means of particle interactions, meant that he was reluctant to propose it as a full-blown theory.

Besides, he was about to face a significant challenge to his explication of the mysteries of Iceland spar from Rømer, based on an observation of Bartholin (Rømer's father-in-law, as it happened) that in certain situations light rays were found to run precisely along the line of the crystal planes.

With peace in prospect between the Dutch Republic and France, Huygens returned to Paris in April 1678. 'My son, the Parisian,' as his father wrote to Stadholder William III, judged that his duties to his French employers outweighed any lingering health considerations. Perhaps all that had been preventing his departure was the fact that the eighty-one-year-old simply desired his continued company, and 'to enjoy his learned and amiable conversation until he might see me die'. He pointedly observed to the stadholder how it had been impossible 'to procure for him a contract of two thousand écus in his fatherland, which he enjoys over there'.

It was a pitch rooted in desperation more than any real hope. The national finances were in a relatively poor state following the war, but even in its famous 'Golden Age' the Dutch Republic could never have supported a centre of intellectual patronage on the scale of Colbert's academies. The country lacked a Paris or a London with the cultural gravity to draw and hold the best thinkers, and the spectacular Dutch successes achieved in fields from botany to medicine, as well as in astronomy, optics and mechanics, came necessarily from the small cities of the provinces, and depended upon the brilliance of relatively isolated individuals, who were reliant in turn on the comparatively modest and haphazard support of burghers who had prospered in an economy where wealth was comparatively widely distributed.

Constantijn found his own income impacted, too, with rents from his estate at Zuilichem 'being reduced by the floodings and other inconveniences of the war' (much of the land had been deliberately flooded as a defensive measure). His difficult son Lodewijk was newly restored to office as the bailiff of the small south Holland city of Gorcum (Gorinchem), having been previously found guilty of corruption, after the father used his influence to ensure the sentence was not enforced. His oldest son, Constantijn, meanwhile, continued his more honourable career in service with the stadholder, accompanying William on the last campaigns of the war. Their sister Susanna wrote regularly to Christiaan with these family titbits and with news of progress of the peace negotiations. She also sent him recipes for 'some little stews' that she thought he might be able to make for himself. She looked forward to the day when it would be easier to receive fashion goods from the city she was desperate one day to visit for herself, but for now she simply exhorted Christiaan to tell her if he noticed 'any exceptional Alteration in the dress of the Ladies or in their Hairstyles'.

Back in Paris, Rømer's theoretical challenge (he did not have his own specimen of Iceland spar to experiment with) drove Huygens to conduct some further tests. By means of a tricky cleaving and polishing of the mineral along new crystal faces, he was able to show that Bartholin's observation of light deflected exactly along the crystal surface was a matter of pure coincidence, and not anything of optical significance. On 6 August 1679, precisely two years after he had performed what he had thought were conclusive experiments, he awarded himself once more a triumphant, and now a little ironical, 'EUREKA'. The Iceland spar had once again served him well in his 'Experimentum Crucis, as Verulamius calls it'.

When Huygens came to describe his new theory of light, he drew repeated analogies with sound and with waves on water. Because light paths can cross, light cannot behave 'like cannonballs or arrows,' he concluded. Rather, 'it propagates, like Sound, by spherical surfaces & waves. For I call them waves owing to the resemblance to those that are seen forming in water when a stone is thrown in.'

Huygens's geometrical analysis showed that a ray of light could be considered as an advancing wavefront. When the wavefront glances onto a mirror surface, for example, each point where the front is incident becomes the source of a new wavelet. The succession of wavelets generated at the mirror surface form a new, reflected wavefront, which advances at an angle equal and opposite to the incident wavefront. In simple refraction, light travelling in air might strike a flat glass surface at a certain angle to the perpendicular. The new wavelets generated at the boundary advance more slowly in the denser medium, which causes the new wavefront to form along a line bent at a reduced angle. The great strength of the Huygens principle (as it came to be known) was that all the laws of optics could now be explained by geometry.

Simple lines on paper offered vivid and persuasive representations of phenomena that had seemed baffling in physical reality.

Lines were mathematical ideals. What was the physical reality? Huygens was not prepared to let go of the Cartesian idea of some form of matter in motion – not cannonballs perhaps, but fine particles of some kind dispersed through an ether, through which light travels as impulses received by one particle and then transmitted in all directions to further particles. He sought to develop the analogy that had been explored by Bacon and Gassendi between light and sound waves, which are transmitted longitudinally by the compression and expansion of the air, but he was unable to do so in a way consistent with the mathematics. We now know that this hypothesis of 'particular waves' is incorrect,

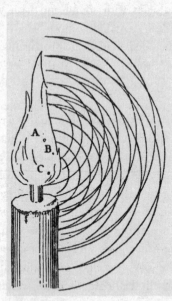

29. Huygens's diagram showing the transmission of light from a candle by means of spherical waves. Light travels in such a wave from each luminous part (A, B, C) of the source.

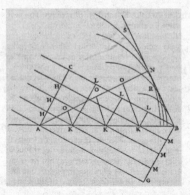

30. Reflection of light from a surface showing how new waves
originate from each point where the light impinges and
then assemble to form a new wavefront.

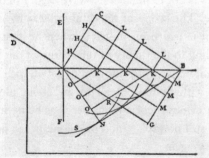

31. Refraction of light showing the geometrical relation between the speed of
light and the angle of refraction according to Huygens's wave theory.

but for Huygens it provided at least a semblance of fidelity to
Descartes while also remaining entirely consistent with his own
rigorous geometrical model of light behaviour. In the end, it was
his successful application of mathematics to the theory of light,
rather than his materialist conclusion, that was Huygens's lasting
achievement and which put his theory of light ahead of rival
theories, including other wave-based theories, such as those of

the French Jesuit Ignace-Gaston Pardies and Robert Hooke in London.

There had been no hint of waves in Huygens's early thinking about optics. The development of his theory depended not only on the ever-present utilitarian drive to understand refraction in order to design better lenses, but also on the arrival on the scene of the curiosity that was Iceland spar, and on Rømer's astronomical observation of Jupiter's moons that led him to deduce a finite speed of light. The combination of these factors was enough to push Huygens to develop the mathematical analysis that could account for these recent observations, and to set aside any worries he may have had about the need to ascertain their ultimate causes as demanded by Cartesian doctrine.

Huygens was not to publish his wave theory of light for more than a decade. The scope of his thinking was by now far more extensive than it had been in 1653, when he first wrote about light, responding to the theory of refraction set out by Descartes, or when Oldenburg was pressing him to get on with his 'dioptrics' in 1669. Yet, with nothing at all to say about physical matter, which most previous theories considered as being integral to the problem, and still no explanation of colour, he no doubt thought it wise to delay again.

In November 1680 a bright comet appeared and lingered in the skies over Europe, sparking fear and excitation. After a few weeks, the eighty-four-year-old Constantijn Huygens could stand the commotion no longer and dashed off this punning quatrain:

> Wat lightgh' en quelt mij vroegh en laet
> Met praetjens van Cometen?

Ick houw meer van Comedi-praet,
En noch meer van Com eten.

What tickles you and torments me early and late
With chatter of comets?
I would rather comedy prate,
And more yet to come eat.

Comets were omens. The great comet of 1618, long-tailed
and reddish in hue, swept through the sky in a year when the
Dutch provinces nearly fell to war among themselves. A comet
was seen in 1651 when the Dutch first went to war at sea with
England. Another appeared as a harbinger of the plague in 1664.

Yet as the seventeenth century progressed, telescopes made
sightings of these astronomical oddities more frequent and more
detailed. Other comets were observed in 1652, 1661, 1665, 1668,
1672, 1677, 1678, 1681, 1684, 1686 and 1689. The 1680 comet
was the first to be actually detected by telescope. The comet by
which Edmond Halley was able to prove the periodic recurrence
of the objects in the solar system appeared two years later. If they
appeared with such tedious regularity, what did it really mean if
another year turned out to be a 'comet year'? Did comets portend
anything at all?

Huygens wrote several more short squibs about the comet
and then, in April, a charming longer poem, *Cometen-Werck*, whose
framing conceit was that everybody had been pestering the old
man whose son happened to be a famous astronomer with their
questions about the new comet 'That each knows so well and
each so little knows'. In answer to these imagined queries, Huygens
emphasized the natural wonder of the phenomenon, and lightly
dismissed its worth as a portent:

A King dies in the East: and over there they mourn,
In the South, they laugh about it: that is the world's way,
The Heavenly sign's not good nor bad for anyone,
But just as sweet for one as sad for another.

Did he believe, then, that the comet meant nothing? 'No, my foot fits not in such a shoe.' He chose instead to read the bright streak in the sky as a symbol of God's comforting rod and staff of the Twenty-Third Psalm. If a comet was a messenger at all, then Huygens for one preferred to 'await the message in full when my soul is brought / To where He lives'.

Most astronomers studied comets without attributing any predictive or mysterious powers to them. But this stance was not without risk. Pierre Bayle, for example, recently appointed as a professor of philosophy in Rotterdam, where he had fled in exile from France, suddenly found his pay withheld following clerical protests provoked by the publication of his entirely rational thoughts about comets in 1682. Constantijn Huygens, too, tended to dismiss superstitious interpretations of unusual natural phenomena, but his poetical mind could not resist putting popular omens to work for fun. In May 1665, attending a ceremony at the old Roman theatre in Orange at the conclusion of his arduous diplomatic mission to wrest back control of the principality from Louis XIV, Huygens saw a corona appear around the sun at the climactic moment. He interpreted this, whether for political advantage or poetic pleasure, in a stanza suggesting that heaven had crowned the occasion with its beneficence. Many years later – when the French had recaptured Orange – he admitted he hadn't believed a word of it, and knew it had been no more than a natural occurrence.

His son was only able to spot the 1680 comet once it emerged from its extremely close approach to the sun in late December. Christiaan wrote to his father:

For some time already there has been talk of a Comet,
but nothing was seen here until yesterday evening
towards 5 hours and a half, when the sky having become
completely clear, it appeared of a surprising size, with a
very long and pronounced tail. I have never seen a
Comet of this strength, and you could tell me if that of
the year 1618 resembles it.

Like many astronomers, Christiaan Huygens had devoted
large amounts of time during the 1670s to largely fruitless speculation about the origin and behaviour of comets. He, too, had
no time for them as portents. 'What does it mean if comets
signify good and bad events equally,' he observed, 'unless they
signify nothing?' He had believed that they were born in the
sun and travelled outward from it in straight lines, but the likelihood that what he saw in December 1680 was the same comet
that his father had seen in November strengthened the possibility
that, as Cassini believed, they in fact occupied orbits around the
sun. In February 1681 Huygens spoke before the French
Academy of Sciences, and explained how the comet could have
travelled through the solar system, describing how the length
and orientation of its tail would vary as it did so. He rejected
Descartes's belief that the tail was no more than an effect of
light refraction, and speculated that it might be composed of a
very thin 'smoke or vapour'. Such bodies could not originate
among the stars, for there would have to be vast multitudes of
them to account for the numbers seen from the Earth. Could
they in fact come from our own sun? He realized now that the
observed trajectory made this impossible, too.

In Paris, though, the omens were not good for Christiaan Huygens. His recurring spells of winter fever were increasingly attended by intense headaches and other complications. Some of his long-time associates, including his greatest ally Chapelain, were now dead. Increasing persecution of Protestants sanctioned by Louis XIV was beginning to produce a Huguenot exodus from France, and the king and the church were also growing more confident in their opposition to mechanistic interpretations of the world, which had been tolerated or ignored during the early years of Colbert's academy of science.

A medal struck in Huygens's honour in 1679 might have offered him some small consolation. The reverse showed the mythological figure of Saturn holding a cycloidal pendulum, with his ringed planet and its first satellite hovering above. But on the obverse, Huygens judged that the 'sculptor has not done at all well as far as likeness'.

He pressed on dutifully with scientific work on all fronts. Answering a request for information from an author planning a history of the early years of the French academy, Huygens compiled a list of the topics it had examined, which, since it corresponds so closely with the range of his personal interests, can be taken as a sign of his centrality within the organization. It included astronomical observations, among them 'new discoveries among the stars', confirmation of the form of Saturn's ring and the discovery of three planetary satellites, as well as studies of eclipses and parhelia (bright spots that appear in the sky on either side of the sun, now known to be due to ice crystals in the Earth's atmosphere), and the measurement of the circumference of the Earth, which came to 20,541,600 French rods (within 0.1 per cent of today's estimate). He also itemized the further development of longitude clocks and their evaluation on long voyages at sea; the construction of models of windmills and fountain

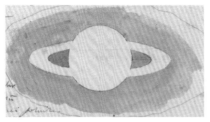

(*above*) 18. One of many drawings of the planet in various phases, this watercolour sketch of Saturn and its ring is to be found in Christiaan Huygens's notes from 1659.

(*right*) 19. Pendulum clock made by Salomon Coster for Christiaan Huygens, c. 1657. The interior shows the pendulum and the brass 'cheeks' devised by Huygens to modify the swing, thereby greatly improving the clock's accuracy.

20. Scheveningen church with the dunes protecting it from the sea, sketched by Christiaan Huygens in 1658, when he was making improvements to its clock.

21. Susanna Huygens, painted by Caspar Netscher in 1669, nine years after her marriage to her cousin Philips Doublet.

22. Henry Oldenburg. Oldenburg was the 'foreign secretary' of the Royal Society in London, doing much to build links between scientists across Europe.

23. A typical scene of music-making in a prosperous Dutch household. The Huygenses' musical evenings must have looked very similar to this.

24. This view of Paris shows Louis XIV crossing the Pont-Neuf. The new buildings of the Louvre Palace erected during the years when Christiaan Huygens lived in the city are visible on the right bank of the river.

25. *Colbert Presents to Louis XIV the Members of the Royal Academy of Sciences Created in 1667*, by Henri Testelin, c. 1675–80. This imagined scene shows many leading figures of the academy, probably including Huygens and Cassini.

26. Jean-Baptiste Colbert. Known as 'the north wind', Colbert believed in the potential of science and arts to increase French power and the prestige of Louis XIV.

27. Christiaan Huygens, by Caspar Netscher, 1671. Huygens had withdrawn from Paris for a short period owing to sickness, and was with his family in The Hague when Netscher painted this fine portrait.

28. Gottfried Wilhelm Leibniz. As a young man, Leibniz studied mathematics with Huygens in Paris, before surpassing his teacher and devising the method of calculus.

29. William III of Orange became the Stadholder of the Dutch Republic in 1672 and King of England, Scotland and Ireland following the 'Glorious Revolution' in 1689.

30. The hanged and disembowelled bodies of Johan and Cornelis de Witt. The slaughter of the de Witt brothers brought an end to the Stadholderless Period and a restoration of the House of Orange in 1672.

32. Replica of one of Leeuwenhoek's microscopes. The tiny bead lens is set in a hole pierced through the brass base plate. The object to be viewed is impaled on the nearby spike, which is attached to an adjustable stage for focusing.

31. Anthoni Leeuwenhoek. Largely self-taught as a natural philosopher, he relied on members of the Huygens family to spread news of his remarkable microscope observations in Paris and London.

33. Baruch Spinoza made his living by grinding lenses while writing some of the greatest works of western philosophy.

34. This painting depicts the great comet of 1680. Once regarded as omens, comets were gradually demystified by means of observations made with ever better telescopes during the course of the seventeenth century.

35. Christiaan Huygens had this painting made shortly after his father's death, when he had given up any hope of returning to his work in Paris. It became known in the family as his 'orphan portrait'.

pumps; 'a most exact map of the surroundings of Paris to some 10 leagues all around'; new observations made using improved microscopes; and experiments concerning refraction, the laws of motion, the vacuum and air pressure. Huygens declared that in addition to the work of the academicians themselves, the Academy had also been busy fulfilling its intended purpose as a national clearing-house for innovation. The inventions of others 'have often been called in to the Academy to be examined, which has been done, with complete fairness, although sometimes, enamoured of their imaginations [and] conceptions, the authors have complained'.

The news of Huygens's persistent ill health caused the family in The Hague to begin to think about ways of bringing him home. Other factors weighed in their calculation. His brother Constantijn had at last moved his own family out of their father's house on the Plein, where they had lived since his marriage in 1668, and the matter was raised of the father's possible loneliness. Christiaan's sister Susanna wrote to him in March 1680:

> I hope, my Brother, that your health might get better and
> better and that the best thing is for you to come and see
> us in Holland. My Father will take a particular Joy in this.
> For though he wants us to think that the solitude will not
> be unpleasant, it seems to me that the Change is too great
> for a man of his Age, having spent all his life in quite a
> large family to find himself alone in such a large House.
> But your Company, my Brother, will remedy
> everything . . .

Christiaan quickly caved in to their wishes, and announced that he would return to Holland in the summer. But Colbert still had projects for him at Versailles and elsewhere, and he was unable to obtain leave. His departure was set back again the following winter when he suffered further bouts of severe head pains. In the spring of 1681 he was joined instead in Paris by Susanna, who had longed to see the city, and her husband Philips and their three children. Christiaan finally set out for Holland with the returning family group in September. In his luggage were the parts of a mechanical brass planetarium that he had begun to construct in order to demonstrate the true paths of the planets by using gear wheels cut with unequally spaced teeth to simulate the planets' varying velocities in elliptical orbits.

Christiaan's brother Constantijn was often in The Hague on secretarial duties. Christiaan hoped the two of them would find the time to resume their work together on telescopes, in which they had found such companionable release as young men, and on microscopes, where he sensed the competition from Leeuwenhoek, 'our bourgeois philosopher of Delft', newly elected to the Royal Society, and now 'the great man of the age'.

One goal was to create telescopes of much greater focal length. Such working instruments were very different from the embossed leather-and-gilt spy-glasses made for presentation to dignitaries. A surviving device made by the Huygens brothers in 1683 consists of a dull tube of thin metal divided into five nested sections which extend to nearly six yards. It resembles nothing more than a drainpipe or a stovepipe, and was in fact made for them by a local tinsmith. But for even longer focal lengths, it was necessary to dispense with the tube altogether, which would be an impractical encumbrance at much greater sizes, and retain only the essential optical parts – the objective lens and the eyepiece. Christiaan had made observations using such a telescope long ago in Paris, but

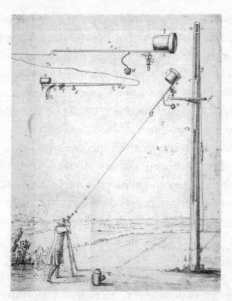

32. Huygens's arrangement of a tubeless long telescope.
The objective lens (*i*) is mounted on a platform raised up a mast, and steered
by means of a cord held by the astronomer.

the instruments he and his brother would erect in their father's
garden in The Hague were larger and more challenging to set
up. The objective lens was placed in a short tube mounted on a
copper ball joint fastened to a small wooden platform. This
assembly was then raised into a remote position by hoisting it up
a mast. A taut silk cord was run from this lens housing down
to the telescope eyepiece on a tripod on the ground. In this way,
the observer could maintain the two lenses in alignment, and
steer the telescope to look at different parts of the sky by means
of the cord while walking, taking the eyepiece on its tripod with
him, to new positions on the ground. In practice, the apparatus
was tricky to operate, especially in windy weather. Huygens
describes how, on one occasion, he had 'a small boy sent to me

by the carpenter climb up to reinsert the rope into the pulley from which it had come out, but this did not happen without his apprehension because of the swaying of the mast'.

Huygens felt himself to be competing in this work with Cassini at the Paris Observatory. Cassini had used lenses supplied by the Roman Campani brothers, Giuseppe and Matteo, when he discovered two new moons of Saturn in 1671–72. Huygens had been unable to match their quality using glass from Paris glassworks, but he hoped that in Holland, with his brother's expertise and better raw materials, he might be able to catch up. These vast instruments demanded objective lenses the size of dinner plates if anything was to be seen through them. Because of the long focal distances, their curvature is hardly apparent at a casual glance, but the lenses nevertheless had to be ground with great precision. Using a local supplier of glass in 's-Hertogenbosch, the two brothers succeeded in increasing the focal distance of their telescopes year by year from 34 feet to 210 feet. But it did them little good. It was Cassini who discovered two more moons of Saturn in 1684. Not long afterwards, Huygens wrote affably to Cassini that he had found his tubeless method 'simpler', but 'less perfect' than his own because of the awkward adjustments it required. He signed off with a cordial greeting: 'Always continue, Monsieur, for the advancement of the sciences.' In private, though, Huygens believed Cassini was jealous of his own record on Saturn. Huygens never saw the fourth and fifth satellites identified by Cassini, and was unable to make any discoveries of his own with his new telescopes.

An important secondary objective for Huygens had always been to devise an automated means of grinding lenses, in order to improve optical quality as well as to speed up production, and in this area he was able to make some progress. Through many iterations, the brothers had found that the quality of the glass

was not the critical factor. Many glasses would produce a satisfactory lens; it did not necessarily require the clearest. In fact,
'usually the best is that which is yellowish, reddish or sea-green
in colour when seen from the side'. The glass for these large
lenses was made in the same way as mirror glass, by blowing a
large glass sphere and then cutting from it a disc, which was then
allowed to relax while still soft into a planar shape on a hot flat
surface. Best results were obtained, Huygens noted, when the
glass was allowed to rest for a couple of days, as over a holiday,
giving time for small bubbles and other flaws to work their way
out of the medium. Each such piece cost three guilders.

The greater challenge was to be able to grind and polish the
disc of glass so that it would produce a clear magnified image at
a predetermined focal length. Grinding by hand was liable to
introduce irregularities if the grinder applied too much pressure
in one place. But a mechanized process risked introducing vibrations. Huygens eventually succeeded in designing a grinding
platform that combined the advantages of both techniques, using
a mechanical arm driven by a treadle to ensure the correct pressure of the grinding medium on the lens, but retaining the
sensitive guidance of the hand. Even so, work was slow. The coarse
grinding stage alone might take five or six hours. The polishing,
with a still finer compound of 'tripoli' powder and copper sulphate
mixed with vinegar, required 3,000 passes in all for a large lens
to be polished to perfect smoothness on both sides. Huygens even
incorporated a mechanical counter into his custom platform to
save manual counting of the score or so passes made in one position before the lens was ready to be adjusted minutely into the
next position for the process to be repeated. Often, a flaw was
found at the end of it all, in which case the lens had to be cleaned,
degreased, reaffixed to its mount and polished some more. 'To
polish by hand would be a very great labour, impossible even in

the case of large lenses of 5 or 6 inches or more,' Huygens explained. 'This is why we have in the first place designed and put into operation an apparatus to hold the glass against the bowl for as long as necessary, in order to be spared this part of the chore.'

Huygens had not intended to stay long in The Hague. He finished work on his 'Automatic Planet Machine' by August 1682, and wrote to Colbert that it was ready, itemizing thirteen aspects in which he believed it to be superior to a similar device already made by Rømer. He proposed to charge 620 écus for the instrument, which, he pointed out, would also make it cheaper than the Dane's device. Driven by an eight-day clock, it replicated the orbits of the planets 'most precisely according to the time of their periods, for average motion and for the inequality which requires them to move more slowly to the extent that they find themselves further from the sun, in which I have conveyed Kepler's hypothesis'.

This was to be Huygens's last communication with his French patron. He made the usual excuses about his ill health, and assured Colbert that he was keen to resume work with the Academy and on the *grands projets* under Colbert's personal direction. But no encouraging word came back from Paris. Instead, it was agreed that he should extend his leave. Huygens busied himself for a few months with other work, including making improvements to his marine clocks, and was thinking about departure again in September 1683 when news reached him that Colbert was dead.

Colbert's successor, François-Michel le Tellier, the marquis de Louvois, who had been Louis XIV's successful minister of war against the Dutch, was less sympathetic towards Huygens.

Dutifully, Huygens wrote to pay his respects to his new master with what optimism he could muster:

> I do not know what change might come to [the Academy].
> But having learned that its care is being put into your
> hands My Lord, who esteems the Arts and useful inven-
> tions, and is even pleased to take note of them, as much as
> your limitless pursuits allow, I am persuaded that our
> matters will go even better than in the past.

Huygens even dared to hope 'that my status might become a little better than it has been in the seventeen years that I have been with his Majesty'.

It was not to be. Constantijn interceded repeatedly on his son's behalf, writing to his own contacts appealing for Christiaan's reinstatement, and, as it became apparent this was too much to hope for, trying at least to ascertain whether 'the good opinion France has had of him still persists'. But Le Tellier seemed deter-mined not to understand Huygens's importance within the Academy of Sciences since its foundation. He revealed his ignor-ance when he addressed a letter 'to Monsr. Huijgens &c. mathematicien', recognizing his apparent trade rather than his noble title. Constantijn took umbrage at this social faux pas. 'He seems to take him for one of his fortification engineers,' he huffed to one of his French friends. 'I had not thought to have an artisan [*gens de mestier*] among my children.' To make the rebuff quite plain, Le Tellier also refused payment for the completed planet-arium that Colbert had commissioned.

The Huygenses continued to plead in an increasingly desperate manner until 1685, by which time Christiaan believed himself to be the victim of the jealous 'intrigues of certain people who do not wish me to return to my post among those of the Academy of Sciences'. His final letter to Le Tellier was abject. Having been unable to ascertain his true standing from any source, he wrote: 'It was as if I had received your very orders and it was the King's will that I not return to which I must acquiesce though completely ignorant of the cause.' When the German mathematician Ehren-fried von Tschirnhaus sent him a copy of his *Medecina mentis et corporis* ('Medicine for Mind and Body'), which was dedicated to the French king, Christiaan commented bitterly to his brother Constantijn: 'Louis has great need of both.'

Huygens was not alone in finding himself unwelcome in France in the years leading up to the revocation of the Edict of Nantes, which finally gave state sanction to the persecution of Protestants. The French Academy of Sciences stopped appointing foreign members. Existing foreign academicians were also affected, leading to a brain drain that decimated French scientific life. Rømer returned to Denmark in 1681. Leibniz had already left to work for the House of Brunswick in 1676, having failed to gain admission to the academy. Only Huygens's fellow Dutchman Nicolaas Hartsoeker moved his family to Paris in 1684, although he, like Leibniz, was not admitted to the academy because of his foreign birth. He became nevertheless the chief beneficiary of Huygens's absence, continuing the work of developing very large telescopes for the Paris Observatory.

In the quiet, *burgerlijk* surroundings of The Hague, Huygens attempted to pull together something of his old life. After the deaths of useful intermediaries, such as Chapelain and Carcavi in Paris and Oldenburg in London, he found new points of contact with the scientific communities in both cities, most notably the Swiss Nicolas Fatio de Duillier, a gifted young mathematician whom he met in The Hague, where they discussed topics as varied as circular geometry and snowflakes, before Fatio moved on to London and made the acquaintance of many fellows of the Royal Society.

His father proposed that Christiaan should succeed to his position on the council of the Prince of Orange, but Christiaan, perhaps recalling his own Ciceronian words on the death of his old friend de Witt, decided he was not cut out for political life. Instead, Constantijn saw to it that a royal stipend was arranged, which enabled Christiaan to pursue his scientific activities in Holland for the remainder of his life, and gave him the independence he needed to produce his great treatises on light and gravity.

Huygens filled much of his time instead designing and supervising the construction of new clocks that would be trialled aboard the *Alcmaer* for the Dutch East Indies Company. The development work with Johannes van Ceulen, the local clock-maker who had also built Huygens's planetarium, lasted four years. Huygens often felt the project was an intrusion on the more enjoyable business of polishing lenses and building telescopes with his brother. He was no more enthusiastic about joining the early trials sailing short runs on the Zuiderzee and off Texel island, finding that the pitching and tossing of the small vessels on the choppy seas made him ill, and he did not accompany the clocks on the *Alcmaer*'s subsequent long voyage to the Cape of Good Hope.

Christiaan never married, and without the responsibilities of dependants or an estate to be looked after there was no compelling reason for him to do so. This was a situation that he shared with most of his professional contemporaries, including Newton, Boyle, Hooke, Descartes and Leibniz. But he never gave up hope of finding lasting female companionship. Although he had promised to remain faithful to her, Marianne Petit's love for him did not survive their separation in 1661. Whether out of 'piety, despair or foolishness', as Huygens put it, she chose a life of devotion and made 'a plan for the cloister'. There were other favourites later, such as the married English beauty Jane Myddelton or her younger sister, who was intrigued by his skill at drawing her likeness. While he was recuperating in The Hague in 1671, he struck up a romance with Haasje Hooft, the young daughter of one of the burgomasters of Amsterdam, but it did not last beyond his return to Paris. During his long, final residency in Paris, Huygens had got to know Suzette Caron, a distant cousin of his, and her husband François de Civille, seigneur de la Ferté, and became godfather to their eldest daughter. The couple separated in 1685, whereupon Suzette was imprisoned in the Bastille for her Protestant beliefs. After three months, she renounced her faith in order to secure her release, and fled penniless to England, and thence to the Netherlands, where she met Huygens again. It was rumoured that they may have begun an affair.

Over the years, Constantijn Huygens noted his growing longevity in ever more woebegone birthday poems. As his seventieth year dawned on 4 September 1665, he wrote: 'One more September, and one more fourth day I have seen appear! / How many Septembers, Lord, and how many fourth days will you suffer me yet?' On his eighty-second birthday it was: 'Cease murderous years, and think no more of me / You come on too late . . .'

He died at last on 28 March 1687 at the age of ninety. It was Good Friday, and perhaps some believed his soul ascended straight to heaven. His body was buried in Sint-Jacobskerk in The Hague. He left behind him more than 800 musical works, 75,000 lines of verse in half a dozen languages and tens of thousands of letters composed during eight decades of loyal service to the House of

Orange. Providing biographical information for the Geneva-born theologian Jean Le Clerc, who was compiling a historical dictionary, Christiaan summarized the positions his father had held with the House of Orange, and drew attention to the restitution of Orange that he had won from Louis XIV. He ended his short note: 'He died in 1687, at the age of 90 years 6 months, being President of the Council of the Prince in service of whom and of his predecessors he had held position during 62 years, having wholly kept his mind to such a great old age.'

There is nothing in Christiaan's surviving correspondence to tell us how he felt about losing the father who had been his lifelong teacher, guide, agent and emotional and financial support. His first letter after Constantijn's death, addressed to the mathematician and astronomer Philippe de la Hire, one of the new intake at the French Academy of Sciences, makes only passing reference to 'the matters that arise on these occasions, and furthermore my own indisposition which has not entirely left me', as he presents his excuses for having been unable to send some promised scientific papers to Paris.

His sense of bereavement is more palpable in a portrait he had done of himself at this time. A note later made by a nephew of Christiaan's describes the work as having been 'done near the end of his life and in such orphan rags that it has stayed greatly impressed upon me'. It is indeed a dour likeness. Gone are the colourful ribbons and satins of the Netscher portrait of 1671. It shows Christiaan dressed in plain black clothes with a lace collar, looking pale and puffy with long dark hair in place of the golden periwig curls of the earlier painting. He still looks, though, younger than his fifty-eight years, and his high, dark eyebrows and wide brown eyes give an impression of alertness and intelligent appraisal.

The house on the Plein was valued at 32,000 guilders. The

father's other estates and their existing contracts came to 80,000 guilders, yielding an inheritance of 28,000 guilders for each of his four surviving children. The house at Hofwijck, valued at 4,000 guilders, was not sold, but retained for the shared use of the brothers, who agreed that Christiaan should take up permanent residence there. After some nighttime agonizing, Christiaan grudgingly accepted the plan. 'I imagine that by doing some building to enlarge the house and install my library, I could live there quite agreeably,' he told Lodewijk. He moved in at the end of April 1688 and took delivery of the major portion of his father's library that was not to be sold. After five days there, he wrote to his other brother: 'I have not gone round The Hague during this time, nor have I received any news from there, and so make my first foray into the solitary life to which I shall have to adjust. What hurts a little is to be alone at dinner and supper, although I do have that in common with the crowned heads.'

The scholar-king was left alone in his moated castle surrounded by his books, with the stars and the planets wheeling silently in the sky overhead and the heavy ticking of the clock echoing from the walls.

14

LIGHT AND GRAVITY

The invasion did not begin well. Lying off Hellevoetsluis, Prince William's fleet of forty-eight warships, twelve fireships and more than two hundred transport ships was at first delayed in its departure for the English coast and then forced back to port by storms with many vessels damaged. The horses taken aboard for the army's march to London had been incorrectly stalled facing the bulwarks, and many of their heads had been stove in by the ships' motion, and so the carcasses had to be dumped overboard. Travelling on the prince's vessel, Christiaan's brother Constantijn alone lost five animals, and the order went out to round up replacements from

33. William III's England invasion fleet assembled at Hellevoetsluis in November 1688.

wherever they could be bought. Fortunately, the 50,000 printed copies of William's declaration of his claim to the English throne that the fleet carried aboard remained intact.

A small pen-and-ink sketch, presumed to be a self-portrait, shows Constantijn Huygens shortly before the 1688 expedition to England, with a smooth, untroubled face emerging from amid the abundant curls of his periwig, wide-set eyes and the ghost of a smile, a man at ease and perhaps even a little smug in his important secretarial office. A medallion profile in ivory made in 1690 while he was gathering maps and equipment in London in preparation for the army's further campaign in Ireland reveals a rather different figure, stout and jowly, tight-lipped and proud, staring coldly ahead. The two years away from home were often testing and trying, but the prince's secretary has the look of one who has carried out his duties well, and is as secure as anybody can be who holds a position at court.

In the frantic days as the fleet was being readied again, and everybody waited for an easterly wind, Constantijn suffered from a fever and bad dreams. The Dutch were on the brink of a momentous action that would, if all went well, secure the safety of the republic from French ambitions. In England, the campaign would come to be known as the 'Glorious Revolution', a term all but coined by Huygens himself as the army waited, when he noted down the words of William's field marshal, the Prince of Waldeck, that they were embarking upon 'a great and glorious enterprise'.

The armada – four times the size of the one Spain had launched against England a hundred years earlier – finally got away on 11 November. The ships flew down the Channel before a 'Protestant wind' passing Dover and the Isle of Wight, and dropped anchor in Torbay four days later. 'We landed at a village called Braxton [Brixham],' Huygens wrote in his diary. 'It is very simple. Its few

34. Possible self-portrait of Constantijn Huygens the younger,
Christiaan's older brother and secretary to Stadholder William III
of Orange, King of England, Scotland and Ireland.

decayed houses are built of stone – hewn from rocks found both
on the coast and inland – and roofed with slate . . . We dined
this evening on a very tough fricassé of lamb. I wrote a letter to
my wife.'

The Dutch army, 21,000 strong including Huguenot, Scots
and English volunteers, set forth nervously at first, uncertain as
to whether they might encounter English troops loyal to King
James II. But they met little resistance. Huygens noted 'the high
mountains and deep valleys' of the Devon countryside.

The roads were incredibly bad, all made of stone and strewn
with loose rocks, and very dirty and muddy to boot. Along
the way we saw country folk everywhere, as on the previous
day. Women and children shouted, 'God bless you,' showering

us with good wishes. They gave the Prince and our men apples, and an old woman with a bottle of mead wanted to pour a glass for His Highness.

In Exeter, they found the people 'polite' but 'extremely frightened' because James had issued a warning not to deal with the Dutch invaders. Huygens had the delicate task of writing to the bishops of Exeter and Bath 'asking – but with threats added – to have our troops billeted in their cities'. Prince William addressed the people and gave reassurances, and what little resistance his men encountered soon melted away. It became clear there would be no counter-attack.

The Dutch sent a message ahead that English troops should remove from London in advance of William's arrival. Rain turned to hail and the mud grew thicker as they marched on through Dorset and Wiltshire, but there were a few highlights on the route. Huygens admired the newly erected Italianate town hall in Abingdon. William made a point of seeing the paintings at Wilton House, although Huygens did not accompany him, preferring to hurry on to Salisbury 'to warm myself'.

They reached the royal castle at Windsor on 24 December. Huygens marvelled at the sight of the River Thames in the splendid weather, and inspected the royal apartments with their tapestries and 'many good paintings by Titian'. Often in England, when William was in the mood to chat, he spoke with Huygens, a *kenner* of art like his father, about the paintings in the royal collection at Windsor and at Whitehall and Hampton Court. The following evening, Huygens called on the aged Isaac Vossius, a scholar and family friend from The Hague who had published a treatise on the nature of light before moving to England and becoming the prebendary at Windsor. He found the old man in his library where he complained about the impoliteness and gruffness of the English, with whom he spoke little, he told me. He said that he was afraid that if His Highness took the crown, he would lose the goodwill of the people. Then he began to talk in his own manner about physics, about the seeming circles on the moon, as he postulates in his book, though without sound reasoning.

Christiaan was relieved to hear of his brother's safe arrival in England, especially after all the difficulties at the outset of the expedition. 'Now your arrival in London is awaited with impatience, and the reception that will be given to the Prince which will doubtless be a marvellous thing to behold,' he wrote. Aware of French interests, Christiaan was keen to hear as soon as possible 'how everything will be set up and run both there and here', and anxious that Dutch soldiers be sent home, 'now that you no longer need them, lest we are the recipients of some great insult as we go about protecting our neighbours'.

Alone in The Hague, Christiaan was clearly missing his sibling's companionship and some back-and-forth with a scientific equal. Cassini's recent claim to have sighted two further new satellites of Saturn at the Paris Observatory still rankled with him while he remained unable to see them with his own equipment. 'When shall we work together again on large lenses?' he asked plaintively. Heedless of his brother's burden of court duties, he urged that when Constantijn reached London he should waste no time in visiting a particular lens-grinder as well as Robert Boyle and other members of the Royal Society. He clearly regarded his brother as the advance guard for the trip he hoped to make himself in the spring. 'I would wish to be in Oxford [he meant Cambridge], only to make the acquaintance of Mr. Newton, of whom I greatly admire the fine inventions I see in the work [the *Principia*] he has sent me.'

Constantijn and the Dutch forces finally arrived at Whitehall Palace on 27 December: 'many bonfires were lit'. Over the next few months he did find time to make some scientific connections. He visited the shop of Thomas Tompion, the leading watch-maker. He had a case made for the Campani telescope that he had carried with him for astronomical recreation during the campaign. He bought books and enquired about the latest *Transactions* of the Royal Society, which was late to the press. The Dutch were horrified to find that the only printing operations in England were in London, Oxford and Cambridge, in sharp contrast with the situation in the Dutch Republic, where every little city had its own press.

Huygens was officially William's secretary for Dutch affairs, but it was often hard to compartmentalize the work or to get the prince's attention. Before they left Holland, he had been entrusted with the English seals, but he had to relinquish these to the secretary for English affairs once they were in England, which caused him some disquiet. At times, there was an absurd dimension to the competitiveness between the secretaries, for example when it came to obtaining a favourable camping pitch while on campaign, or on one occasion that Huygens ruefully records when a rival's coach pushed ahead of his own in order to move closer to the prince.

Sometimes Huygens was able to use the minor distinctions in court duties to work his way out of an awkward situation. In February 1689 he was approached by men seeking to put off the execution of George Jeffreys, James II's Lord Chancellor, the notorious hanging judge of the Bloody Assizes held after the Monmouth Rebellion against the Stuart king. He was offered 500 pounds, which he declined to accept, saying it was an English matter. He primly told the men that in Holland such behaviour would be seen as corruption.

That same month saw the first session of Parliament held before Prince William. 'The Lords and Commons came to His Highness to thank him for his care and trouble in delivering them from popery and slavery, and asked him to continue governing,' Huygens wrote. The joint coronation with his English wife, Mary, the Protestant daughter of James II, took place on 21 April. All that spring Huygens suffered from colds and gout. He grew dissatisfied with his cramped quarters at Hampton Court and with his salary, which he felt would be insufficient to support his family when they followed him over. He cannot have been so badly off, however, as he consoled himself by buying art for his master and for himself, including a book of human movement studies that he took to be by Leonardo da Vinci. His relationship with William remained companionable, though, and occasionally the prince would ask after his secretary's family or his health. When Huygens complained about the damp weather, William commiserated: 'It is the air here . . . Do you not long for home sometimes?'

Huygens was clearly on good terms with William, but he was not as favoured as some. Rumours of homosexuality at court arose in England after the 1688 invasion, fuelled by Jacobite partisans. Although William and Mary produced no heir, it is known that Mary suffered many miscarriages. Their relationship, which had been cool at first upon their marriage in 1677, grew into one of true love. But William did have male favourites. He was distraught when, upon arriving in London, he learned of the death of one close friend, Gaspar Fagel, who had helped to orchestrate anti-Jacobite propaganda in advance of the invasion: 'I am losing the greatest friend that I can have in this world,' he wrote.

However, William's most faithful companion was Hans Willem Bentinck, a page who was soon promoted to chamberlain and thereafter to more senior advisory and diplomatic positions. When the prince contracted smallpox in 1675, Bentinck lay in bed with him, on doctor's advice, in an effort to cure him; the prince recovered but Bentinck caught the disease in turn. This selfless act earned William's lasting gratitude, and Bentinck was showered with favours and later given the title of Duke of Portland. In England, however, another page, Arnold van Keppel – charming, vivacious and twenty years Bentinck's junior – caught William's attention when his horse broke its leg. When Keppel was promoted, Bentinck felt rejected and eventually asked to be relieved from the service of the court.

William's sexuality had never been an issue in the Netherlands (where sodomy was a capital offence, but prosecutions were very rare). Nor was it a great scandal at court or even sufficiently noteworthy for Huygens to write about in his diary, where it seems that the subject simply held no interest for him. The evidence in any case amounted to no more than a habit of effusively emotional greetings revealed in hand-kissing and letters – William signed his letters to Bentinck with endearments such as 'I am always till my last sob yours'. Besides, other factors may have guided the king's judgement. Huygens observed that Bentinck sometimes came across as 'too Dutch', whereas Keppel could be relied upon to behave like an Englishman on difficult occasions.

Others experienced more severe pressures, which contributed further to the stresses of Huygens's job. One of the new king's first actions was to appoint John Temple (the son of Sir William Temple, who had been Charles II's special ambassador to the Netherlands) to be his secretary of war. But Temple had gone to the king to confess that he did not feel up to the job. Huygens recorded the rest in his diary:

The king had encouraged him and told him that if he hired
the old clerks, he would learn the trade very soon. Then
Temple left in a pair of oars on the river, and had himself
rowed to London Bridge. When he arrived there, he took
out a paper or a letter, together with a shilling for postage.
When he was under the bridge, he said, *Adieu, watermen*, and
threw himself in the river and drowned.

According to other sources, Temple had stopped off on his
final journey downriver to call on various friends, who found him
'very Melancholy and Discontented, or at least somewhat disturbed
and troubled of Mind'. He had then switched boats and ordered
the new skipper to shoot the bridge, where he leapt out into the
rough water. His body was retrieved two days later a little way
downstream at Pickle-Herring Stairs. The suicide was a reminder
– hardly needed in Huygens's case – that the service of Prince
William was onerous in many ways.

On 11 June 1689 Constantijn's wife and son arrived in Harwich
along with his brother. Christiaan had felt abandoned in The
Hague, not only by family but also by his nation now that the
stadholder was in England. He had prophesied to Constantijn:
'It is only England that will profit in the end from this great
revolution, and the only advantage that will redound to us
[Dutch] is, as I think, that we should otherwise have fallen into
greater misfortunes.' Above all, perhaps, he regretted that Prince
William had 'so little fondness for studies and the sciences'. If
matters had been otherwise, he might have had the hope of
royal patronage such as he had enjoyed in France, and a little
more scientific companionship. As things stood in The Hague,

there was 'not a single soul to talk to about things of that nature'.

His principal reason for going to London, then, was not to bask in the aura of the Dutch court abroad, and he was not sorry to miss William's coronation, the second such occasion that he had managed to avoid in London. Instead, he wished 'to see some old friends, apart from those who have recently passed away, and what is being done in the way of sciences, in London and in Oxford and Cambridge or wherever I am sufficiently known'.

It was to prove a short but significant visit. He travelled to Greenwich, where John Flamsteed showed him the observatory and his instruments. He shared a laugh with Robert Boyle, who told him a story about a man who claimed to have made an ounce of gold out of lead by using various powders, and who had subsequently been arrested in France. On 22 June he attended a meeting of the Royal Society at Gresham College, where he read parts of his works on light and gravity. Among the audience was Isaac Newton.

Newton was in his pomp, forty-six years old, with his masterwork, *Principia Mathematica*, published just two years before. But Huygens, at sixty, was still the acknowledged senior figure, the most respected physicist in Europe. Newton had sent a personal copy of *Principia* to Huygens upon publication (it was handdelivered by Edmond Halley). This was not just good manners. Newton knew that Huygens was one of very few recipients of the work in a position to offer useful comments on it. Furthermore, the ground had been prepared by Fatio de Duillier, who, as an informed acquaintance of both men, was in an ideal position to mediate a correspondence between them.

Huygens had devoured the *Principia* at his Hofwijck retreat, weighing the merit of its new ideas against his own, still unpublished, theory of gravity influenced by Cartesian vortices. He duly responded with various suggestions and a list of errata, which he hoped might be incorporated into a second edition. Newton graciously acknowledged the comments, but they were not such as to cause him to revise his basic thinking. In his private note-book, meanwhile, Huygens wrote: 'Vortices destroyed by Newton. Vortices of spherical motion instead.'

Huygens felt unable to embrace Newton's work *in toto*. He was uneasy about the Englishman's dynamical approach, based on unseen forces, including the strange centripetal force that held celestial bodies in orbit. Surely any force had to be accompanied by matter in motion. How could the force of gravity be transmitted through empty space, as Newton's theory stipulated, without the aid of vortices? He was critical, too, of Newton's reliance on empirical evidence. For him, it was not enough to say that a hypothetical force explained all kinds of motion, and that it was consistent with rigorous mathematical exposition. He still wanted

35. Isaac Newton in 1677, fellow of the Royal Society
and of Trinity College, Cambridge, and Lucasian Professor
of Mathematics at Cambridge.

to know where the force actually *came from*. Even Newton had no answer for this, and to Huygens the omission seemed a step backwards, perhaps even towards occultism. On the other hand, the *Principia* offered persuasive explanations of many phenomena that had troubled natural philosophers. For instance, Newton's deduction that the Earth was ellipsoidal in shape – an idea that Huygens had entertained independently – neatly explained the pendulum anomalies that had dogged the sea trials of his marine clocks. Furthermore, Huygens recognized the mastery of Newton's (still unpublished) method of calculus, and it was this above all that convinced him that his theory of gravitation must be essentially correct. For Huygens as for Newton, mathematics was the ultimate arbiter of truth.

Despite the two men's differing philosophies, it was an affable meeting. If there was any sense of *noblesse oblige* on Huygens's side, or of triumphalism on Newton's, nobody thought to note it down. In his Gresham College lecture, Huygens had set out his explanation of the double refraction of Iceland spar, and it was this novelty that drew them into conversation afterwards, causing Newton to concede that the *Principia* was not yet complete. It was a meeting of minds rarely equalled even under the auspices of the Royal Society. The men met at least twice after this and began to correspond directly.

The relative standing of the two great physicists at this time is illustrated by something that happened the following month. Though he had long been Lucasian Professor of Mathematics at Cambridge and a fellow of Trinity College, as well as having been recently elected as one of the two members of Parliament for the university, Newton nurtured an ambition to be the provost of King's College. This required a petition to the king himself. Through his brother, the king's secretary, Christiaan Huygens was able to provide access for his new friend. On 9 July Newton called on Christiaan, who was with Constantijn at Hampton Court, and the following day Christiaan, Newton and Fatio de Duillier were granted an audience with William. The king announced that he was prepared to support Newton's nomination, but the idea was later rejected by the fellows of the college – an unprecedented action, after which royal prerogative was never again used in this way. Though it may not have had the desired outcome, this moment of scientific internationalism shows that it is Newton who is put in Huygens's debt, and not the other way round as modern observers might assume.

A few weeks after this episode, Christiaan set sail from Gravesend on a slow but eventful crossing of the North Sea.

24 Wednesday. Embarked in the morning on the ship
Briel, the same which had carried the King of Holland and
England . . . Saturday 27, sighted since morning 3 vessels
approaching close to us thought to be French. All
prepared for combat, the passengers except 2 or 3 each
taking a carbine and a bandoleer. The women went down
in the rope hole and were there for more than 2 hours.
Finally with my telescope I began to discern Orange,
white and blue banners, and it was realized a little later
that these were vessels from Amsterdam.

He arrived safely in The Hague on 30 August.

Constantijn continued to exchange scientific news with Christiaan, although he was now mainly preoccupied with the forthcoming military expedition to Ireland, which was now 'only too certain'. That autumn and winter, they discussed Leeuwenhoek's microscope observations and his notion of what Christiaan called the 'infinite inclusion of animals and plants', that is, the idea that each creature contains the seed of its successor and so on *ad infinitum*, so that no truly new living creatures are made in the world. Constantijn visited lens-grinders and watch-makers, and heard Thuret speak at the Royal Society, claiming to have invented a new marine clock. On one occasion, he was amused to find himself accosted by a Scottish milord in the royal antechamber who wanted to know if it was his father or his brother who had invented the pendulum clock. But the political upheaval made life difficult both in Holland and in England. Christiaan sometimes thought it would have been better for both countries if James had been allowed to rule his kingdom in peace. In Holland he noted 'a great cooling towards all the finer things'.

In June 1690, Constantijn Huygens rode with the king's army to Chester, from where they sailed via the Isle of Man to Carrickfergus Bay. On a long stone bridge on the way to Belfast, he was 'disconcerted to see a large crowd of poor and miserable people, men, women and children, bad-tempered and looking very ugly and unhealthy'. In Belfast itself, the king took lodgings at a grand house 'which had very bad paintings' while Huygens lodged in an alehouse. He noted that the people 'do not want to admit to being Irish, and say that they are Scots who came over'.

The progress that had been largely bloodless through England was to turn remarkably bloody in Ireland. William's army marched south from Newry to Dundalk for Dublin, and on 10 July they spotted James's troops, which they estimated to be up to 20,000 strong, assembled on the far bank of a bend in the River Boyne. An exploratory salvo from the Jacobite side found its mark. Huygens wrote: 'The King was hit by a cannonball, which stripped his coat, waistcoat and shirt and scorched his skin. He was bandaged in a hollow-way, making no appearance of discomfort from the first, and saying only, "That should not have gone further".'

The following day, Huygens rose early, uncertain about the king's intentions. 'Nobody spoke with any assurance about the passage of the river,' he observed despondently. Nevertheless, a plan was hatched. William's men waited for low tide and then waded across the river, firing as they advanced. James's troops appeared to retreat, but gathered on a hilltop and rushed down in a counter-attack. This manoeuvre was repeated until Dutch reinforcements arrived, whereupon the Jacobites fled 'so fast that our cavalry could not overhaul them'.

James fled to France and his army of mostly untrained soldiers deserted. William's men continued their march south through the ramshackle villages, plundering as they went. Huygens noted that 'our English soldiers committed various atrocities to women and wretches', while a Huguenot commander freely told Huygens that he had 'chopped off both of a girl's hands, gouged out her eyes, and left her lying there'. Protestant Dubliners rushed out of the city to greet the advancing troops. Huygens was pleasantly surprised by Dublin. 'The houses are fairly good, compared with most cities in England . . . The shops are rather poorly stocked and the people clad worse than in England, but very much in their own way. The womenfolk are reasonably pretty.'

William's Irish campaign would drag on for another year yet, with some far bloodier battles ahead. But Constantijn Huygens's war was over. He sailed from Dublin to Chester on 13 August, and was back in London just over a week later.

Christiaan Huygens's many-times postponed work on light finally appeared in print at the beginning of 1690. Published together with the (also delayed) *Discours de la Cause de la Pesanteur* ('Discourse on the Cause of Gravity'), the *Traité de la Lumière* ('Treatise on Light') was his first major work written in French rather than Latin, an indication of the increased prestige of French science. He made the change from his decades-old working title of *Dioptrics* to the more ambitious *Traité de la Lumière* just before the work went to press, at the urging of his French colleagues. This was an acknowledgement of the expanded scope of the final text, but also a signal that Huygens was ready to move on from Kepler and, especially, from Descartes, both of whom had produced works called *Dioptrics*. The contents dealt with the sun and other sources of light, the speed of light, reflection, refraction and the design of optical media, and yet still gave half its pages over to the new discoveries related to Iceland spar.

The preface to the treatise was notably modest. 'One may ask why I have been so slow in bringing this Work to light,' Huygens wrote with disarming candour. He answered his own question by explaining that he had begun to write in Latin, in which language he felt he could describe the science better, before switching to French. 'I finally judged it better that this text should appear as it is than to let it run the risk, in waiting longer, of being lost.' He confessed that some of the experiments he described had not produced results quite as crisp and clear as the geometrical analysis implied, and referred, with obvious pain, to 'where I leave difficulties without solving them'. But this was a new, relaxed and almost carefree Huygens, who was at last prepared to publish what he had and let others be his judge.

Perhaps he was emboldened by Newton, who could not give an ultimate explanation of the cause of colours, and was not especially concerned to do so. Huygens summarized his theory of light, still largely based on Cartesian principles of collisions occurring in the ether, with the almost offhand statement that 'light expands successively in spherical waves, & how it is possible that this expansion happens at such high speed, that experiments, & celestial observations must investigate'. In fact, Huygens had long been comfortable in principle with such uncertainties. In 1673 he had written to Pierre Perrault with this eminently quotable axiom: 'I do not believe we know anything with complete certainty, but everything probably and to very different degrees of probability.' What was new was that he had overcome his fear of the effect it might have on his reputation to commit uncertain findings to cold print. Cajoled by Leibniz, Huygens also abandoned the classical method of argument of earlier works such as *Horologium Oscillatorium*, based on a succession of theorems and proofs, in favour of a more modern approach, which made the *Traité* in some respects more forward-looking even than Newton's *Principia*.

Huygens sent out the treatise in his customary fashion to his international scientific colleagues. He sought to use the offices of his brother serving William III in London to distribute copies in England, but although his letter of instruction arrived, at first the books, sent along with a consignment of horse-blankets, did not. This was eventually discovered to be because the emissary entrusted with the task, a royal table-dresser named, appropriately enough, L'Orangeois, 'was found having tried to smuggle through Customs a piece of Holland cloth and some lace which he had hidden all round his naked body'. Soon enough, though, copies were dispatched to Newton, Halley, Boyle, Fatio de Duillier and John Flamsteed, the first Astronomer Royal. 'You will see that I have in several places marked the divergence of my feelings from those of Mr. Newton,' Huygens noted to Fatio de Duillier, 'as I have found myself obliged to do in order to support my Theory, but I have not made them of such a kind that I think he will take them in bad part.' On the continent, Leibniz, Pierre Bayle and Johannes Hudde also received copies of the treatise. Another recipient was the French Protestant inventor Denis Papin, who had collaborated with Huygens on the design of a gunpowder-operated engine when they had both been living in Paris. Following the revocation of the Edict of Nantes, he had fled to Marburg, where he was about to build the first prototype steam engine.

Newton responded warmly to the *Traité*, calling it 'perfectly beautiful and worthy of the author', in a comment passed on to Huygens by Fatio de Duillier. Clearly, much of the tension between the two great optical scientists dating from the time of their argument about colour had dissipated with their meetings

36. Nicolas Fatio de Duillier. The young Swiss mathematician
made himself a useful conduit for Huygens,
Newton and Leibniz.

in London. Newton was at work on what would become his own *Opticks*, but where their theories differed, they sought to avoid conflict. Huygens was opposed to all colour theories, and had purposely omitted to include anything in the *Traité* about colours, concerning which both Newton and Hooke had such strong ideas. He told Leibniz: 'I have said nothing about colours in my Treatise on Light, finding this topic very difficult; above all because of the many different ways in which colours are produced.' For his part, Newton took Huygens's discussion of Iceland spar as a cue to begin his own experiments with the material, but he made no mention of them – or of Huygens's substantially correct analysis – when his *Opticks* finally appeared in 1704, some years after Huygens's death.

Huygens sent a further copy of his treatise to the philosopher John Locke, who had returned to England with the 'Glorious Revolution' following five years living in exile in Amsterdam and Rotterdam. Locke had made contact with Huygens during that time, seeking his confirmation that the mathematics in Newton's *Principia* was sound. (Huygens assured him it was.) Locke and Huygens found common ground in disagreeing with Newton's model of gravitational attraction, in which neither was able to accept the notion of force required to act at a distance through a vacuum. In return, Locke sent Huygens his own newly published *An Essay Concerning Human Understanding*. Huygens read it 'with much pleasure, finding there a great sharpness of mind, with a clear and agreeable style, which not all those in that country possess'.

Huygens's theory of light took its place in a lineage of theories of this puzzling yet most apparent of natural phenomena stretching back to antiquity. Geometers such as Euclid had been unable to decide whether the eye was the originator or the receptor of the light signal. Sometime after 1000 CE Al-Hazen in Cairo separated reflected and refracted rays into vertical and horizontal components, which greatly assisted in the geometric analysis of light. In the fourteenth century William of Ockham and his followers speculated that light might travel like a wave. Leonardo da Vinci also proposed a wave mechanism, while Huygens's closer contemporaries Francesco Grimaldi and Evangelista Torricelli conducted early experiments on light diffraction, a wave-like behaviour easily observed in water.

In 1801 the British physician Thomas Young demonstrated that interference patterns produced by the diffraction of light through closely spaced slits were analogous to the patterns generated by the combination of water waves from multiple sources. This development at last provided conclusive evidence in favour of Huygens's wave theory, with Young adding the correct surmise that colours corresponded to light of different wavelengths. Later, Young suggested that the wave action was transverse to the direction of travel of the light rather than longitudinal like sound in air. A few years later, in 1815, a French engineer named Augustin Fresnel, apparently unaware of the work of either Young or Huygens, independently developed his own purely wave-based model of light. The Huygens Principle is now often known as the Huygens–Fresnel Principle.

The long hiatus has played out badly for Huygens's lasting reputation. As Newton's stature grew, wave theories of light were soon forgotten. Young even experienced abuse in Britain for daring to suggest that Newton might not be right about everything. Confirmation that Huygens's theory was correct had to await the advent of optical instruments suited to experiments at scales close to the wavelength of light. Huygens himself did not even live long enough to see his theory effectively ousted by Newton's misleading corpuscular theory.

Huygens drew. He drew to help guide his scientific thinking and to express new concepts and give form to inventions; he drew to communicate his ideas to others; and he drew for amusement. Though never quite so much or quite so well as his older brother Constantijn, Christiaan sketched and tinted and painted throughout his life: designs for telescopes and microscopes and what he saw through them; the shifting shapes that eventually resolved themselves into the ring around Saturn; Venus in its phases; striped Jupiter; and Mars, recording the first impressions of surface features, including the south pole and the dark blotch of Syrtis major.

He drew designs for lens-grinding equipment and carriages, fountains and pumps. Some drawings reveal the essence of unrealized inventions – for example, a shoe with a sprung sole for a more comfortable walk, or the same idea scaled up as a device to be fitted to ships' bottoms to float them if they ran aground.* He drew pendulums and the gear mechanisms of clocks. Sometimes a hand emerges from a ruffled sleeve to grip some manually operated device. He drew harps and recorders and keyboards. The margins of his paper and the interstices between the strings of numbers, staves of music, anagrammatic codes and scribbled notes, often running vertically as well as horizontally on the page,

---

* Huygens presented this idea to the French Academy of Sciences in September 1678, a few months after the loss of a dozen French ships on rocks off the Caribbean island of Bonaire, which the French had tried to attack, having previously captured the Dutch colony of Tobago. 'We do not see that Huygens makes any allusion to the fact that this accident was to our advantage,' write the Dutch editors of the *Oeuvres Complètes*. 'Considerations of this kind were not his department' (OC22 707).

are filled with calligraphic doodles and shading exercises. Occasionally the face of an attractive girl emerges from the chaos of data. He drew the Dutch landscape – trees, cottages, windmills, canal boats, the beach at Scheveningen, the house at Hofwijck. He even drew Death, in the series of comically posed skeletons to be projected using his magic lantern.

It was inevitable that the Huygens brothers should grow up to become at least competent draughtsmen. The Dutch Republic was soaked in ink and paint. Mathematicians such as Simon Stevin developed painters' sense of perspective into a science of optical projection. In turn, optical accuracy became an 'ethical imperative' for many artists, a means of exciting in the viewer both wonder and a sense of insight. At the same time, advances in printing technology made it easier to include detailed illustrations in books. One of the most alluring and persuasive images in a scientific text was the representation of vortices produced by an unknown woodcut artist for Descartes's *Principia Philosophiae* of 1644. The ethereal vortices, so essential to Descartes's theory of matter in motion, but conceptually hard to grasp, were given compelling visual form in a set of illustrations showing the solar system with the sun and the planets in orbit around it. With its key features labelled with letters of the alphabet, the illustration was clearly diagrammatic, but it also seemed to be a natural image – a honeycomb of cells, each carefully shaded with pinpoint dots arranged like lines of longitude to give an impression of actual presence and bulk, and all of them together filling space like a mass of soap bubbles or onions packed in a box. Significantly, it was the elder Constantijn Huygens who advised Descartes to use woodcuts rather than copper engravings, because they could be placed directly in the text, where they could best amplify the Frenchman's theory.

Christiaan Huygens's facility made drawing an important part of his scientific thinking. He often articulated his ideas visually, and understood, like Descartes, that a visual image could be a useful weapon in his rhetorical armoury. This was especially the case with his many drawings of Saturn's ring, which not only resolved the blurry shapes seen around the planet into a convincingly pure and simple form, but also explained its appearance and disappearance during different phases of the planet's orbit. His colleagues realized the persuasive potential of his diagrams very early on. In August 1656, for example, when theories still abounded concerning the nature of what surrounded Saturn, Gilles de Roberval urged Huygens to divulge his interpretation of the latest observations, adding: 'a figure will assist the imagination; and, if need be, you will do it easily'.

When Huygens employed a simile, it was often visual, too. Light is compared to cannonballs and arrows on the one hand and to water waves on the other. When he discussed the idea of the relativity of motion, arguing that not only objects might be in motion but also the space around them, he introduced the analogy with men in moving barges, even deleting the word 'space' from his description and replacing it with 'boat' to make his argument more convincing. When describing the double refraction of Iceland spar, he asked his readers to imagine light falling on a 'toothed surface' like an array of prisms and thus being split off in two directions.

Huygens's visual mentality also helps to explain his abiding fondness for the geometrical exposition of physical phenomena over algebraic methods, even when, during the latter part of his career, the latter became more potent with the advent of the differential and integral calculus of Newton and Leibniz. Unlike Huygens, Newton and Leibniz had relatively little training in Greek geometry, and were always more comfortable using equations to express line and form, although Newton did admire Huygens's proofs enough to want to find his own ways of representing the methods of calculus in geometrical terms.

This is not to say that Huygens avoided more abstract methods. His wave theory of light surpassed those of Hooke and others because of the sophistication of its mathematical description. But even when algebra was his working tool, geometry would often remain his preferred mode of presentation, because it gave an instant sense of physical relationships that many were unable to discern from algebraic formulae alone.

All this puts Huygens firmly on the side of the visualizers in a perennial debate in physics. This most mathematically expressible of sciences has long nursed a division between those who think visualization is a helpful means of interpreting physical concepts that lie beyond the realm of the visible, and those who prefer abstract mathematical methods, and fear that visualizations may be erroneous in themselves and, worse, might lead scientists further astray in their thinking. The split widened at the beginning of the twentieth century with the advent of quantum mechanics, whose weird revelations of the invisible world of the atom led some to demand new visual images to assist interpretation.

There is no doubt that mathematics has the upper hand in the development of physical theories, but images remain hard to resist. Albert Einstein's famous 'thought experiments', involving a train speeding through a storm with its lightning flashes and observers here and there, gave his abstruse ideas about relativity an approachable face that he found helpful not only for himself, but also when explaining his concepts to other scientists and the broader public. It is no surprise to learn that Einstein praised Huygens for coming so close to the concept of relativity himself when he presented his very similar image of bargemen on the Holland canals.

Painters need light and astronomers and microscopists need light. Is it more than coincidence that these disparate pursuits prospered together in the Dutch Republic? Of course, the wealth and confidence of the new country, and its peculiar, self-imposed requirement to display that wealth, if at all, then in quiet, unobvious ways, must remain the major impetus for its astounding art and science. But all this work has its media and its materials, and light is chief among them. Not light in the fundamental physical sense, of course, which we know to possess certain absolute and universal properties, but the ambient light, light that has been filtered through a local atmosphere, light that illuminates a local geography, light that belongs to the place.

The abundance of sand for glass-making in the dunes of Zeeland and Holland may have provided the essential raw material for the lens-grinding innovators of the telescope and microscope. Was there also some special quality of the Dutch sky, as frustrated Italian astronomers muttered when Huygens saw Titan and they didn't? Is there – was there – a Dutch light?

The artists of the Dutch Republic had no thought of being bathed in a special light. For them, light was more usually discussed as a representation of the force of moral rectitude derived from Calvinist theology. The idea of light and optical devices such as mirrors were frequently invoked in the titles of pamphlets issued for the purpose of illuminating heresies and corruption. In 1668, for example, the Amsterdam brothers Adriaan and Johannes Koerbagh, one a physician, the other a preacher, wrote a polemic, 'Een ligt schijnende in duystere plaatsen' ('A Light Shining in Dark Places') attacking Reformed Church dogma; Adriaan was put in the workhouse, where he perished the following year. An antinomian tract with the same title by one 'Christianus Constans' distributed in 1710 was banned by the States of Zeeland. The author was in fact one Grietje van Dijk, a scholar of Hebrew and cult leader who rejected the authority of the church. Even men like Huygens, whose concern was to understand and exploit light in a strictly physical sense, would have been aware of the importance of this other light.

Dutch painting lapsed from fashion during the eighteenth century, its moralizing intent no longer well understood, its quotidian subjects and pedestrian landscapes ill-matched to new decorative tastes. The English portraitist Joshua Reynolds in his *Journey to Flanders and Holland*, recording travels made in 1781, made no particular observation about the quality of light in the country he was passing through, and entirely neglected to mention many of the artists we prize today for their handling of the light, such as Jan Vermeer, Jacob van Ruisdael and Meindert Hobbema. His attitude towards the Dutch masters was notably equivocal. He did acknowledge Rembrandt's handling of light, but he compared other Dutch artists unfavourably with their Renaissance forebears and with recent painters of the French baroque, and occasionally expressed the wish that they had been born in Italy, where they might have exploited their talents to better effect.

Many French artists, including Corot, Courbet and Manet, visited the Netherlands during the middle years of the nineteenth century in order to see for themselves the great paintings of the 'Golden Age'. Claude Monet made the journey too, but told his friend Camille Pissarro he would have no time for museums because he meant to paint. In June 1871, on the outbreak of Franco-Prussian War, he travelled to Holland on his way to London, and then back through the country again on his return, when he stayed at Zaandam a few miles north of Amsterdam. There, over a period of four months, he produced some two dozen canvases, mainly river scenes with windmills, bridges and sailing vessels. In his *Archives de l'impressionisme*, the critic Lionello Venturi wrote: 'This contact with reflections from the waters, which suggested to him the analysis and reconstitution of tones, renews his manner. This is already the definitive style of the Impressionist period: the effect of light is complete and perfect.' Monet wrote to Pissarro that Holland was more beautiful than people had said, and that in Zaandam alone there was enough to paint for a lifetime. Henry Havard, a French art historian whom Monet accompanied for a time in Zaandam, observed more precisely: 'The sky above and the water below which remirrors the air are both silvery white or of an extremely pale azure.' In Monet's canvases, the houses dotted along the rivers provided contrasting oranges and reds. Monet returned to Holland twice, visiting Amsterdam in 1874 and painting the bulb fields in bloom round Rijnsburg and Sassenheim in May 1886.

Another Frenchman, the writer Edmond de Goncourt, visited Amsterdam ten years before Monet, in September 1861. He found, as he noted in the celebrated journal that he kept with his brother, 'a country at anchor, an aqueous sky: rays of sun that appear to have passed through a jug of brackish water'. He saw the same light shining forth from Rembrandt's *The Night Watch*, 'a warm, vibrant ray of sun' that the artist had seized and twisted so as to fall vertically from above and splash across the assembled figures.

The sun shines from above but the weather comes from the west over the long, low coast fronting the pettish North Sea. The land is constantly exposed to the prevailing westerly winds and the damp Atlantic airstream. The atmosphere is cleansed at regular intervals by showers, its 'daily deluge', as Andrew Marvell put it. An endless parade of clouds populates the sky. The illumination is alternately soft when the air is laden with moisture or razor-sharp after rain. The low northern light etches the line of the horizon across flat fields and the rooflines in the town. Sometimes the sun is so low that it catches the underside of the clouds with a fierce luminosity, and a bonus radiance is reflected groundward.

No light is constant, however. Because of the unstoppable breeze, any sky is soon transformed. It is surely a defining characteristic of Dutch landscape painting that so many works capture this sense of the weather having just changed, or being about to change, in a way that simply never occurred to artists labouring under the cerulean skies of southern Europe. The rain-washed bricks in Vermeer's *View of Delft* show such a moment. So do the ragged hems of clearing black clouds in Ruisdael's winter landscapes. The popular subject of a breach in a dyke was inevitably also accompanied by departing storm clouds – the damage done, man's lesson hopefully learned, and in the light in the sky, God's vague promise never again to visit the same terror on the land.

Are we going to be scientific about this? Any theory must be testable. Is Dutch light truly unique? Is it demonstrably different from other light? It is easy to believe that it is different from the harsh light of Florence or Madrid. But what of other centres of painting? Venetian painters developed a softer style with a greater emphasis on colour and less on outline. Is that connected with the abundance of water there? A similar case might be made for other schools, such as St Ives on the Cornish coast or the Norwich school on the edge of the Norfolk Broads, a waterland created by the same peat-digging practices that carved out the 'hol-land' or hollow country that now lies well below sea level behind the protecting dunes.

Perhaps the test is not to think about other places but about other times. In the 1970s, the German artist Joseph Beuys provocatively suggested that the characteristic light that suffuses Dutch painting had been lost forever owing to the poldering of the Zuiderzee, once a substantial bay of the North Sea, during the middle decades of the twentieth century. His underlying supposition was that it is not just the impinging radiant light of the sun that makes the light of the country. Consider the path of a ray of sunlight. It passes through the atmosphere, where it encounters any condensed moisture. The tiny water droplets that make up this aerosol scatter a fraction of the light in all directions. The rest of the light carries on and strikes the ground. If it strikes water, much of this incident light may be reflected upward to pass a second time through the moist atmosphere, where yet more of it will be scattered by the aerosol of water droplets. The exceptionally high proportion of *diffuse* light produced by the sun's rays taking this longer path, so the thinking goes, was once the secret ingredient of the Dutch painters. In the riverine landscapes of Albert Cuyp, for example, a luminous haze hanging low over the water is a characteristic feature, and is often the brightest band of colour lying across the canvas.

No people have physically manipulated their topography as much as the Dutch. From the beginning of the seventeenth century, the provinces set about systematically claiming land from the sea, using dykes and drainage pumps to create polders. Thousands of square kilometres of land were created, displacing the large inland seas that once lay around the major cities, such as the Haarlemmer Meer, the Leidse Meer and the Zoetermeer near The Hague (which were already the product of human activity, having been enlarged by digging peat for fuel during medieval times). Constantijn Huygens the elder played a part in this. In August 1633 he bought a parcel of land recently drained from a former lake known as De Waert near Alkmaar. Even though a network of new canals was created at the same time, these would not have compensated for the loss of shining water elsewhere. These works, industriously pursued in many provinces, and continuing into modern times, have undoubtedly had the effect of reducing the area of water within Dutch borders. Perhaps the ambient brightness of the air has fallen in parallel.

However, there are difficulties with Beuys's theory from an optical and geographic point of view. The Zuiderzee mostly did not lap the shores of those parts of the Netherlands most celebrated for their painters. The province that was most greatly affected by these changes during the seventeenth century was Holland. And here, the cities that were home to the greatest artists – Delft, Leiden, Haarlem, Amsterdam – had already lost many of their surrounding waters by the time that they were working. So even here, in the critical place at the critical time, the quality of the light cannot have been suddenly altered from what it always was by changes made to the watery landscape.

Identified and celebrated only long after these old masters' work had been rediscovered, it seems that the phenomenon of 'Dutch light' is more the product of this art than its vital stimulus, and it is the admiration of later generations of artists seeking to copy their techniques that has embedded the idea in the public consciousness.

15

OTHER WORLDS

In the first wintry months of 1691, Christiaan Huygens was unwell at his home in The Hague, missing the summer pleasures of Hofwijck and fondly recalling the days spent with his brother Constantijn, taking turns at the telescope there, wondering about the celestial bodies they saw above them. 'I recall with pleasure those times of our diverting work in devising and polishing those lenses, inventing new methods and devices, and always striving for greater things,' Christiaan wrote. Constantijn had been away for two years serving – feet and eyes very much on the ground – as secretary to the stadholder in England and Ireland.

Christiaan was alone now, but still pushing forward, and his thoughts wandered to the possible inhabitants of the planets and stars. In 1686 the French writer Bernard de Fontenelle had published a wildly successful speculation on the nature of other worlds, *Entretiens sur la pluralité des mondes* ('Conversations on the Plurality of Worlds'). An early example of popular science, it took the form of a dialogue between an ingenuous marquise and a wise philosopher. Written in plain French so that it might appeal to those without any scientific knowledge, and specifically to women readers, it offered a primer to then current astronomical theories. Thirty years earlier in Rome, the German Jesuit Athanasias Kircher had published a more mystical dialogue on the same subject, *Itinerarium exstaticum* ('The Ecstatic Journey'), which

Huygens read, but found to omit all that he considered probable about other planets while including 'a company of idle unreasonable stuff'. As for those genuine scientists who had discussed the possible plurality of habitable worlds – Huygens counts Nicholas of Cusa, Tycho Brahe, Giordano Bruno and Johannes Kepler – they had, wisely no doubt, ventured little detail as to the forms that life might take. They were merely the latest names to join a list that included Democritus, Epicurus, Lucretius and Aristotle in ancient Greece, and Thomas Aquinas, William of Ockham and Nicole Oresme among medieval scholars who approached the subject from a Christian point of view.

CHRISTIANI
HUGENII
ΚΟΣΜΟΘΕΩΡΟΣ,

SIVE

De Terris Cœleſtibus, earumque ornatu,

CONJECTURÆ.

AD

CONSTANTINUM HUGENIUM,

Fratrem:

GULIELMO III. MAGNÆ BRITANNIÆ REGI,
A SECRETIS.

HAGÆ-COMITUM,
Apud ADRIANUM MOETJENS, Bibliopolam.

M. DC. XCVIII.

37. Title page of *Cosmotheoros*, 1698. Following a lifetime of observations of Saturn and other astronomical features, Huygens wrote this speculation about life on other planets, ensuring that it would only be published after his death.

So there was clearly scope for a new work informed by recent knowledge of the planets as observed by telescope. This was what Huygens attempted in *Cosmotheoros*.

Two great revelations, both so vast in their implications that it took them more than a century to sink in, largely explain the fad for such books in the seventeenth century: Copernicus's heliocentric theory of the solar system that demoted the Earth to a status equal with the other planets; and the European discovery of the Americas, where in 1621 the Dutch established New Netherland, a short-lived settlement centred on today's New Jersey and New York states, squeezed in among territories claimed by other European colonial powers.

But there were reasons for authors to be cautious, too, since raising the prospect that there were inhabited planets beyond our own risked running foul of Christian doctrine, which cherished the unique earthly paradise into which Adam and Eve had been placed by God. Huygens himself hesitated to go to press, fearful of censure by 'those whose Ignorance or Zeal is too great'. But eventually, tired out by his illnesses and knowing he would not live long, Christiaan asked his brother to supervise the publication of his own extraterrestrial speculation. In the event, *Cosmotheoros* did not appear until 1698, three years after Christiaan's death, and a year after Constantijn's.

'It is hardly possible,' Christiaan began, 'that a follower of Copernicus considering the Earth we inhabit as one of the Planets, in orbit around the Sun and receiving therefrom all its light, does not sometimes think that it is not unreasonable to allow that, the same as our Globe, others also might not be devoid of culture and ornament, or perhaps of inhabitants.' The key phrase here is 'not unreasonable', recalling Huygens's investigations of statistical likelihoods. For, as he warned his readers: 'We advance nothing here with complete confidence (could we do it?), contenting ourselves with conjectures the truth of which everyone is free to judge.'

Huygens assembled a detailed probabilistic argument. In the early 1690s, Galileo's discovery of four of Jupiter's moons was already more than eighty years old. Huygens had made his own discovery of the first moon of Saturn nearly forty years earlier. In the 1670s and 1680s Cassini discovered four more moons of Saturn. This profusion of new worlds, together with telescopic surveys of the 'mountains and plains of the Moon', invited comparison with the Earth. Huygens also demonstrated that other stars are comparable with our own sun, providing an estimate of the distance from Earth to one star, Sirius, based on a comparison of its diameter, observed by telescope, with that of the sun. He reasoned that, though we cannot see them, such suns may have their own planets in orbit around them, pointing out that if our solar system were viewed from a sufficient distance, it would not reveal its planets even to the most powerful telescopes. Finally, marvelling at the richness and fitness of plants and animals so well adapted to life on Earth, he argued more sentimentally that if we were to deny this abundance to other planets, then 'we should sink them below the Earth in Beauty and Dignity; a Thing very unreasonable'.

What form might this life take? Based on new information that American species are enough like those of the Old World, Huygens presumes a general similarity with terrestrial species. But he does give some consideration to the different physical conditions that may prevail on other planets. The atmosphere might be thicker, for example, which would suit a greater variety of flying creatures. Gravity might be different, too, although he does not provide estimates of the comparative gravitational force on each of the planets, and in any case he rejects the notion of a simple correlation between the size of a planet and the scale of its flora and fauna. '[W]e may have a Race of Pygmies about the bigness of Frogs and Mice, possess'd of the Planets,' he says, although he thinks it unlikely.

For Huygens, however, 'the main and most diverting Point of the Enquiry is . . . placing some Spectators in these new discoveries, to enjoy these Creatures we have planted them with, and to admire their Beauty and Variety'. Remarkably, these need not be 'Men perhaps like us, but some Creatures or other endued with Reason'. Huygens imagines there might be planets capable of accommodating several species of 'rational Creatures possess'd of different degrees of Reason and Sense'. The nature of reason and morality would be the same as on Earth. The creatures would be social, and they would have houses to shelter them from the weather. Huygens struggles with their appearance, though. He wants to indicate that they might not be humanoid, and yet, he says, surely they must have hands and feet, and stand upright. Perhaps they would have exoskeletons, like crustaceans, for example. After all, ''tis a very ridiculous opinion, that the common people have got among them, that it is impossible a rational Soul should dwell in any other shape than ours'.

They would have science, especially astronomy, the study of which was thought to have arisen as a consequence of the fear of eclipses, which would also be a familiar occurrence on other planets. They would doubtless have some of our inventions, too, 'yet that they should have all of them is not credible'. In particular, Huygens cannot believe that they would possess telescopes, since he considers those which he has used himself as being so fine that it would be impossible for other intelligences to produce their like. Instead, he invests the denizens of the planets with far superior natural eyesight.

In 1600, Giordano Bruno was burnt at the stake in the Campo de' Fiori in Rome by the Inquisition for many heresies, including his insistence on the plurality of potentially inhabited worlds. A century later, Huygens was safe from such a fate. Nevertheless, he attempted to pre-empt criticism from the church by making the semantic point that the heaven and earth referred to in scripture must apply to the totality of the universe and not to the planet Earth exclusively. His suggestion that the inhabitants of other worlds, though rational, might not take human form may therefore be read as his provisional defence against a theological challenge based on the grounds that the Earth was made exclusively for man.

Cosmotheoros marked the 'first attempt to mount a rigorous scientific case for life on other worlds, without doing harm to Scripture', according to the science writer Philip Ball. Upon its publication in Latin in 1698, three years after Huygens's death, it was immediately translated into Dutch, and into English, with the shamelessly overselling title *The Celestial Worlds Discovered*. French and German editions followed a few years later.

It has never been entirely respectable for scientists to speculate about life beyond Earth. The famous astrophysicist and author Carl Sagan, who did much to popularize the idea of extraterrestrial intelligence in the twentieth century (he coined the acronym CETI, for Communication with Extraterrestrial Intelligence, later amended to SETI; Search for Extraterrestrial Intelligence), spoke happily in public and on television about the prospect of life on Mars, for example, but he expressed far greater scepticism when he was with other scientists. His apparently unqualified speculations irritated many of his peers, and in 1992 Sagan was turned down for membership of the American National Academy of Sciences.

Sagan was in fact acutely conscious of the many probabilities involved in the question. Although his enthusiasm for the idea of extraterrestrial intelligence dated from his teenage years, his thinking was shaped at a conference in 1961, when he learned of an equation devised by the radio astronomer Frank Drake. Drake's formula for the number of extraterrestrial civilizations (or the number capable of interstellar communication, since the SETI project was above all about contact) was simple. It involved multiplying the total number of stars by a series of fractions governing the various factors that needed to be satisfied for extraterrestrial intelligence to exist – the proportion of stars with planets, the proportion of those planets considered habitable, and so on, as well as more subtle factors such as the duration of existence of an alien civilization during which it might be capable of sending out detectable signals.

Huygens, who did more than anyone before him to define probability and apply it as a meaningful concept in the natural sciences, had a similar sense of the problem. Halfway through *Cosmotheoros*, he anticipates his critics: 'some will say, we are a little too bold in these Assertions of the Planets, and that we mounted hither by many Probabilities, one of which, if it chance to be false, and contrary to our supposition, would, like a bad Foundation, ruin the whole Building, and make it fall to the ground'.

The value of each fraction in the Drake equation is still debated, but recent discoveries concerning the atmospheric composition of some of the planetary satellites in our own solar system appear to have shortened the odds on life being found.

Mr. des Cartes had found the way to have his conjectures
and fictions taken for truths. And to those who read his
[Les] Principes de [la] Philosophie, he arrived at something
similar to those who read Novels which please and make
the same impression as true stories. The novelty of the
figures of little particles and vortices made a powerful
appeal. When I read this book of Principles for the first
time, it seemed to me that all went well with the world,
and I thought, when I found some difficulty, that it was
my fault for not understanding his thinking. I was only
15 or 16 years old.

Christiaan Huygens made these notes in 1693, nearly fifty
years after he had read Descartes's *Principes* as a boy. He was
commenting on the first published biography of Descartes, by
Adrien Baillet, in which, among other errors, the author 'constantly
confuses me with my father'. By this time, Huygens had found
much to disagree with in the substance of what Descartes had to
say, especially concerning the nature of the physical world. He
had long since reached a position where, though he might wish
to be faithful to the principles set out by Descartes, he had no
compunction about abandoning them when they did not accord
with his scientific observations or experimental results.

Was Huygens ever really a 'Cartesian'? Since the ancient Greeks,
only a few philosophers have achieved the cultural penetration that
has made it necessary to coin an adjective to record their impact
in the world beyond philosophy: Confucius, Machiavelli, Marx.
Descartes is one of these. To be a Cartesian, especially for anglophone writers after the rise of Newton, is to align oneself with a
dogmatic and sometimes dubious philosophic programme, and even
to mark oneself out as in a sense anti-science. It is a characterization that still colours some accounts of Huygens's life.

To be a Cartesian in the Dutch Republic of the mid seventeenth century was hardly less problematic. The Frenchman's mechanistic view of the world had brought him into increasing conflict with the universities and religious authorities, which detected a creeping atheistic agenda in his pursuit of the ultimate causes of physical events without invoking a proof of the existence of God as the prime mover. This split Dutch science into Cartesian and Aristotelian factions, and rippled out beyond into the stormy seas of church and state relations such that one's position on Descartes became a benchmark of one's political ideology more broadly.

Christiaan had every reason to follow Descartes. His father had assisted the Frenchman in his lens-making, and interceded on his behalf when he was branded an atheist by an influential Calvinist pastor, obtaining protection for him from the stadholder, Frederik Hendrik.* He was also responsible for arranging his sons' academic instruction from the mathematician Frans van Schooten, who was a strong proponent of Descartes's ideas. Descartes was radical and new, and his scientific philosophy embraced the universe. It is hardly surprising, then, that the teenage Christiaan found himself enthralled.

If there were forces leading Huygens towards Cartesianism early on, then there were others later to drive him away. Descartes's ideas were problematic in the Netherlands, but even more so in Catholic France, where Louis XIV forbade any public ceremony when his remains were brought home from Sweden after his death in 1650. His philosophy was banned from French universities, and his followers were among the perceived ideologues excluded from the French Academy of Sciences. Yet it is clear already that

---

* Constantijn Huygens told Descartes: 'Theologians are like swine; when you pull one by the tail, they all squeal' (Descartes, vol. 3 676–9).

Huygens's Cartesianism was not so rigidly held as to cause him difficulties personally when he took up residence in France. He may also have been influenced against Descartes by the literary scholar Isaac Vossius, a Dutchman who had served as the librarian at the Swedish court when Descartes was there, before coming to Paris and being appointed by Colbert as an academician at the same time as Huygens.

But the principal factor guiding Huygens's thinking was always what he saw before his eyes. As early as 1652 it was his own experimental work that led him to doubt what Descartes had written about the collision of bodies of unequal mass or velocity. He was never inclined towards dogmatic belief in Cartesian precepts, but such revelations of his own caused him to reassess the extent of his loyalty each time. A few years later, for example, Huygens saw that Cartesian vortices offered a tolerable interpretation of the tendency of bodies in circular motion to move away from the centre of rotation. It was not blind allegiance to Descartes, though, but rigorous mathematical analysis that enabled him to develop from this a more complete quantitative model of centrifugal force.

However, there were limits to how far Huygens was prepared to depart from Descartes. Although the mathematics was convincing when it came to explaining the action of force at a distance, Huygens could never fully accept Newton's theory of gravitation, remaining true to the Cartesian notion that force can be transmitted only through a medium. And although his own mathematics allowed him to rationalize much of Descartes's theory of lenses, he was not able to leave behind Cartesian ideas sufficiently to state his wave theory of light boldly in his own terms in a way that subsequently might have tipped the balance in its favour against the corpuscular theory of Newton. Because he was certainly no Newtonian either, it later became easy for others to portray – or dismiss – Huygens as a Cartesian. It would be more accurate to say that he was simply unideological. Even Newton was able to acknowledge that Huygens was his own man enough not to have fallen foul of the 'false taste' for Descartes.

Huygens was never less Cartesian than in his practical dealings as a scientist. Though not as empirically minded as some of his English peers, he prized experiment and understood the value of proven experimental results. Whereas Descartes had no hesitation in ignoring the findings of others, preferring to find his own 'truths', Huygens readily acknowledged the social principle at the heart of emerging modern scientific practice of respecting experimental results and subjecting them to comparison and replication. The successes of Dutchmen such as the microscopists Leeuwenhoek and Swammerdam, who were not philosophers or even great scholars, showed further that science could advance by observation alone, without necessarily having a grand philosophical framework in place to guide it.

Huygens offered a rare insight into his own temperament when he noted down his mature impressions of Descartes in response to Baillet's biography. He surmised that the Frenchman must have been 'very jealous of the fame of Galileo': he had wished to be the author of a new natural philosophy, but unfortunately hardly any of his scientific laws turned out to be correct. Furthermore, in daring to think that he might be able to explain everything, he plainly lacked Galileo's fundamental humility.

Huygens likewise had no burning desire to explain everything. He was at ease with the thought that some things might only ever be known to a degree of probability. Like most scientists, he was not a revolutionary by temperament, and did not seek to overturn the existing order. His disputes with other scientists and philosophers, and especially with Cartesian ideologues, were enough to put him off that mad course. He had seen the harm that could ensue.* Strenuous proponents of new theories in physics, wrote Huygens feelingly to Leibniz in 1692, 'do much injury when they wish to pass off their conjectures for truths, as Mr. des Cartes has done, because they prevent their followers from seeking anything better'. He for one did not want to erect a universal system, or to leave behind him a Huygenian 'school'.

Huygens was, rather, a synthesist. He built upon the most robust elements of theories that might seem to some to represent irreconcilably opposed ideological systems. Heir to Galileo and Bacon, judicious follower of Descartes, he nonetheless surpassed all of these in his mathematical prowess, which put him on a par with Newton and Leibniz. His pragmatic urge to pull together the best of scientific knowledge and methodology from all available sources – irrespective of nationality as well as ideology – perhaps also owes something to observing his father's work as a diplomat. But it leaves an indistinct impression of Huygens's scientific personality, which may help to explain why some historians have felt it necessary to inject a philosophical programme into the career of a man who appears otherwise as 'a mere problem solver'.

Yet it is surely enough that Huygens be remembered for what he was, a mere problem solver indeed: pragmatic, eclectic and synthetic, sceptical and ready to settle for the most probable rather than hold out for the absolutely certain – in other words, what we expect a scientist to be today.

---

* Perhaps he even believed the harm could be physical. When Leibniz complained to Huygens that he could find no Cartesians interested in the philosophy of medicine, Huygens replied: 'There are enough doctors who think they follow Cartesian philosophy, but they are the ones I would call last if I were in need' (OC10 55–8).

Is this a vision of what might have been? It is midsummer 1693. Christiaan Huygens is sitting out in the garden at Hofwijck with Suzette Caron and her children, including her eldest, his goddaughter, under the pines and elms planted by his father fifty years ago. Though Suzette has been 'greatly changed' by her ordeals, her presence there lifts his spirits. And, as members of his family would observe, she remains 'not ill-disposed towards him'. 'The key to my heart is the one to this garden,' his father had written.

Huygens found Hofwijck pleasant enough during the summer. But it was lonely in the winter. In September 1689, wishing to avoid passing a second winter confined within its labyrinth of icy channels, he wondered whether he might be permitted to take a room or two with one of his brothers in The Hague. This would spare him the inconvenience of having to catch the last boat home on days when he ventured into the city.* His brothers were not keen on the idea, however, and he took lodgings near the Noordeinde Palace instead.

His sense of isolation was not only geographic. It was also intellectual. There is 'nothing to aspire to here', he groaned to Constantijn, who was still in London with William III of Orange. He wished he had stayed longer in London. Before he died, his father had proposed that Christiaan take his place on Orange's council, but Christiaan had turned down the suggestion. However, made aware of the taxes he was now obliged to pay as a man of property, he began to see the advantages of a secure appointment. When another member of the council died suddenly, he tried to revive the idea, writing repeatedly to Constantijn with requests

---

* A canal-based public transport network linking many of the cities of Holland had been in place for several decades (de Vries 2006 14). Huygens tells us that the last barge from The Hague that stopped at Voorburg departed at 6.30 p.m. (OC9 346–7).

to have his name put forward. Knowing that to do so would put him in an acutely difficult position, Constantijn did what he could to dissuade his brother from the idea, but Christiaan would not be deterred. 'With a second letter, in which brother Christiaan pestered me to demand of the King a place on his council, I did speak with him,' Constantijn noted in his journal,

> and he said through his teeth that he did not know if he would fill the position. When, a little later, I said again that I thought that he would not be badly served by my brother, he being of a penetrating intelligence and good application, he replied that he thought he had higher concerns than to dally (or some such word) with administrators, at which I insisted no longer.

Both Constantijn and the king could see, as Christiaan at that moment could not, that neither the position nor the working environment would be at all suited to Christiaan's talents. Constantijn told his brother honestly how the conversation had gone, and tried to make it clear to him that he might achieve a better result by approaching the matter more circumspectly. Christiaan seems not to have heeded the advice, however, for he then pursued the idea through another contact, while continuing to nag Constantijn about it. No council appointment was forthcoming.

Huygens's frustration was especially acute in his practical work. In September 1689 the directors of the Dutch East Indies Company asked him to prepare pendulum clocks for a new sea trial on board the *Brandenburg*, bound for the Cape of Good Hope. Despite the poor performance of his instruments aboard the *Alcmaer* three years earlier, Huygens continued to be held in high regard by the company, which was constantly assailed by claims from hopeful and sometimes fraudulent inventors. Making the clocks, installing them aboard and calibrating them on short voyages off the Dutch coast occupied most of 1690. The *Brandenburg* reached the Cape in June 1691, but was greatly delayed on its return by sickness among the crew. During this time, the sequence of clock readings was interrupted, which once again compromised the overall results. While Huygens waited for the ship to return, the East Indies Company asked him to give his assessment of other longitude proposals received, including from 'some ignorants' lured by the prospect of prize money.

The captain's report of the expedition made disappointing reading as he revealed that he had had to resort to astral navigation. Huygens's own analysis of the readings that had been taken showed that the clocks had performed well on the outward journey. He pointed out 'that from S. Jago [Cape Verde Islands] to the Cape the distance is most perfectly measured by the clockwork according to the newest Maps of the Globe, as your honour yourself will acknowledge, and doubtless rejoice that all your honour's faithful work and multiple observations have not been fruitlessly spent'. But the readings on the return journey led him to suspect that some kind of systematic human error must have crept in. The *Brandenburg*'s captain bridled at this suggestion, and responded with a description, complete with a sketch diagram showing exactly how the clocks had been mounted, and adding that parts of the mechanism had broken and they had stopped so frequently that he ceased taking readings. In his final report to the directors of the East Indies Company, Huygens nevertheless robustly defended the performance of his clocks, observing that despite the captain's 'misunderstanding' and 'miscalculation', 'they very well and precisely accomplished the Longitude measurement'. In private, though, Huygens acknowledged that there remained fundamental disadvantages to the pendulum clock at sea, and he renewed his experiments with alternative isochronous mechanisms that would continue to function irrespective of a ship's motion.

Astronomy proved equally unrewarding. Christiaan and Constantijn worked on telescopes and microscopes occasionally when they found time together in The Hague, and corresponded on the subject when Constantijn was away with King William in London, and then, later, in the field when war erupted once more with France. They were spurred on by Cassini's spectacular success in discovering the fourth and fifth satellites of Saturn, but hampered by equipment that was by now plainly inferior to that available in Paris. In London, too, astronomers struggled; even Halley was unable to see Cassini's new moons for himself. In September 1693, Christiaan found himself busy constructing a square pine box to house a forty-five-foot lens, which he intended to put to use for showing the Moon and planets to his visitors, recognizing that they, as inexperienced observers, would find it hard to operate the tubeless telescopes that he was accustomed to using for scientific astronomy. But the exercise seems to have provoked an epiphany for Christiaan, who wrote bitterly to his brother: 'For since Mr. Cassini claims that he sees all 5 Satellites of Saturn with lenses of shorter length, why do I not see them too? I regret not having used a tube for 6 years, for it would surely be better than the method I have invented.'

No such impediments applied to paper-based investigations. In mathematics, Leibniz tried to induct Huygens into the intricacies of his differential and integral calculus, demonstrating how it could reveal the properties of complicated curves, such as the catenary, and Huygens's favourite, the cycloid, without recourse to geometry. '[W]hat I like most about the calculus,' Leibniz told

him, 'is that it . . . liberates us from working with the imagination.'
Their informal correspondence course – Leibniz was employed
in Hanover, ostensibly writing a history of the House of Brunswick
– continued for some time, but Huygens always remained sceptical
of the new method with its fancy notation. 'What you tell
me about the effect of your differential calculus in enquiries
concerning the Cycloid, to tell the truth, seems to me hardly
believable,' he wrote on one occasion.

Huygens preferred not to be liberated from working with the
imagination in any case, and sought when he could to make the
mathematical visual or physical. For example, his tutorials with
Leibniz touched on another curve, known as the tractrix, which
is the path taken, in the illustration Leibniz gave, by a dog on a
taut lead as it wanders off the straight-line path of its owner and
is then pulled back into line. In geometric terms, the tractrix is
also related to the catenary. Huygens's investigation of this curve
included devising various mechanical means of generating a tractrix, such as one based on a wheeled shuttle with a stylus attached,
which is dragged through a layer of syrup, whose resistance causes
the stylus to trace the curve.

Huygens felt a similar disconnection from physical reality
when he read *Principia Mathematica*. He acknowledged that
Newton's work explained the motion of bodies such as comets
much better than Cartesian vortices did, but he was still troubled
by aspects of his theory, such as what happened when the tidal
oceans, moved by gravitational force, met the solid matter of the
terrestrial Earth, and the apparent implication of the inverse-
square law that gravitational attraction would have to be infinite
at the centre of the Earth. Huygens found the *Principia* erroneous
in places and often obscure, but his new young friend Fatio de
Duillier, a fellow admirer of geometric methods, was able to
interpret Newton's analysis for him, and tried to enlighten him

as to the relative merits of Newton's and Leibniz's versions of calculus.

Fatio paid a brief visit to Huygens in April 1690, and returned to London with a list of Huygens's suggested amendments for Newton to consider. Fatio, too, read the work closely, and made his own list of errors, keen that Newton should prepare a second edition of the work. Thinking on a continental scale, Huygens encouraged Fatio in this project:

> Monsieur Newton should be most happy if you should undertake the second Edition of his work, which would be something else with your clarifications and additions, which it presently lacks. But it should not be that this labour impairs your health. As for the expense, it seems strange to me that there are not in that country printers who would wish to risk the cost of printing books of this importance. They would certainly be found here. This method of subscriptions [Edmond Halley paid for *Principia* to be published, the Royal Society having used up its periodic book budget] is not suited for works which ought to be sold throughout Europe; for England and this country will not supply enough.

In the end, a revised edition of the *Principia* did not appear until 1713, made possible by the support of the Master of Newton's Cambridge college, Trinity. By then, Newton's relations with Fatio had long cooled.

With his mathematical skill and willingness to find the merit in competing theories, Fatio earned the respect of both Huygens and Newton, positioning himself as an international intermediary much as Henry Oldenburg had once done. However, Fatio lacked Oldenburg's tact, and was careless in passing Newton's confidences on to Huygens, from whom they could easily reach his rival, Leibniz. Fatio thus trapped himself – and Huygens, too – between Newton and Leibniz as they descended into a long-running dispute over the invention of calculus.*

Notwithstanding the notorious dispute over calculus, this international quadriga – Dutch Huygens, English Newton, Swiss Fatio and German Leibniz – can be seen to display some mature features in their scientific discussions. Compared to the situation only a few decades before, cooperation and collaboration were on the rise, bringing with them a greater willingness to expose new ideas to criticism and to compare methods and results with others. Even Huygens now fully recognized the importance of announcing new discoveries as early as possible. In 1691 he felt able to chide Leibniz, who was then involved in another priority dispute, this time with the Swiss mathematician Johann Bernoulli: 'You would have been able to prevent all these doubts . . . by communicating your inventions under cover of a code, as I have advised you more than once.'

The greatest change is in the tone of the discussions. Gone is the quickness to feel insulted and take umbrage that characterized so many exchanges – domestic as well as international – in the early days of the French and English academies of science.

---

* When Huygens learned – from another source – that Newton had experienced a sustained attack of 'phrenzy' (a mental breakdown) in 1693, he carelessly shared the news with Leibniz (OC10 615–16). 'The English, as it seems, had thought to conceal this accident but in vain,' he wrote to his brother Constantijn. 'Aside from his too earnest studies, it could be that the misfortune of a fire which he had, which carried off his Laboratory and some papers, has contributed to so disturb his mind, which is the worst of all ills that can come upon a man' (OC10 617–18). Newton was, in fact, fully recovered by the time that Huygens received this intelligence, and Leibniz wrote back cordially: 'It is to men like you, Monsieur, and him, that I wish a long life and good health, in preference to others, whose loss would not be so considerable speaking comparatively' (OC10 639–40).

Instead, divergent opinions coexisted, such as those favouring the wave (Huygens and Leibniz) or corpuscular (Newton and Fatio) nature of light. The template of civil exchange so painstakingly hammered out by Mersenne, the elder Constantijn Huygens and Oldenburg among others had become the norm within the scientific community of Europe.

The exchanges are newly generous in the range of their content as well as in their tone. Leibniz made good use of Huygens as a sounding-board to test out his philosophical ideas about matter. Huygens did his best to answer these queries, though he grew doubtful as the questions became more esoteric. A discussion of the nature of material hardness in one letter, for example, led Leibniz to wonder what it would mean if this property were variable; to this hypothetical poser, the empirically minded Huygens simply put a baffled 'Why?' in the margin of the letter. However, Leibniz clearly felt that it would do no harm if Huygens were to open his mind further to a more ecumenical way of thinking. He signed off the letter: 'I shall try to remember occasionally what you say in your letter with respect to Descartes, that it is useful that people of great reputation give their conjectures on all sorts of topics in order to stimulate others. It is what I would wish you to do yourself.'

The winter colds and fevers to which Huygens had always been susceptible grew more frequent in these years, and were often exacerbated by persistent headaches, insomnia and what he himself called his 'tristitia'. But in July 1694, while he was at work on *Cosmotheoros* and still engaged in lively correspondence with Leibniz, he experienced a new pressure in the area of the heart, and felt that his pulse was becoming irregular. Diederik Lieberghen, the long-standing family doctor, advised Huygens to work less and walk more, and to avoid raw food and vegetables. 'I think however that the refreshment of the blood that comes from eating cherries and redcurrants has done me good,' Christiaan wrote to Constantijn. If these too were part of Lieberghen's prescription, it implies a worrying reliance on the ancient humoral 'doctrine of signatures', which largely passed out of fashion during the course of the century.

In London, Constantijn received regular bulletins on the state of Christiaan's health from their sister Susanna and her husband Philips Doublet, and from his wife Susanna Rijckaert. Their letters sometimes raised his fears, not least because by the time they arrived, they described events of a week or more earlier. In the spring of 1695, Christiaan's condition worsened. In the last letter that he was to write to his brother, dated 4 March, he asked him to buy him some books available in London, one on anatomy, another dealing with infinite numbers. The letter showed no dimming of his intellectual curiosity, touching on the usual matters of clocks and telescopes, but also on architecture, archaeology, geology, Congreve and the weather. However, on the twenty-third, a notary was summoned to take down the details of his will. Christiaan added a personal codicil a few days later.

On 16 April Constantijn received three letters from his wife in the same post. Their import was the same, 'that brother Christiaan was still very ill, and in the last that he could not sleep, and feared that he would go mad'. She told him that Christiaan was keeping it 'very dark in his room, and nobody was allowed to come in to him, because if he spoke much, it would definitely make him worse'. She added – this was perhaps her own idea – that he greatly longed to see his brother and, as Constantijn transcribed it into his diary, 'thought that the joy of seeing me would do him good'.

By the twenty-sixth, Christiaan's condition had deteriorated still further. There was, however, no question of Constantijn leaving the king's side. Susanna reported that he 'was getting the most deplorable ideas about the world'. Three days later, Constantijn received more news. 'My wife writes that brother Chr. had a grievous sickness, which looked not a little like insanity, since he sometimes raved and spoke such desperate things, that one became afraid of it. That Liebergen said it was the black bile.' This was the same diagnosis of hypochondriac melancholia that the doctor had made at a distance in 1670 when Christiaan was in Paris. He prescribed the same remedies of goat's milk and baths.

Early in May, rumours began to circulate in the English court that their Dutch king would shortly be sailing for home. The rumours were well founded, and he departed with his entourage from Gravesend on 17 May, arriving at Rotterdam a week later, from where the royal party, including Constantijn, travelled on to The Hague, arriving there at six o'clock in the evening. Constantijn found his brother in a desperate state, suffering from violent intestinal pains and hallucinations. 'Real madness seemed to underlie his sickness.' The family had removed anything with which he might harm himself, because he had begun to cut his skin with broken glass and stick pins into his body. He even attempted to put a marble down his throat; his manservant heard him choking, ran to him and clapped him on the back to release the blockage. He imagined he was hearing voices, cried out 'godless things', and feared that people would tear him apart if they heard his thoughts on religion. 'The only hope in this was that it not lead to his being taken for a madman,' Constantijn wrote. A few days later, Constantijn was recalled to service. When he attempted to call in on Christiaan to take his leave, Christiaan's servant came to the door with the instruction that 'if I wanted to see him in the utmost misery, I could come in, but that otherwise he wished me a good journey'. Constantijn did not insist.

In June Christiaan's condition deteriorated again, and he refused to eat, believing his food was being poisoned. Susanna Rijckaert tried to persuade him to allow a minister to come to him, but again 'he began to curse and rage'. He continued to lose weight for three more weeks until, on 7 July, a minister was brought in at the family's insistence, a man somewhat known to Christiaan. Constantijn wrote in his diary what Susanna told him, that the minister 'spoke with him for a long time and said a prayer or two, but that he answered him in the same manner as when I had last heard him speak, and that, whatever was said or not said to him, he was not going to change his opinion'. His wife was greatly dismayed by these words. After that, Christiaan spent a troubled night. When Susanna checked on him at 3.30 a.m., she found him unconscious. He remained so until later that morning 'when he very softly passed away'. He was sixty-six years old.

For a long time, Huygens had considered mortality as if it were a problem of probabilities, like one of his calculations of dice. In 1670, when he experienced his first major health crisis and considered leaving his papers to the Royal Society, his brothers discussed his apparently flippant and irreverent attitude towards human existence. Constantijn told Lodewijk:

> I can well forgive him at his age and full of thoughts of
> the great plans he could carry out if God prolonged his
> days. But being in the state where he finds himself, which
> he ought to regard as close to immortality [i.e., his own
> Christian death], he amuses himself in arguing as if it were

a problematic question for and against, as I have learned in truth with great sadness. It seems to me this does not pay the respect it should to the word of the good Lord who has taught us what to believe about the state of the soul after this life and its immortality . . . The loss of the good brother would be much worse if we came to lose him with these feelings.

This divergence of attitudes explains the family's insistence, against Christiaan's own wishes, that a minister of the church be called to his sickbed. Christiaan's religious views did not soften as death approached. A few of his considered opinions are set out in some Ciceronian meditations on glory and death that were probably written after his return from visiting his brother in London in 1689. 'De Gloria' observes that we survive in what we produce, and that natural talent should be directed towards good effects and useful ends. 'De Morte' begins with the idea that we are only ourselves by virtue of the combination of our memory of the past with present experience, and goes on, relying on the Cartesian notion of the separateness of body and soul, to contemplate the theoretical survival of the soul, with its memories, and without the body. In the real world, the wish to live on is tempered by the knowledge that the two are connected and that the body will inevitably deteriorate. This thought brings different considerations to the fore that reveal something of Huygens's state of mind during his final weeks. 'Who would not wish to be immortal?' Huygens demands to know. Anybody would opt for immortality, given a healthy body and mind and everlasting youth; 'but with the certainty of approaching death, with bodily woes and physical decay, and loss of memory and intellect, who would not strive to escape from these evils, even at the expense of life, when they close in?' Recognizing that the misery of disease strikes at random,

Huygens adds: 'Whoever happens to die without this torture, he should consider how much more unhappy are those to whom this euthanasia [the word is spelled out in Greek in the otherwise Latin text] is not granted.'

Huygens was hardly alone among his peers in harbouring unorthodox ideas about religion. Leibniz formulated a doctrine of monadism in which non-physical elements of being, or monads, each amounting to a 'soul', constitute all substance. Newton was an antitrinitarian, for which he was considered a heretic in his lifetime, as well as being interested in occultism and alchemy. Fatio de Duillier eventually joined a millenarian sect. In this context, Huygens's loss of faith is unsurprising. However, he was different from some others in that he neither found a ready-made alternative doctrine to turn to, nor did he make much progress in formulating a coherent philosophy of religion of his own. Perhaps the closest he got were his lively speculations about other planets and the prospect of there being new forms of life on them, which induce an idea of the closeness, and even the oneness, of Creator and Creation, God and Nature. This vision must have been reinforced by the strange new forms of life revealed under the microscope, and by the exotic plants and animals brought back from far-flung colonies, captured in the precision brushwork of painters, so that it is tempting to suggest that it was an idea crystallized by the Dutch 'Golden Age'. Certainly, it was one that found concise philosophical expression in the words of Huygens's sometime neighbour, supplier and co-worker Baruch Spinoza, whose *Ethics* drew God and Nature into an identity.

Huygens's own writing was more tentative. In five numbered paragraphs headed 'What to think about God?' made around the year 1686 or 1687, he noted first of all that 'it far surpasses man to have an idea of God'. Undeterred by his own injunction, he issues this exhortation: 'Let us seek to prove that there is a supremely intelligent creator, but with an intelligence quite unlike our own.' With regard to the mysteries of the solar system, such as the motion of the planets and the puzzling inclined axes of the Earth and Saturn, he asserts confidently that 'although God has so disposed these things, it is however certain that he acts according to the immutable laws of nature'.

Christiaan Huygens was buried in Sint-Jacobskerk in The Hague, the church in which his father had been laid to rest only eight years earlier.

Leibniz learned of Huygens's death less than two weeks after the event. '[I]t is fatal for me to write letters to friends who cannot reply,' he observed to another friend, listing three others among his correspondents recently deceased. '[T]he loss of the illustrious M. Huygens is inestimable. Few people know as well as I that he has in my opinion equalled in reputation Galileo and Descartes and, helped by what they have done, he has surpassed their discoveries. In a word he is one of the prime ornaments of the age.'

Christiaan was survived by his three siblings – Constantijn, Lodewijk and Susanna – and their many children. The Lordship of Zeelhem, which Frederik Hendrik had conferred on his father in 1647, had passed to Christiaan upon his father's death in 1687. (Constantijn became the Lord of Zuilichem.) In his will, Christiaan instructed that this title should pass to one of the two eldest sons of his brothers, to be decided by casting lots; it went to Lodewijk's son, another Constantijn, who was Christiaan's godson. He also inherited one of Christiaan's silver wall sconces 'because he had no christening present from me'.

The surprise to the family was that the major legatees were Suzette Caron and 'her oldest daughter who I held at the Baptism', to whom Huygens left 2,000 and 500 guilders respectively, with another 300 guilders added in a codicil. Smaller amounts, totalling 1,300 guilders, were split among various servants, with some extra going to the Hofwijck gardener's wife and her mother, who had cared for him in his final illness. The remainder of his estate was to be divided equally between his nine nieces and nephews, which came as a disappointment to Constantijn, who had only the one son. His papers were mostly bequeathed to the University of Leiden, with instructions and money to arrange for the publication of the major works on optics and mechanics that he had always been reluctant to call finished. Only the preparation of *Cosmotheoros*, his most controversial work of astronomy, was left to its dedicatee, his brother and fellow observer of planets Constantijn, to oversee.

AFTERWORD

Christiaan Huygens was the greatest scientist alive in Europe during the period between Galileo and Newton, whose lifespans are separated by nearly eighty years. Though fewer people may know Huygens's name today, he was no mere interloper, and this period was no thumb-twiddling interlude, but coincided exactly and completely with what historians sometimes call the Scientific Revolution, whose new ways of understanding the world prepared the ground for the Enlightenment.

Huygens's discoveries and inventions were central to this shift. His continually improved telescopes brought new aspects of the solar system into focus for the first time, while his unerring ability to analyse what he saw, informed by his visual sensibility and mathematical dexterity, enabled him to resolve these features correctly as the first satellite and the ring around Saturn. His lifelong work in optics led him to a substantially correct wave theory of light which, but for the baleful influence of Newton's corpuscular model, might have advanced understanding in that field by more than a hundred years. His pendulum clocks were more accurate than their predecessors, even if he, like so many of his contemporaries, was unable in the end to solve the longitude problem. This practical project nevertheless yielded a body of theoretical work of great importance in general mechanics, including the description of centrifugal force, the analysis of the centres of oscillation, percussion and flotation, and the principle of the conservation of kinetic energy, formulated in some of the first scientific equations. Huygens's work on moving bodies of all kinds also led him to reject the idea that motion was absolute. He realized that the concept of motion made no sense without reference to an environment in which the motion is taking place. He described this for linear motion, such as that experienced by the man in the barge passing by the man on the canal side. From here, he might have gone on to assert more daringly that there is no such thing as absolute space – in the sense of the imaginary mathematical grid imposed on the universe by Descartes and Newton – and that all frames of reference are equivalent.

Newton several times recorded his admiration for Huygens's inventions, discoveries and investigative methods, as well as gratitude for Huygens's contribution to his own work. Two centuries later, Albert Einstein acknowledged his debt to Huygens, and praised him for his intimation of relativity.

Huygens was not motivated by the quest for a grand universal theory. His early realization that Descartes was not a god must have disabused him of such notions. A demonstrable social need, or at least a patron's whim, drove his work on clocks and telescopes, magic lanterns and ornamental fountains. At other times, his inventiveness was the product of a kind of leisured opportunism. He always had financial support – from Colbert, from his father's allowance, from estate incomes – even if it was not always as liberal as he would have liked. This gave him the freedom to investigate the topics that most greatly interested him. Without such support, he might not have dived so deeply into matters such as the geometry of curves and the rules of probability.

But there is another important aspect to Huygens's scientific contribution. Where Newton sought seclusion, Huygens sought connection. Whenever he was sequestered in The Hague, he longed for nothing so much as to be back with his friends in Paris or London. He knew that his scientific work depended on the expert knowledge of colleagues. He maintained a lively correspondence with scholars across Europe, and relied on them to be kept apprised of the latest thinking. He worked closely with others on many occasions, making refinements to optical arrangements and conducting mechanical experiments. Huygens's telescopes drew upon what Galileo and other Italians had done, and were perfected in dialogue with Danish, German, English and Scottish astronomers. His work on the air pump confirmed and furthered the ideas of Irish and English physicists. His mathematical work was stimulated above all by his long engagement with French mathematicians. Through this informal international network, Huygens and his colleagues were more readily able to replicate their experiments and confirm their results, and to accelerate the dissemination of their discoveries to the wider world.

As the first foreign member of the Royal Society of London, and an effective founder of the French Academy of Sciences, Huygens saw the value of formalizing this dialogue within dedicated institutions. The natural philosophers who comprised the membership of these early academies incubated the modern scientific method. Informed by the civility that was the watchword of educated people in the seventeenth century, and guided by patient interlocutors such as Mersenne, Oldenburg and Fatio de Duillier, Huygens and his scientific peers learned that it was in general advantageous to talk, to write and to share and compare results with one another. For most of these scientists, most of the time, their nationality was of little consequence. Indeed, it is remarkable how little the virtually constant wars between the countries most active in science affected this intellectual exchange, even when the subjects under discussion were longitude clocks or telescopes, which might easily be regarded as military secrets. Except for the occasional practical inconvenience of a lost letter or the border seizure of some strange-looking piece of scientific equipment, scientists were generally able to carry on unmolested with their esoteric work.

Why does this story matter now?

Science depends ever more on international connections, and so do our lives. As we peer further out into space and deeper into the heart of matter, the sheer physical dimensions of the telescope arrays and particle accelerators require territory that outruns the capacity of any individual country to accommodate them. The vast numbers of scientists involved in their research teams make internationalism a necessity. This cooperation is essential, too, in more down-to-earth but no less vital projects such as minimizing the risk to global public health from infectious diseases or understanding and reversing the crisis of nature that is increasingly evident from our changing climate and from the gathering pace of extinction of plant and animal species. It requires continental or global organization to coordinate this work, and national academies have now been supplemented by international agencies to reflect this.

Networks of science have blossomed since they began with a few personal friendships and convenient collaborations in the seventeenth century. Every step of the journey has demanded patience, tolerance and willingness to reach out across the divide. Yet how easy it is, in moments of tribal wantonness, to endanger what has been painstakingly built by figures such as Huygens for the greater good of all.

And what of Huygens the man?

What must a scientist do to be remembered for himself? By

any measure, his are major achievements: new laws of physics, new heavenly bodies, new devices. Yet the man fades from view.

The burials of Constantijn and Christiaan Huygens, father and son, are commemorated on a modest plaque in the east end ambulatory of Sint-Jacobskerk in The Hague; the church is usually locked to visitors. A notice placed nearby explains: 'They never received a tombstone, and the location of their grave had already been forgotten 80 years after the funeral.' It was only identified during the course of restoration work in 2007.

There have been fitful and belated attempts to commemorate Christiaan Huygens in other ways. Historians of science received satisfaction when his treatises, papers and correspondence were finally collected and published in a massive international editorial project lasting more than sixty years, coordinated by the Dutch Society of Sciences. The last of the twenty-two volumes of these *Oeuvres Complètes*, running to more than 12,000 pages in French, Latin and Dutch (and occasionally English, German or Italian), appeared as recently as 1950. Several universities in the Netherlands have a Huygens Laboratory or a Huygens Library. The European Geosciences Union awards an annual Christiaan Huygens medal. The Royal Dutch Academy of Sciences gives out a Huygens award, rotating each year among Huygens's several fields of interest.

Older Dutch citizens will recall that Huygens's periwigged face once gazed out from the twenty-five-guilder banknote, with Hofwijck in the background and Saturn hovering overhead. Here and there in Dutch cities is the odd Huygensstraat or Huygensweg, but it is usually the father Constantijn who is intended to be commemorated.

The most ambitious, if not the most visible, memorial to Christiaan Huygens is probably the sandstone monument that stands crumbling gently in a neglected corner of the garden of the Dutch Society of Sciences in Haarlem. The sculpture – a steeply tiered affair like a multistage rocket – is a clumsy melange of classical and Gothic styles. From a niche on one side of the lower section emerges a statue of Huygens, elaborately draped and clutching a staff. Directly above his head is a relief of a seated man demonstrating a cycloidal pendulum. Four robed figures (Archimedes and other great Ancients?) stand at each corner of this lower part, from which rises an octagonal column crowned with a circular frieze of the zodiac (an ill-chosen device to celebrate the rationalist Huygens). Above this, four winged angels clad in copper hold aloft a bronze armillary sphere. The whole work is some four metres tall.

The sculpture, unveiled in 1909 in the presence of Hendrik, Queen Wilhelmina's prince consort, is in fact a one-third-scale maquette for a much grander monument. A few years earlier, the Society of Sciences had celebrated the completion of the first ten volumes – those containing his complete correspondence – of Huygens's *Oeuvres Complètes*. During the formalities, a legacy was read out from one of the society's recently deceased members, who had left a sum of 40,000 guilders for 'the raising in The Hague of a statue of Christiaan Huygens, the famous Dutch physicist of the seventeenth century'. The legacy stipulated that the monument should be erected on 'The Plaats, The Vijverberg, Lange or Korte Voorhout or near surroundings', in other words, in the ceremonial heart of the city, close to the seat of the national government, the Binnenhof, and the spot where Johan de Witt and his brother were so bloodily slaughtered.

However, when a model of the design was temporarily put up facing the Hotel des Indes on the Lange Voorhout, the burghers of The Hague showed themselves to be less than keen on the project. Their principal objection was that the design was more suited to a graveyard than a public space. Surely, said others, a monument to such a versatile and progressive scientist as Huygens

should be executed in a more modern style. It is hard to disagree
with their judgement. The Hague city council voted against
proceeding, and the legacy was spent instead on the acquisition
of new property for the Society of Sciences. In 1999, the Society
made enquiries to see if the town of Voorburg would like to take
the sculpture off its hands and position it in front of the Hofwijck
house, but even this desperate last bid to honour Huygens in the
public arena failed when the town turned down the offer.

Compare this fate with that of his English colleague.

Perhaps nobody was ever more smitten by Newton, or by the
idea of Newton, than the Parisian neoclassical architect Etienne-
Louis Boullée. He summons up his god in *Architecture, essai sur
l'art*: 'Sublime mind! Prodigious and profound genius! Divine
Being! Newton! Deign to accept the homage of my feeble talents!'
In this textbook-cum-confessional, unpublished in his lifetime, he
continues like a celebrity stalker: 'you have defined the shape of
the earth; I have conceived the idea of enveloping you with your
discovery'.

Boullée's architectural projects typically displayed the highest
degree of symmetry and made copious use of perfect geometric
solids – sphere, drum, cone, cube – which he claimed were derived
from nature and were symbolic of transcendence. His cenotaph
for Newton, designed in 1784, was conceived as a massive hollow
sphere 150 metres high, resting on a circular fortress-like base
girded by belts of cypress trees. Taller than the pyramids or St
Peter's in Rome, it was intended to be expressive of the whole of
nature and the cosmos and its fundamental indivisibility and
oneness. Inside, it was to be utterly empty but for Newton's tomb,
and the vault would be pierced with holes to give an illusion of

the stars and planets. Visitors (worshippers?) would be transported 'as if by magic' through this vast airy space to be brought ultimately into the presence of the tomb, placed at the centre of gravity. 'I wanted to give Newton that immortal resting place, the Heavens.'

Unlike his colleague, Christopher Wren, who did in the end truly succeed in surrounding himself with his own project of St Paul's Cathedral (*Si monumentum requiris circumspice*, reads his epitaph there), Newton never got the monument that his ardent admirer designed for him. In 1727 Voltaire exulted to find himself present in Westminster Abbey for the funeral of the man 'who destroyed the Cartesian system'. Newton, he wrote, 'was buried like a king who had done his subjects much good'. Four years later, the architect William Kent and the Flemish sculptor Jan Michiel Rijsbrack produced a more conventional, though still extravagant, memorial in grey-and-white marble, in which Newton reclines majestically on a pile of books and gestures offhandedly towards a geometric pattern drawn on a scroll held out by two putti. Behind him – Boullée would have been pleased – stands a pure pyramid supporting a globe with the female figure of Astronomy lying somewhat uncomfortably across the top. Another neoclassical statue adorns the antechapel of Newton's Cambridge college, Trinity. Here, the French sculptor Louis-François Roubiliac presents a robed Newton standing, prism in hand, emphatically making his point.

Thereafter, as the historian of science Patricia Fara has revealed, Newton swiftly 'passed into mythical realms' to complete his transformation into the 'legendary figure' that he is still held to be today, a 'secular saint endowed with supra-human capacities'. A man who, unlike Huygens, had not been widely known even in his own country during his lifetime, became an icon of rationality and progress, the embodiment of Enlightenment ideals and a symbol of Britannia's intellectual spirit. The most famous poet of the day, Alexander Pope, produced a suitable (though not to the Abbey) epitaph: 'Nature and Nature's Laws lay hid in Night: / God said, "Let Newton be!" and all was Light.' Rijsbrack and Roubiliac both capitalized on their commissions by knocking out replica Newton busts to adorn the salons of Britain's country estates, available in marble, terracotta and plaster versions to suit all pockets.

Although he was slower to gain acceptance in France, which remained faithful to the scientific ideas of Descartes, Newton's reputation there grew even more lustrous than in England. After returning from his British exile, Voltaire produced a popular gloss on Newton's ideas, while his lover Emilie du Châtelet translated the *Principia* with explanatory notes, thereby creating a work that many said was superior to the original English translation (from Newton's Latin). Later, Pierre-Simon Laplace was able to provide a mathematical explanation for certain observed anomalies to do with Newton's law of gravitation, such as periodic irregularities in the planets' orbits, that Newton had been happy to leave to divine arbitration. Laplace even went so far as to style himself as the 'French Newton'. The world was ready – almost – for Boullée's crazy *hommage*.

The construction of Newton as the first genius of science continued apace throughout the nineteenth century. Casts of his death mask used by the sculptors of his statues were widely traded. The cast of his skull was subjected to phrenological analysis, which, needless to say, confirmed his unique talents. 'In the whole course of his life he never exhibited extraordinary aptitude for anything save mathematics,' opined a correspondent in the *Phrenological Journal*; 'in later days, his mind appeared to be singularly devoid of passion and pride; and he had none of that eager and restless hungering for popularity which rendered almost abortive the wonderful acuteness of his contemporary, Hooke.' All this might have gone on even longer but for the arrival on the scene in the following century of Einstein, whose achievements in the same field as Newton at last challenged his primacy in the popular imagination.

For Newton to attain this exalted status in the first place it was necessary that others be forgotten. Newton himself did what he could to demolish the reputation of his principal rival in English physics, Robert Hooke, after the latter's death in 1703. The cult of Newton also effectively extinguished Descartes's once considerable reputation as a scientist, even in France. The effect was even more devastating on Huygens. He makes no appearance at all in Bill Bryson's *A Short History of Nearly Everything*, William Bynum's *A Little History of Science*, or Patricia Fara's revisionist *Science: A Four Thousand Year History*. (Newton, of course, features abundantly in all of these works.) In *The Character of Physical Law*, Richard Feynman glides from Copernicus to Brahe, Kepler, Galileo and Newton without mentioning Huygens once, either in his coverage of forces or in his discussion of the relation of mathematics to physics.

The omission is not quite universal, it must be said: Huygens is adequately covered by both John Gribbin and David Wootton in their general histories, although Wootton – or his indexer – seems perplexed, and perhaps even a little revolted, by Huygens's internationalism, and includes incredulous subentries for him: 'French by choice', and 'moves country'. Nevertheless, the imbalance is clear. A search in the library catalogue at Cambridge (Newton's university) turns up 656 results for 'Isaac Newton' and 53 for 'Christiaan Huygens'. At Leiden (Huygens's university), the situation is only partially redressed, with 612 results for Huygens and 455 for Newton.

There is no question that Huygens and Newton were effectively equals in their own time. Although Newton often resisted the opportunity to correspond directly with many of his peers, he was eager to compare notes with Huygens. It was Huygens who offered the most sustained and informed critique of Newton's theories, placing himself as one of very few natural philosophers whose opinions Newton felt bound to take seriously. It was Huygens, thirteen years Newton's senior, who then enjoyed the greater established reputation. And it was Huygens who, in 1666, while Newton secretly experienced his *annus mirabilis*, was already being wooed by Louis XIV. As Newton's biographer Richard Westfall has observed:

> The parallel between Newton and Huygens in natural philosophy is remarkable. Working within the same tradition, they saw the same problems in many cases and pursued them to similar conclusions. Beyond mechanics, there were also parallel investigations in optics. At nearly the same time and stimulated by the same book, Hooke's *Micrographia*, they thought of identical methods to measure the thickness of thin coloured films. No other natural philosopher even approached their level.

And there lies the problem. In popular perception, it seems, the science of any era can only accommodate a single 'genius'. In the generation before Newton and Huygens, it is today Galileo who shuts out Kepler, who in fact made the greater contribution to astronomy and physics. By the end of Huygens's life, it was no longer Padua or Prague, nor even Paris, that was the major centre of science in Europe, but London, driven on by the ferment of activity of the Royal Society and England's economic expansion. From there, despite the critical comments of continentals such as Huygens and Leibniz, the *Principia* 'took Britain by storm', provoking gushing admiration even from those who couldn't understand it.

After that, there was then nothing to sustain widespread esti-
mation for Huygens, even though there was still much that would
later prove to be correct about his ideas to do with force and
light. Huygens never spawned an adjective analogous with *cartésien*
and Newtonian (the former coined just fifteen years after
Descartes's death, the latter while Newton was still relatively
young). 'Huygenian' was used only in a technical context and as
a last resort: Titan was the 'Huygenian satellite' before it was
named in its own right, for example.

The Système Internationale unit of force is the newton; there
is no 'huygen'.

Huygens leaves something else behind him, though, something
more in keeping with his commitment to science and suspicion
of honours. This is the extraordinary Dutch contribution to
astronomy. Though the Netherlands is famously flat, with big
skies, it is still a surprise to find that this often cloud-covered
land has produced so many notable astronomers. When yet
another Dutchman appeared at his door looking for work, Harlow
Shapley, the twentieth-century director of the Harvard College
Observatory who calculated the size of the Milky Way, is said to
have called it 'the place where they grow tulips and astronomers
for export'. Jan Oort, Willem de Sitter and Gerard Kuiper are
among many who have given their names to features of the solar
system and universe.

Of course, Christiaan Huygens is not directly responsible for
this comparatively recent pre-eminence. In the view of Carl Sagan,
who once sat with Kuiper observing the planets and speculating
about extraterrestrial life much as Christiaan and Constantijn
Huygens had done, it is, rather, its history as an outward-looking,

exploratory nation that accounts for 'the fact that Holland has, to this day, produced far more than its per capita share of distinguished astronomers'. But there again, Huygens's discoveries do surely have an intoxicating appeal. As another American science writer has suggested, haloed Saturn is, for many amateurs, the 'gateway drug' to their astronomical habit.

On 14 January 2005, after a seven-year journey, the European Space Agency *Huygens* spacecraft parachuted through the thick nitrogenous atmosphere of Titan and settled onto the satellite's oozy surface. It might have pleased Christiaan that the craft was French-built. It might have pleased him, too, that many space scientists have felt, ever since the two NASA *Voyager* space probes passed by Saturn and its moons in 1980 and 1981, that it might be *his* moon, Titan, of all the bodies in the solar system, that would reveal itself to be the most suited to the formation of biogenic molecules and perhaps even capable of supporting life.

During its two-hour descent and in the few minutes left before it expired on the surface, the probe beamed measurements of temperature, density, electrical activity and chemical composition up to the *Cassini* orbiter for relaying back to Earth. Visual images showed river channels, probably generated by flows of liquid methane, cutting through jagged mountains of water ice. Isotopic analysis of the carbon found on Titan indicated, however, that it was not compatible with a biological source. Huygens's moon is far too cold to support life now, and there is no sign that it ever has done. But there is a chance that it could do so in billions of years' time, when the sun has swollen so that its heat causes the methane to evaporate and the ice mountains to melt, forming oceans of water rich in organic compounds and minerals. Then,

Titan might enter the 'Goldilocks zone' presently occupied by the Earth, where it is neither too hot nor too cold but just right for life. For this reason, Titan is now regarded as a potential 'simulation' of the very early Earth. In 2026 NASA is planning to launch a new probe, which will drop a drone into Titan's atmosphere, able to fly from place to place to search out conceivable habitable environments.

Dozens of fly-bys of the *Cassini* orbiter during the months before and after the *Huygens* spacecraft touched down took it close to many of Saturn's sixty-two major moons and even through some of the gaps between the planet's rings. Data sent back confirmed that the rings are composed chiefly of ice, in pieces varying in size between snowflakes and bergs, spread in a layer just ten metres thick. The orbiter also resolved a mystery as to why the rings were bright enough for Christiaan Huygens to discern them with his primitive telescope at all. The quantities of space dust detected during the mission suggested that the rings should be dirty and much less reflective. The fact that they are bright suggests that they are much younger than the planet itself, perhaps as little as ten million years old. It was found, too, that the rings are being drawn slowly into the planet itself and may be gone again in as little as a hundred million years.

When it flew past Titan, *Cassini* revealed lakes of methane near the pole the size of some of the inland seas on Earth, which confirmed that the rugged terrain of the satellite was due to methane rivers that once flowed there. *Cassini* also flew close by the smaller, ice-covered moon Enceladus on several occasions. Here, it found geysers of ice particles thrust up high into the atmosphere from underground oceans of salty water. This water was found to contain silica and organic molecules, strongly suggesting that it was issuing from thermal vents under the moon's seas, and that the core of the moon was therefore hot. Such warming is not achievable so far from the sun by solar radiation; on Titan and Saturn, surface temperatures are a chilly $-180°$ Celsius. Instead, it is thought that Enceladus is heated by friction generated by the constant movement of its solid core produced by the fluctuating gravitational field of its massive parent planet, much as a squash ball becomes warm in the hand with repeated squeezing. By chance, this vigorous massage has raised temperatures to levels conducive to the emergence of life.

In a touching remark at the beginning of *Cosmotheoros* concerning the mysteries of the solar system, directed at the beloved older brother whom he had asked to see the work into print after his death, Christiaan lamented of their days spent at the telescope together: 'we were always apt to conclude, that 'twas in vain to enquire after what Nature had been pleased to do there, seeing there was no likelihood of ever coming to the end of the Enquiry'. It would surely have delighted Huygens more than any stony memorial to know that a space probe bearing his name is now proving him wrong.

Main Personalities

ADRIEN AUZOUT, 1622–91. Astronomer and instrument-maker active in the scientific circles in Paris during the 1660s, he was a frequent correspondent of Huygens on optical matters.

FRANCIS BACON, 1561–1626. Philosopher, statesman, Lord Chancellor of England, First Viscount of St Albans, thus also known as Lord Verulam or Verulamius. Bacon held that science should be based on systematic investigation rather than blind acceptance. He set out these views in works such as *Novum Organum* and *New Atlantis*, a utopian novel depicting an ideal scientific community, which served as a template for the later Royal Society.

ISAAC BEECKMAN, 1588–1637. Copiously self-educated Middelburger, sometime candle merchant, Latin teacher and church minister, he corresponded with Snel and Mersenne and befriended Descartes. His later years devoted to mathematics, mechanics and optics make him the successor to Stevin and the forerunner of Christiaan Huygens.

WILLEM BOREEL, 1591–1668. Dutch ambassador to France, 1649–68, he had a lively interest in telescopes.

ISMAËL BOULLIAU, 1605–94. Born in Loudun to Calvinist parents, he converted to Roman Catholicism and became a priest. He moved to Paris, where he pursued many scholarly interests, publishing significant works on astronomy, geometry and the nature of light.

ROBERT BOYLE, 1627–91. Irish natural philosopher, his studies mainly in Oxford established him as the father of modern chemistry. He also investigated the pneumatics of air and the vacuum; Boyle's law states that pressure is in inverse proportion to volume for a given quantity of gas.

ALEXANDER BRUCE, c.1629–80. Earl of Kincardine and a wealthy Scottish landowner, he collaborated and later quarrelled with Huygens in early attempts to make seagoing pendulum clocks.

SUSANNA CARON, 1652–? Known to Huygens as Suzette, she was a cousin and close friend. She married a French nobleman, François de Civille, becoming Madame de la Ferté.

GIOVANNI DOMENICO CASSINI, 1625–1712. Professor of astronomy at Bologna, he moved to Paris in 1669 to run Louis XIV's new observatory. He made calculations of the rotational periods of Mars and Jupiter and its four Galilean satellites. Using ever longer and more powerful telescopes, he discovered four moons of Saturn in addition to the first, Titan, discovered by Huygens.

JACOB CATS, 1577–1660. Poet and statesman, grand pensionary of Holland, 1636–51. His often humorous and moralistic verse was more conservative than that of Constantijn Huygens and the Muiden Circle.

MARGARET CAVENDISH, 1623?–73. Writer and philosopher. Of a staunch royalist family, she served at the court of Henrietta Maria, wife of Charles I, but later lived in exile in Paris and Antwerp, where she became interested in scientific questions.

Jean Chapelain, 1595–1674. Poet, critic, and a founding member of the Académie Française. He turned down prestigious appointments offered by Cardinal Richelieu and other powerful figures, and remained without ambition and loyal to his many friends, including Huygens.

Jean-Baptiste Colbert, 1619–83. Minister and controller of finance under Louis XIV. *Colbertisme* demanded the increase of industry, trade and regulation, all directed towards enriching the French state in the person of the king.

Salomon Coster, *c.*1620–59. Huygens's first clock-maker in The Hague.

Pierre De Carcavi, *c.*1600–84. French royal librarian and mathematician, friend of Pascal and Fermat.

Pierre De Fermat, 1607–65. Mathematician and lawyer, his analysis of curved lines was a precursor to calculus. He also made advances in number theory and, with Pascal, founded the field of probability. He was not overgenerous in leaving proofs of his theorems.

Jacques (Jacob) De Gheyn II, 1565–1629. A Flemish dynasty of painters and engravers, the de Gheyns were neighbours of the Huygenses in The Hague. De Gheyn II found favour with the House of Orange and made an engraving of Prince Maurits aboard Stevin's sand-yacht.

Leopoldo De' Medici, 1617–75. Born at the Pitti Palace in Florence into the ruling family of the Grand Duchy of Tuscany, he was made cardinal and governor of Siena. Inspired by Galileo, he became an enthusiastic supporter of the sciences and established the Accademia del Cimento.

HENRI LOUIS HABERT DE MONTMOR, *c.*1600–79. Royal councillor, mathematician and cheerleader for Descartes. The 'académie Montmor' that he hosted at his house in Paris from 1657–64 was an informal forerunner of the French Academy of Sciences.

GILLES PERSONNE DE ROBERVAL, 1602–75. Eager to be considered the equal of his colleagues, he made innovations in fields from mathematics to mechanics and optics. He claimed other discoveries and inventions, but was often reluctant to publish, poor at taking criticism and quick to argument.

JOHAN DE WITT, 1625–72. As grand pensionary during much of the Stadholderless Period, 1650–72, he attempted to introduce a more liberal, tolerant and devolved form of government to the Dutch Republic. But following the invasion of the country in 1672, he and his brother Cornelis were lynched by an angry mob.

RENÉ DESCARTES, 1596–1650. Though born in France, he spent most of his working life in the Dutch Republic, where he found the quiet and tolerance he sought to develop his philosophical ideas. His famous *Discourse on Method* found a Dutch publisher. He was a friend of the elder Constantijn Huygens, with whom he worked on lens development, and became a profound influence on his son Christiaan. It was not always to his advantage that he preferred the power of reason to empirical tests, and while his reputation as a philosopher still stands, he is no longer seriously regarded as a scientist.

JOHN DONNE, 1572–1631. Considered the foremost English metaphysical poet, he was also an Anglican priest, becoming Dean of St Paul's, and a member of Parliament.

CORNELIS DREBBEL, 1572–1633. An inventor of wild imagination, Drebbel was born in Holland but moved to London, where he was employed at court, putting on masques and presenting an array of miraculous devices, from perpetual-motion machines to the camera obscura. Beneath the showmanship lay a fine understanding of mechanical and optical principles.

NICOLAS FATIO DE DUILLIER, 1664–1753. Swiss-born, he lived mainly in England and Holland, where his facility with the new mathematics brought him into contact with Newton, Leibniz and Huygens. He did not fulfil his early promise, however, and later joined a religious sect.

FREDERIK HENDRIK, Prince Of Orange, Stadholder, 1584–1647. The youngest son of William the Silent, he played a decisive role in the second half of the Eighty Years War. In 1625 he married Amalia van Solms and succeeded to the stadholderate.

GALILEO GALILEI, 1564–1642. A physicist and astronomer of unprecedented range and achievement, he was the giant upon whose shoulders Christiaan Huygens stood. Lamps swinging in Pisa cathedral sparked his investigation of pendulum motion. In 1610 he used a telescope to observe four moons of Jupiter, the first bodies in the solar system that clearly did not orbit the Earth, contradicting both Aristotelian and Church doctrine.

NICOLAAS HARTSOEKER, 1656–1725. Astronomer. Passionate about the sciences from childhood, he gained instruction in microscopy from Leeuwenhoek. In 1684 he settled in Paris, where he built large telescopes at the observatory.

THOMAS HOBBES, 1588–1679. Best known as the author of *Leviathan*, the philosophical fig leaf for authoritarian governments ever since, he was well travelled and well connected, and his interests were all-encompassing. He was an enthusiastic but incompetent mathematician, and his efforts were derided by Huygens and others. Descartes's writings excited his interest in optics, but he feared the vacuum was the work of the devil.

SUSANNA HOEFNAGEL, 1561–1633. Constantijn Huygens's mother came from a wealthy Antwerp trading family. Her brother Joris was renowned as a miniature painter, and other members of the family also made their name as artists.

PIETER CORNELISZOON HOOFT, 1581–1647. Poet, playwright and historian. As magistrate of Muiden, he hosted a famous literary circle at the castle where he resided.

ROBERT HOOKE, 1635–1703. While studying at Oxford, he assisted Boyle, becoming the best experimenter of his day. He made several discoveries, including the law of elasticity that bears his name, but jealously guarded his work, which often led to conflict with his peers and denied him due recognition. His major published work was *Micrographia*, a magnificently illustrated work on optics.

JOHANNES HUDDE, 1628–1704. He studied law at Leiden, but became more interested in mathematics, Cartesian geometry and probability, about which he corresponded with Spinoza, de Witt and Huygens. Later, as mayor of Amsterdam, he made improvements to the hygiene of the city's canals.

CHRISTIAAN HUYGENS, 1629–95. *Passim.*

CHRISTIAEN HUYGENS, 1551–1624. Christiaan's paternal grandfather. Born in Brabant, he became secretary to William the Silent in 1578, continuing in that position to the council of state appointed after William's assassination in 1584, thus inaugurating a lineage of Huygens service to the Dutch Republic.

CONSTANTIJN HUYGENS, 1596–1687. *Passim*, especially chapters 2–4. Christiaan's father. He married Susanna van Baerle in 1627.

CONSTANTIJN HUYGENS, 1628–97. Christiaan's older brother. The pair collaborated in lens-grinding and astronomical observation, when his secretarial duties to William III did not take precedence. He married Susanna Rijckaert in 1668. One son.

LODEWIJK HUYGENS, 1631–99. Christiaan's younger brother and black sheep of the family, he got into trouble for duelling at college and improperly accepted gifts as the magistrate of Gorinchem. He married in 1674. Four sons.

PHILIPS HUYGENS, 1633–57. Christiaan's youngest brother, he became ill and died in Marienburg (now Malbork in Poland) while on an embassy mission.

SUSANNA HUYGENS, 1637–1725. Christiaan's sister. Although her accomplishments might perhaps have equalled those of her brothers, her father did not wish his daughter to become an intellectual. She married a cousin, Philips Doublet, in 1660. She had six children, three of whom survived to adulthood.

ZACHARIAS JANSEN, *c.*1580–*c.*1630. Middelburg lens-maker, unreliably claimed as the maker of the first telescope.

ANTHONI LEEUWENHOEK, 1632–1723. Work as a draper's bookkeeper in Amsterdam may have introduced him to lenses. He returned to his hometown of Delft and began making microscopes with tiny bead lenses, through which he was able to observe numerous protozoa for the first time.

GOTTFRIED WILHELM LEIBNIZ, 1646–1716. Employed by a series of electors of Germany, he was able to find time for philosophy and mathematics among routine diplomatic duties. He developed the form of differential and integral calculus employed today, leading him into dispute with Newton, whose method of 'fluxions' was devised earlier but not published.

JAN LIEVENS, 1607–74. Hailed as more brilliant than Rembrandt when a young man in Leiden, he travelled widely in search of work, coming under the influence of van Dyck in England, after which his output became more conventional in both subject matter and style.

HANS LIPPERHEY, ?–1619. German-born spectacles-maker of Middelburg, he demonstrated his telescope to Prince Maurits at The Hague in 1608.

JOHN LOCKE, 1632–1704. English philosopher and physician regarded as the founder of modern liberalism. His empiricism owed much to Bacon and was opposed to Cartesian *a priori* reasoning. He spent five years in exile in Holland during the reign of Charles II.

LOUIS XIV, 1638–1715. King of France for all but the first five years of his life, he took steps to increase French military might, centralize power, and embellish the state through the arts.

MARIN MERSENNE, 1588–1648. French theologian, mathematician and music theorist, he was one of the principal scholarly figures of his age, counting Galileo and the elder Constantijn Huygens among his many correspondents.

JACOB METIUS ?–1628, and ADRIAAN METIUS, 1571–1635. Sons of an Alkmaar military engineer and cartographer, Jacob was wrongly believed by contemporaries to have invented the telescope. Adriaan was a surveyor and mathematician.

ROBERT MORAY, 1608/9?–73. A Scottish army officer, he served widely in France. Upon the Restoration in 1660, he settled in London and was one of the founding members of the Royal Society.

ISAAC NEWTON, 1642–1727. England's greatest scientist, he unwove the rainbow, to the dismay of poets from Blake to Keats. His experiments with light, the laws of motion and his theory of gravitation are set out in a 1672 paper, *New Theory of Light and Colours*, and in *Principia Mathematica* (1687) and *Opticks* (1704). His ideas nullified Descartes's science and laid the scientific foundations for the Enlightenment. A disputatious and unpleasant man, he was a fellow of Trinity College, Cambridge and master of the royal mint.

HENRY [HEINRICH] OLDENBURG, *c.*1619–77. Born in Bremen, he settled in England in 1653. During his continental travels, he had made the acquaintance of many leading scientists, and subsequently proved an invaluable 'foreign secretary' of the Royal Society, promoting an international scientific discourse between sometimes reluctant correspondents.

DENIS PAPIN, 1647–1713. French physicist and inventor, he worked with Huygens, Leibniz and Boyle on experiments with air pressure, and later invented a steam digester, a forerunner of the steam engine.

BLAISE PASCAL, 1623–62. French mathematical prodigy and philosopher of religion, he developed the theory of probability with Fermat. He participated in physical experiments on pressure, and invented a calculating machine, to which Huygens later made modifications.

PIERRE PETIT, 1598–1677. Military engineer and surveyor, he frequently worked with Christiaan Huygens on inventions and made instruments for him in Paris.

REMBRANDT HARMENSZOON VAN RIJN, 1606–69. Painter and print-maker considered by many to be the greatest artist of the Dutch Golden Age. The elder Constantijn Huygens secured early commissions for him, although his style subsequently diverged from courtly tastes.

WILLEBRORD SNEL [SNELLIUS], 1580–1626. He succeeded his father as professor of mathematics at Leiden and independently discovered the law of refraction that bears his name.

BARUCH SPINOZA, 1632–77. Philosopher disowned by his Sephardi Jewish community, his work on ethics, consciousness, God and nature, and biblical truth made him a figurehead of Enlightenment thinking. He made a humble living as a lens-grinder, living near Leiden and in Voorburg, near The Hague, which brought him into contact with the Huygens brothers, who held him in high regard.

SIMON STEVIN, 1548–1620. Born in Bruges, he moved to Leiden, where he made a European reputation as a highly imaginative inventor, while also teaching mathematics and serving as superintendent of finance to Prince Maurits of Orange. He was a prolific writer, producing works – in Dutch, unusually for the period – on everything from bookkeeping to the design of fortifications.

JOHANNES SWAMMERDAM, 1637–80. He studied medicine at Leiden, but was early on captivated by the possibilities of the microscope, discovering red blood corpuscles. His compendious observations of insects laid the foundations of modern entomology.

SUSANNA VAN BAERLE, 1599–1637. Her parents died when she was young, and she inherited a share of her father's trading fortune. She wrote verse and painted nature-based still lifes, but her work does not survive. She married the elder Constantijn Huygens in 1627 and died following the birth of their fifth child.

JACOB VAN CAMPEN, 1595–1657. Painter and architect of Amsterdam's city hall (now used as a royal palace) and, with Pieter Post, of the Mauritshuis in The Hague. He had lived in Italy, from where he introduced to the Dutch Republic a severe version of Palladian classicism.

JOOST VAN DEN VONDEL, 1587–1679. Playwright and poet born in Cologne, he became the most important poet and dramatist in the Dutch Republic, writing verses in celebration of the House of Orange and plays reflecting the political and religious disputes of the time.

FRANS VAN SCHOOTEN (THE YOUNGER), 1615–60. Followed his father as professor of mathematics at Leiden. He lived for a time in Paris and met Mersenne and Descartes. He later inducted de Witt, Hudde and Huygens into Cartesian geometry.

ANNA VISSCHER, 1584–1651, and MARIA TESSELSCHADE VISSCHER, 1594–1649. Muses to the poets of the Muiden Circle, the Visscher sisters were highly accomplished in many arts and well able to hold their own in verse.

JOHN WALLIS, 1616–1703. England's leading mathematician until Newton. Ordained in 1640, he acted as code-breaker for the Parliamentarian side during the Civil War, but later became the king's chaplain and a founding member of the Royal Society.

WILLEM [WILLIAM] I, PRINCE OF ORANGE, STADHOLDER, 1533–84. He led the Dutch Revolt against Spain in 1568. For his shrewd political judgement and military skill, he earned the sobriquet of William the Silent and, in the Netherlands, the father of the fatherland.

WILLEM [WILLIAM] III, PRINCE OF ORANGE, STADHOLDER, KING OF ENGLAND, SCOTLAND AND IRELAND, 1650–1702. Grandson of Frederik Hendrik, he was appointed as stadholder in the 'disaster year' of 1672 and crowned king of England in 1689 by virtue of marriage to his cousin Mary twelve years earlier.

Acknowledgements

My first debt of gratitude is to the Nederlands Letterenfonds (Dutch Foundation for Literature) for awarding me a writer's residency in Amsterdam, which immeasurably broadened and deepened my understanding of places known to Christiaan Huygens and his peers. Thank you to Maaike Pereboom, Fleur van Koppen, Hanneke Marttin and Greetje Heemskerk, as well as to Tim Visser, Hilde de Weerdt and Fabian Kraemer at the Netherlands Institute for Advanced Studies in the Humanities and Social Sciences, for their generous support of my work.

I am grateful to Ruth Scurr, Patricia Fara, Michael Pye, Felicity Henderson, Max Porter and Karolina Sutton, with all of whom I enjoyed productive conversations during the early stages of development of my ideas.

A book like this, though not written for an academic audience, depends for its progress on the kindness of academics, whose only reward perhaps for the gift of their time, expertise and contacts is seeing their beloved field of specialization traduced for the questionable gain of appealing to the general reader. In this case, it was Anne Goldgar, professor of early modern history at King's College London, who opened the door for me. I am immensely grateful for her help. Without it, I would not have benefited from the uniquely informed insights of many others, and in particular Rudolf Dekker, Ad Leerintveld, Pamela H. Smith and Sven Dupré.

I am grateful to Clina Tuijtjens, who helped me to obtain the Museumkaart without which I would surely have omitted to look at so many of the paintings and objects related to the seventeenth-century Dutch Republic. Possession of this card made me feel properly Dutch. My rudimentary Dutch was greatly augmented in informal conversation classes with Sia van Deutekom and Claire Kokelaar: *bedankt voor de moeite*. I also benefited from the Early Modern Dutch Course offered by the Centre for Languages and International Education at University College London, where An Castagnia proved a doughty instructor. In my efforts at translation, especially of Dutch poetry, I was emboldened by the advice of Amanda Hopkinson, visiting professor of literary translation at City University London.

In the Netherlands, I gained insight into the optical sciences in seventeenth-century Holland from Tiemen Cocquyt at the Rijksmuseum Boerhaave in Leiden, which has excellent displays of some of Huygens's scientific instruments. I was fortunate also to see an exhibition on the life and work of Christiaan Huygens at Huygens' Hofwijck in Voorburg while the Boerhaave Museum was closed for renovation. Paul Junger kindly gave me a practical introduction to lens-grinding in the traditional fashion as it is still taught at the Leidse Instrumentmakers School.

Rijk-Jan Koppejan of the Philippus Lansbergen Observatory led me round the display he has created about the invention of the telescope in Middelburg, while Peter Louwman showed me the early telescopes that are the pride of his collection at the automotive museum that bears his name in The Hague. I enjoyed further discussions concerning Christiaan Huygens's astronomy with Vincent Icke and Charlotte Lemmens, and his innovations in clocks with Kees Grimbergen and with Pier van Leeuwen at the Uurwerk Museum in Zaanse Schans.

Ineke Huysman at the Huygens Instituut voor Nederlandse

Geschiedenis told me about the women in the Huygenses' lives. Gary Schwartz answered my questions about the family's intersection with the life and works of Rembrandt. Quentin Buvelot at the Mauritshuis Museum in The Hague told me more about the paintings of the Huygens family in that collection, while Sjors Dekkers at the Koninklijke Nederlandse Akademie van Wetenschappen (Royal Netherlands Academy of Arts and Sciences) in Amsterdam showed me the 'orphan portrait' of Christiaan. Timothy de Paepe's virtual-reality recreations of Huygens's house on the Plein in The Hague provided a compelling adjunct to written sources on this architecture.

Diederik Burgersdijk at the University of Amsterdam, Saskia van Manen at the Koninklijke Hollandsche Maatschappij der Wetenschappen (Royal Holland Society of Sciences and Humanities) in Haarlem, and René van Rijckevorsel of the *Elsevier Weekblad* informed me about the unbuilt monument to Christiaan Huygens.

Many others helped me by answering miscellaneous queries. They include: Maaike Doorenbos at the Zeeuwsarchief and Caroline van Santen at the Zeeuwsmuseum in Middelburg; Huib Zuidervaart of the Huygens Instituut voor Nederlandse Geschiedenis; Majo Slosser at the Stadskasteel in Zaltbommel; Boudewijn van Calseijde at the Koninklijke Militaire Akademie in Breda; Thijs Weststeijn and Dirk van Miert at the University of Utrecht; Nadine Akkerman and Suze Zijlstra at the University of Leiden; Klaas van Berkel at the University of Groningen; Alan Moss at Radboud University in Nijmegen; Helmer Helmers at the Netherlands Institute for Advanced Studies; Wim Euwe; Mariëlle Hageman; André Klukhuhn; Leonoor Broeder and Jan Paul Schutten. Thank you to all.

In Paris, my questions were answered by Isabelle Maurin-Joffre at the Académie des Sciences and Suzanne Débarbat at the

Acknowledgements

Observatoire de Paris. Elsewhere I would like to thank Sebastian Kokelaar; Clare Beck; Henry Jelsma; Peter Strengers; Jane Partner; John Parker; Simon Schama; Hugh Bowden; Deborah Hamer; Taylor Skerritt; Sally Holloway; Steven Nadler; Spike Bucklow; Matt Weiland; Laura Snyder; Virginia Mills; James Nye and Sebastian Whitestone.

I would also like to thank the staff of the Cambridge University Library and its constituent libraries; the British Library; the Royal Society; Jeroen Vandommele and staff of the Koninklijke Bibliotheek in The Hague; Leiden University Library; the Rijksmuseum Research Library; and Amsterdam Public Library. The creators of the magnificent online resources that I have been able to use also deserve thanks, including especially the Digitale Bibliotheek voor de Nederlandse Letteren, as well as others listed in the Bibliography.

Philip Ball, Rudolf Dekker, Felicity Henderson, Ineke Huysman, Rijk-Jan Koppejan, Martin Kemp, Ad Leerintveld, Arthur Miller, Steven Nadler, Laura Snyder and Sebastian Whitestone kindly read portions of the text. I am grateful to them for their suggestions and corrections. Any remaining errors are, of course, mine.

I am indebted to my wonderful agent Patrick Walsh, who immediately saw the merit of my subject. Without his patient and exhaustive efforts in helping me to develop the proposal, as well as encouragement during the research and writing, this book would have come to nothing. I thank him and his assistant, John Ash, who was also hugely supportive in ways that went far beyond the call of duty.

George Morley was my inspiring and delightful editor at Picador. She very gently ushered me away from some of the traps I had set for myself, and helped me to plot a more logical course forward, while still finding time to point out some scenic possibilities along the way. Trevor Horwood's copy-editing saved me from many smaller embarrassments. Marissa Constantinou made light work of the search for illustrations. My thanks also go to Laura Carr, Nicholas Blake, James Annal, Lindsay Nash and all at Macmillan. At my Dutch publisher, Thomas Rap, I thank Arend Hosman, Henrike de Goede and Catharina van Schilder, as well as Haye Koningsveld and Anne Kramer at Thomas Rap's sister house, De Bezige Bij, for their early encouragement. Translating my work for Dutch publication, Ineke van den Elskamp and Gertjan Wallinga kindly set me straight on some of the finer points of that language.

My family once again accompanied me on the long journey of creating this work. I am forever indebted to Moira for her historical insights, as well as for bringing me back to the twenty-first century, and to my son Sam. If it were not for his dauntless engagement with European languages, I would never have dared to attempt Dutch. This book is dedicated to him.

Hugh Aldersey-Williams
Norfolk, April 2020

Notes

OC   Christiaan Huygens, *Oeuvres Complètes*

HUG Christiaan Huygens, manuscripts in the Leiden University Library

KA   The elder Constantijn Huygens, manuscripts in the Koninklijke Bibliotheek, The Hague

RS   Archive of the Royal Society, London

INTRODUCTION

p. 2   'as if by night': Huygens ed. Zwaan, *Hofwijck*, lines 127–8.

p. 7   'The Art of Painting': Hoogstraeten 24.

p. 10   'The country is flat': Nooteboom 8.

p. 11   'the greatest': Gribbin 2002 118.

p. 16   'The Lord's benevolence': Huygens ed. Worp, *Gedichten*, vol. 6 36.

I SAND, LIGHT, GLASS

p. 19   according to a French observer: Dijksterhuis 1943 208–10.

p. 19   'celebrated sailing chariot': Sterne 87.

p. 21   'the lack of': Prak 233.

p. 22   *duytsche mathematycke*: van Berkel, van Helden and Palm 34.

p. 22   Stevin's greatest legacy: Dijksterhuis 1943 110.

p. 22   'to stargazers': Stevin, Simon, *De Thiende* (Leiden: Christoffel Plantijn, 1585) 3.

p. 23   an imagined *wijsentijt*: Dijksterhuis 1943 128.

p. 23   The words he gave: Struik 53.

p. 23   Stevin also invented: Dijksterhuis 1943 308–9.

p. 24   *spiegeling*: van Berkel, van Helden and Palm 372.

p. 24   The population of Middelburg: Israel 328.

p. 26   **nearly half the population**: ibid. 219.

p. 26   **one of them was the tulip**: Goldgar 20–6.

p. 27   **Middelburg became the first city**: Buckley 66.

p. 27   **Govert van der Haghen**: Zuidervaart 2017.

p. 27   **'rude and coarse'**: Buckley 66.

p. 27   **in Low Countries glassworks**: Davids 513.

p. 27   **The Murano glass-makers**: McConnell 8–9.

p. 28   **painters and glass-makers were members**: Tummers 259.

p. 28   **'so characteristic'**: Klukhuhn 13.

p. 28   **kiln on the Kousteensedijk**: ibid. 79.

p. 29   **patent for glass-making to van der Haghen**: Zuidervaart 2017.

p. 30   **Lipperhey constructed a tube**: Willach 95; Zuidervaart 2017.

p. 30   **'sights involving glasses'**: van Helden et al. 11.

p. 31   **'From now on'**: ibid. 13–14.

p. 31   **the States General gave Lipperhey**: Zuidervaart 2017; Willach 91.

p. 32   **De Drie Vare Gesichten**: van Helden et al. 14–15.

p. 32   **'concerning certain "lunettes"'**: McConnell 46.

p. 32   **the French ambassador took**: Zuidervaart 2017.

p. 32   **Zacharias Jansen**: ibid.

p. 34   **In 'Cupido brilleman' et seq. quote**: Cats, vol. 2 624.

p. 34   **Jacob Metius of Alkmaar**: van Helden et al. 85.

p. 34   **It is possible that Metius**: Ploeg 22.

p. 34   **The suggestion that Metius**: van Helden et al. 21.

p. 34   **Descartes was nevertheless troubled**: Vrooman 129.

p. 35   **A firmer connection**: Harris 130.

p. 35   **Around 1600 he moved to Middelburg**: Tierie 3.

p. 36   **believing Metius to have been the inventor**: Ploeg 22.

p. 36   **One of his pupils was Isaac Beeckman**: Porter and Teich 127; van Berkel 68–9.

p. 36   **'looked like a Dutch farmer'**: quoted in Smit 87.

p. 36   **He impressed the young diplomat**: Huygens ed. Heesakkers 127.

p. 37   **similar exercises in England, Italy and France**: Meyer 229.

p. 37   **inaccuracies in his earlier measurements**: Struik 49.

p. 38   **translating the work of Simon Stevin**: van Berkel, van Helden and Palm 33.

p. 38   **his eponymous law of refraction**: Dijksterhuis 1961 389–90.

p. 39   **The English mathematician Thomas Harriot**: Dijksterhuis 2004, *Lenses and Waves* 35.

p. 39   **It seems that Descartes discovered:** Dijksterhuis 1961 390; Scott 36–7; Jorink and van Miert 170; OC13 741.

p. 39   **'that the laws of refraction':** OC10 398–9.

p. 39   **by the time that Descartes published:** Dijksterhuis 2004, 'Once Snell Breaks Down'.

p. 41   **In England around this time:** Godfrey 241–3.

p. 41   **A popular joke in Holland:** Dekker 2001 20.

2 THE *KENNER* OF ALL THINGS

p. 43   **'between ten and eleven o'clock':** Huygens ed. Heesakkers 8.

p. 45   **'Religious Peace':** Israel 195.

p. 45   **Willem Janszoon van Hooren:** Smit 14–15; Hooft 778.

p. 45   **the first assassination . . . with a handgun:** Jardine 2005 x.

p. 46   **'attractions of children's games':** Huygens ed. Heesakkers 22; Smit, 26.

p. 47   **a flowering of art:** Israel 550.

p. 47   **'I would have achieved':** Huygens ed. Blom, vol. 2 67–71.

p. 47   **His earliest education at home:** Smit 25.

p. 48   **even trying ivory-carving:** Bergvelt 66.

p. 48   **It is hard to translate:** Goldgar 23.

p. 48   **who were disparaged by real artists:** Tummers 283.

p. 49   **the Flemish painter Gonzales Coques:** ibid. 203.

p. 49   **'kennelyke':** ibid. 105.

p. 49   **'Cast a compassionate eye':** Huygens ed. Worp, *Gedichten*, vol. 4 51.

p. 50   **to learn the art of painting watercolours:** Smit 34.

p. 50   **he took up the guitar:** ibid. 295.

p. 50   **fretted that too much prowess:** Stoffele 17.

p. 50   **He commissioned portraits:** Sluijter.

p. 50   **'embarrassment of riches':** Schama 1987; see also, for example, Sir William Temple's comment in Jan de Vries, 'Luxury in the Dutch Golden Age in Theory and Practice', in Berg and Eger 41–67.

p. 51   **Huygens had not neglected the sciences:** Smit 38.

p. 51   **a fancy for perfumery:** Ploeg 83.

p. 51   **many of his fellow students:** Smit 42.

p. 52   **boils and carbuncles on his feet:** Bachrach 133.

p. 52   **to sample the ripe cherries:** Huygens ed. Blom, vol. 2 64; Smit 63.

p. 52 'from which you' et seq. quote: Huygens ed. Worp, *Briefwisseling*, vol. 1 32–3.

p. 53 'When water, air': Huygens ed. Worp, *Gedichten*, vol. 1 145.

p. 54 By January 1621: Smit 83–7.

p. 55 On the ride back to London: ibid. 101.

p. 55 influence of the aged Caron: Hofman 65.

p. 55 'He is an honest Spy': Huygens ed. Worp, *Gedichten*, vol. 2 1.

p. 55 'an ambassador is an honest man': Wotton's original Latin leaves scope for ambiguity in translations, which exist in several variants.

p. 56 'always possessed a kind of': Huygens ed. Heesakkers 124.

p. 56 'not to be outdone' et seq. quotes: ibid. 126.

p. 57 fared better with Cornelis Drebbel: Tierie 5.

p. 57 He only ever obtained: Harris 131.

p. 57 Drebbel's most astonishing: ibid. 166; Jaeger 75.

p. 58 'One invention to set against': Huygens ed. Heesakkers 129.

p. 58 It was said that the vessel: Harris 166.

p. 58 'that deservedly' et seq. quotes: Boyle 363–5.

p. 59 the life-sustaining gas: Harris 180; Szydlo 207.

p. 59 'pipe with quicksilver': Dekker 2013.22.

p. 59 'an instrument by which': Harris 146.

p. 59 in Alkmaar with Jacob Metius: ibid. 130.

p. 59 process to produce standardized lenses: Tierie 48.

p. 59 introducing two convex lenses: Cook 289.

p. 59 projected images on a wall: Alpers 13.

p. 59 struck by his great knowledge: Tierie 27.

p. 60 'very probable': ibid. 48.

p. 60 'I laughed at your last': Huygens ed. Worp, *Briefwisseling*, vol. 1 89.

p. 60 'lightly constructed instrument': Tierie 52.

p. 61 'It is impossible for me': Huygens ed. Worp, *Briefwisseling*, vol. 1 94.

p. 61 'Nothing will exhort us' et seq. quote: Huygens ed. Heesakkers 132–3.

3 ENCOUNTERS WITH GENIUS

p. 63 'roof of leaves' et seq. quotes: Huygens tr. Davidson and van der Weel 43–63.

p. 64 'may be read just as well': quoted in Smit 97.

p. 64 the Muiden Circle: Mak 78.

p. 64   'to mould manners and emotions': Zumthor 218.

p. 65   his two beautiful daughters: Smit 83–4.

p. 65   They could ride, and had even learned to swim: Worp 1976 xvii.

p. 65   Anna's speciality was diamond-point engraving: Brüderle-Krug 14; van Elk 91.

p. 66   her byname, Tesselschade: Worp 1976 xvi.

p. 67   'Amstel-nymphen': ibid. 16.

p. 67   'O Scheveningen dune': ibid. 26.

p. 67   'Foe-friendly hand': ibid. 60.

p. 67   'his best prose and poetry': Hofman 97.

p. 67   A journey to Zierikzee: Smit 57.

p. 67   He favoured the iambic: Hermans 175–6.

p. 68   Acquaintance with scientific topics: Bots 434n.

p. 68   He heard Donne deliver sermons: Huygens tr. Davidson and van der Weel 202.

p. 68   Huygens picked out two elegies: ibid. 203, 208.

p. 69   'Translations fall as short': Worp 1976 118.

p. 69   'preserved the English fruit': van Tricht, vol. 2 516.

p. 69   'The British Donne': Worp 1976 120.

p. 70   The technical challenges of translation: van Dorsten.

p. 70   His translation of 'The Flea': Daley 178–9.

p. 71   'fabric was so good' et seq. quote: quoted in Smit 161.

p. 71   'How solid': Huygens ed. Worp, *Gedichten*, vol. 2 268.

p. 71   Rembrandt overpainted the largish panel: Schwartz 1985 64.

p. 72   'prince among painters': Huygens ed. Heesakkers 79.

p. 72   'If I say that' et seq. quote: ibid. 84.

p. 72   'more boy than young man' et seq. quotes: ibid. 85–9.

p. 73   'set against all Italy' et seq. quotes: ibid. 86.

p. 74   just what Rembrandt was seeking: Schama 1999 267.

p. 74   'movements of the soul': Tummers 223.

p. 74   knew that his protégés: Schwartz 1985 76–7.

p. 74   Not only did they shun: ibid. 81.

p. 74   'pensive expression' et seq. quote: ibid. 76.

p. 75   Rembrandt, dishevelled and bareheaded: Schwartz 2006 151.

p. 75   to paint a pair of small panels: White 54–5.

p. 75   But royal subjects were perhaps: Schama 1999 291.

p. 76   Rembrandt's only surviving correspondence: White 56; Sass 4.

p. 76   'mutual dissatisfaction': Schwartz 1985 106.

p. 76   'very diligently' and 'skilfully': Huygens ed. Worp, *Briefwisseling*, vol. 2 150.

p. 76   'nine-sixteenths': Schama 1999 435.

p. 77   'but I should be content' et seq. quote: Huygens ed. Worp, *Briefwisseling*, vol. 2 151–2; Schwartz 1985 117.

p. 77   'because these two are' et seq. quote: Huygens ed. Worp, *Briefwisseling*, vol. 2 425.

p. 77   The effect that Rembrandt was drawing: Schama 1999 443.

p. 78   'no less than a thousand guilders': Huygens ed. Worp, *Briefwisseling*, vol. 2 427.

p. 78   However, the artist: Crenshaw 2006 46.

p. 78   'since His Highness in goodwill': Huygens ed. Worp, *Briefwisseling*, vol. 2 433.

p. 78   repeatedly invoked their 'friendship' et seq. quote: ibid. 434.

p. 79   The wealthy Amsterdam official Jan Six: Schwartz 2006 194; Mak 121.

p. 79   'that will do honour': Huygens ed. Worp, *Briefwisseling*, vol. 2 425.

p. 79   'against my lord's wishes' et seq. quotes: ibid. 428–9.

p. 80   Many sources accept this version: for example, Gerson 10.

p. 80   not in the inventory: Broekman 2005 41.

p. 80   admiring a friend's copy: Hofman 208; White 60; Schama 1999 423.

p. 80   Others believe that Rembrandt: for example, Schwartz 1985 130.

p. 80   'Were this at all': Huygens ed. Worp, *Gedichten*, vol. 2 245.

p. 81   An eighth couplet: ibid. 246n2.

p. 81   'On Jacques de Gheyn's': ibid. 245.

p. 82   *Satire on Art Criticism*: Crenshaw 2013.

p. 83   Christiaan's highly competent rendering: Schwartz 1985 64–6.

p. 83   'you can hardly see the difference': OC1 12.

p. 83   In the spring of 1632: Smit 167.

p. 83   'wonderful Gaul': Huygens ed. Worp, *Briefwisseling*, vol. 1, 348.

p. 83   'There are qualities which occasion': Descartes, vol. 1 314–16.

p. 84   Descartes came to the Dutch Republic: Vrooman 43.

p. 84   Descartes fell in with Isaac Beeckman: ibid. 48.

p. 84   One Père François: ibid. 31.

p. 84   Although the attraction of such a move: ibid. 75; Porter and Teich 129.

p. 87   Descartes was distressed: Vrooman 129.

p. 87   He also, characteristically: Scott 34.

p. 87   Their relationship was one of: Gaukroger 7.

p. 87   'if you were to take': Descartes, vol. 1 53–69.

p. 88   Descartes reasoned that another: Ruestow 18.

p. 88   Descartes soon fell out with both: Smit 188; Roth 3–8; Bots 352.

p. 88   Between them, they agreed: Burnett 60.

p. 89   'out of my Hyperbola': Huygens ed. Worp, *Briefwisseling*, vol. 2 139.

p. 89   The two men issued new instructions: Roth 19, 40.

p. 89   with Descartes remaining convinced: Burnett 18–19.

p. 89   Descartes had held back: Vrooman 84.

p. 90   'I devoured your *discours*': Roth 40.

p. 90   He persuaded van Schooten . . . to provide the figures: Dupré and Lüthy 109.

p. 90   Huygens was able to use: Joby 156.

p. 90   the Frenchman may not: Bunge 44.

4 AT HOME

p. 93   Dorothea van Dorp: Jardine 2015 65–83; Keesing 1987 23; van der Vinde 34.

p. 94   'Although the D': Huygens ed. Worp, *Gedichten*, vol. 1 122.

p. 94   Susanna van Baerle visited: Smit 102–110; van der Vinde 39.

p. 94   'Walk into your garden': Huygens ed. Worp, *Gedichten*, vol. 2 136–42.

p. 95   'Jan could read': Huygens tr. Davidson and van der Weel 79.

p. 95   His modest salary: Hofman 219.

p. 96   As a gift, she sent: Frans R. E. Blom and Ad Leerintveld, 'Vrouwen-schoon met Mannelicke reden geluckigh verselt', in *De Zeventiende Eeuw* 2009 97–114.

p. 96   She brought 80,000 guilders: Hofman 221, 211.

p. 96   As he sat for the picture: Smit 149.

p. 97   In their first months: ibid. 142–8.

p. 97   Their first son was born: Eyffinger 89.

p. 98   'so somebody said': ibid. 110.

p. 98   The fourth son, Philips: ibid. 113–14.

p. 98   Susanna did finally: ibid. 115–21.

p. 99   As his own father: ibid. 89–93.

p. 99   'Christiaan our second': HUG 30:1; Eyffinger 103; Worp 1913.

p. 100   The demonstrator: Haeseker 42.

p. 100 'Nevertheless, it appeared': HUG 30:1; Eyffinger 103; Worp 1913.

p. 101 When the stadholder marvelled: Smit 155.

p. 101 Huygens bought the estate of Zuilichem: Hofman 211–12.

p. 101 A fine sketch: Heijbroek.

p. 101 the Zuilichem arms: Frankenhuis-van Scheijen.

p. 101 He acquired further: Hofman 211; Schinkel 15.

p. 101 he sat for Antoon van Dyck: Sass 28; Smit 165.

p. 102 Hatfield Chace: Schinkel 69.

p. 102 seventy-two houses at Zeelhem: ibid. 22.

p. 103 In all, although: Hofman 211–19; Schinkel 64–8.

p. 103 'modern building, but truly' et seq. quotes: quoted in Worp 1894.

p. 104 He needed a larger home: Hofman 219.

p. 104 Frederik Hendrik made available: Smit 179.

p. 104 painter and architect Jacob van Campen: Huisken 43.

p. 105 The two friends worked together: Bos et al. 42; KA 47 f539–42.

p. 105 Dutch translations of Vitruvius: Huisken 66.

p. 105 'mis-builders': *Hofwijck*, lines 612–20.

p. 105 in a letter to Rubens: Huygens ed. Worp, *Briefwisseling*, vol. 2 132.

p. 106 Built of brick: Huisken 163.

p. 106 Huygens having laboriously compiled tables: KA 48 480–9.

p. 106 In practice, it was Susanna: Frans R. E. Blom and Ad Leerintveld, 'Vrouwen-schoon met Mannelicke reden geluckigh verselt', in *De Zeventiende Eeuw* 2009 97–114.

p. 106 Huygens, meanwhile, paid regular visits: Huisken 21.

p. 106 In October 1635, when plague broke out: Smit 185.

p. 106 'lazier than sleeping sickness': quoted in Frans R. E. Blom and Ad Leerintveld, 'Vrouwen-schoon met Mannelicke reden geluckigh verselt', in *De Zeventiende Eeuw* 2009 97–114.

p. 106 Despite taking the economical step: Smit 179–89.

p. 107 'Blessed are the faithful rays': Huygens ed. Worp, *Gedichten*, vol. 2 305.

p. 107 By the end of 1636: Smit 191.

p. 108 'soet mockeltje' et seq. quotes: Frans R. E. Blom and Ad Leerintveld, 'Vrouwen-schoon met Mannelicke reden geluckigh verselt', in *De Zeventiende Eeuw* 2009 97–114.

p. 108 'My beloved has dedicated': Huygens tr. Blom 138–40; Frans R. E. Blom and Ad Leerintveld, 'Vrouwen-schoon met Mannelicke reden geluckigh verselt', in *De Zeventiende Eeuw* 2009 97–114.

p. 109 service for royal entertainments: Stoffele 47–9.

p. 109 **Its value was hardly less:** Hofman 210.

p. 110 **'the single most imposing':** Israel 867.

p. 110 **He bought the land:** Smit 206.

p. 110 **'pretty village':** Huygens ed. Zwaan, *Hofwijck*, lines 139–40.

p. 111 **In other directions:** ibid., lines 2802–4.

p. 111 **Huygens swiftly acquired:** van Strien and van der Leer 80; Hofman 215.

p. 111 **'reason-rich mind' et seq. quote:** *Hofwijck*, lines 616–17.

p. 111 **'crooked corners' et seq. quote:** ibid. 981–90.

p. 111 **'pseudo-aristocrats':** Mak 159.

p. 112 **The Vliet had become:** de Vries 2006 14–30, 339; Parker 623.

p. 112 **In the middle was the 'stomach':** Christiaan Huygens itemizes the fruit grown at Hofwijck in OC4 179–80.

p. 112 **The entire estate:** Huygens tr. Davidson and van der Weel 186.

p. 113 **Trees were planted:** van Strien and van der Leer 81.

p. 113 **'like tapers in church':** Huygens ed. Zwaan, *Hofwijck*, line 341.

p. 114 **'a tame wilderness':** ibid. 145.

p. 114 **Concerned about the cost:** van Strien and van der Leer 82; Huygens tr. Davidson and van der Weel 187.

p. 114 **'the white wall of dunes':** Huygens ed. Zwaan, *Hofwijck*, line 2800.

p. 114 **'The key to my heart':** ibid. line 1387.

p. 114 **'mushroom revealed':** ibid. line 128.

p. 114 **Hofwijck is in effect:** Hermans 286; Hunt 82.

p. 115 **'I ban the whole':** Huygens ed. Zwaan, *Hofwijck*, lines 1499–500.

p. 115 **'as if our yesterday':** ibid. line 130.

p. 115 **'Hofwijck as it is':** ibid. lines 25–7.

p. 115 **'opulent alchemy':** ibid. lines 2707–12.

p. 116 **'. . . the lowest thrust up':** ibid. lines 770–4.

5 ALMOST A PRODIGY

p. 117 **The family took their meals:** Dekker 2000 28–9.

p. 118 **A possible early sketch:** reproduced in van Gelder 1957 plate 17.

p. 119 **'the image of his mother':** OC1 1–3.

p. 119 **a cousin, Catharina Zuerius:** Smit 192.

p. 119 **her latest antics:** see for example Constantijn to Christiaan in OC3 145–6.

p. 119 **'Here lies Auntie Catharine':** Huygens ed. Worp, *Gedichten*, vol. 8 248.

p. 120 **Although the boys:** Andriesse 68; see for example OC1 354–5.

p. 120 **Tien and Tiaen, as they were affectionately abbreviated:** Keesing 1983 7.

p. 120 **They were not sent to school:** Hooijmaijers 7.

p. 120 **Young Constantijn was the quieter:** ibid. 7–8.

p. 120 **'dog-like':** OC1 1–3.

p. 120 **It is possible:** Andriesse 43.

p. 120 **Lodewijk was always:** Dekker 2000 28.

p. 120 **Philips was also:** Huygens ed. Blom 196.

p. 120 **Susanna, on the other hand:** Dekker 2000 30.

p. 121 **'a worn old man':** ibid. 29.

p. 121 *Juvenilia pleraque,* **it reads:** HUG 17.

p. 121 **Their father was able:** Andriesse 52–8.

p. 122 **Teaching both boys together:** ibid. 60.

p. 122 **'he might almost be called' et seq. quote:** OC1 552–3.

p. 122 **He had regarded:** Dekker 2000 73.

p. 122 **Other tutors came:** Andriesse 65.

p. 122 **His reading list:** OC1 5–10.

p. 123 **One challenge nearly went too far:** Bots 378–81.

p. 123 **'first a good intelligence':** OC22 399.

p. 124 **On 11 May 1646:** Andriesse 70; OC22 403–4.

p. 124 **They were encouraged:** Stoffele 69.

p. 124 **By the 1640s:** Israel 572.

p. 124 **In 1633, Leiden became:** van Berkel, van Helden and Palm 53.

p. 124 **Encouraged by their father:** OC22 403.

p. 124 **Frans van Schooten (the younger):** van Berkel, van Helden and Palm 149; Andriesse 72.

p. 125 **Christiaan perhaps did not find:** OC22 412.

p. 125 **Van Schooten had a rare ability:** Dupré and Lüthy 89–111.

p. 126 **'most eager about':** OC2 547–50

p. 126 **'I don't doubt' et seq. quotes:** OC1 47–9, 83–6, 50–6, 64–6.

p. 126 **put paid to plans:** OC22 463.

p. 127 **Indeed, Christiaan, who busied himself:** OC12 241–7.

p. 127 **The catenary seemed to be:** OC1 34–5; Andriesse 74.

p. 128 **attention to another special curve:** OC14 200–7; Yoder 1988 11.

p. 128 **the House of Orange college at Breda:** Andriesse 80–1; Smit 225; OC22 410–17.

p. 129 **'Today we burn the Victory':** OC1 99.

p. 129 **verse in Spanish:** Joby 195.

p. 130 **Although many accounts suggest:** see for example OC16 3 and Dijksterhuis 1961 408, both of which hold that Christiaan 'without doubt' and 'more than once' met Descartes 'at his parents' home'.

p. 130 **What is certain is that:** Westfall 1971 185.

p. 130 **Descartes's philosophy was exciting:** Dijksterhuis 1961 406.

p. 131 **Huygens was struck by the fact:** Fokko Jan Dijksterhuis, 'Huygens's Dioptrica', in *De Zeventiende Eeuw* 1996 117–26.

p. 132 **faithful all his life:** Dijksterhuis 1961 414–16; Hall 1981 121.

6 REVERSALS AND COLLISIONS

p. 133 **'That in Sweden' et seq. quotes:** OC1 127.

p. 133 **'Nature, take to mourning':** OC1 125.

p. 133 **'galliardement':** OC1 115–16.

p. 133 **'powerfully flavoured' et seq. quote:** OC1 127–8.

p. 134 **'the greatest I have seen':** OC1 118–19.

p. 134 **Later that year, on 6 November:** Israel 608–9.

p. 135 **there was such heavy snowfall:** Parker 6.

p. 135 **This was a great hall of paintings:** Smit 234.

p. 135 **The projects he took on:** Andriesse 87–94.

p. 136 **new value, accurate to nine decimal places:** ibid. 119.

p. 136 **'He praised it abundantly':** Huygens, Lodewijck 75.

p. 136 **He first raised his concerns:** OC16 4–5.

p. 137 **the 'quantity of movement' is conserved:** OC16 24.

p. 137 **This new theory of collisions:** Andriesse 112.

p. 137 **Van Schooten, however:** ibid. 113; OC16 9.

p. 138 **The man in the barge:** OC16 29.

p. 138 **Setting the balls:** Dijksterhuis 1961 374.

p. 139 **Huygens believed that the visual image:** OC16 12.

p. 139 **'In the case of two':** OC16 73.

p. 139 **Galileo was the first:** Farmelo xiii, 259n11.

p. 139 **But Huygens appears to have been:** OC16 95–8; OC22 459.

p. 140 **'It requires ingenuity':** Feynman 44–5.

p. 141 **Later in 1652:** Dijksterhuis 2004, *Lenses and Waves* 11–14.

p. 141 **In his workshop in the attic:** OC9 12; OC22 454.

p. 141 **Huygens's mathematical analysis:** OC13 iii–clxvii.

p. 142 **Like so much of Huygens's work:** Dijksterhuis 2004, *Lenses and Waves* 24.

p. 142 **If he had published:** OC13 xxiii.

p. 142 Towards the end of his life: van Berkel 66.

p. 143 Having studied the literature: ibid. 68–9; McConnell 76–80.

p. 143 The flat piece of glass: Beeckman 380–3; McConnell 76–80.

p. 144 'One must grind lightly': Beeckman 380.

p. 144 Even a careless final wipe: ibid. 383.

p. 144 'Up until this 25th of July': ibid. 386.

p. 144 But his persistence had its reward: van Berkel 69.

p. 145 In November 1652: van Helden and van Gent; OC1 191; Dijksterhuis 2004, *Lenses and Waves* 57.

p. 145 Gutschoven responded helpfully: OC17 252–3.

p. 145 Huygens copied many: McConnell 80.

p. 146 '*Always think about*': OC17 293–4 and n6.

p. 146 he pined for these placid times: see for example OC9 308.

p. 146 Which of the brothers: Keesing 1983 39.

p. 147 Parliament passed the Navigation Act: Israel 719.

p. 148 'as but the off-scouring' et seq. quotes: Marvell 112–16.

p. 149 the play that Dutch writers often made: Meijer Drees 131–4.

p. 149 especially when tensions flared: Israel 726.

p. 149 Christiaan thought it advantageous: OC2 371.

p. 150 'to the renowned philosopher Hobbius': Huygens, Lodewijck 74–5.

p. 151 Although the diplomatic effort: Smit 243; Huygens, Lodewijck 3.

p. 151 'one of the most uproarious farces': Mak 69.

p. 151 The play was notable for its use: Hermans 189.

p. 152 'a dilettante in the best sense': Eyffinger 19.

p. 153 'wide-open large and bulging': Huygens tr. Kan 100.

p. 153 'more blind . . . than we' et seq. quote: Huygens ed. Worp, *Gedichten*, vol. 4 83–119.

p. 153 'our eyes are bows' et seq. quote: quoted in Bots 388; see also de Kruyter 16–20 on moral blindness in *Ooghentroost*.

p. 153 'Painters call I blind': Huygens ed. Worp, *Gedichten*, vol. 4 100.

p. 154 Among those contributing heroic scenes: Broekman 2005 69.

p. 154 he must have quailed: Tummers 203; Vlieghe 1987.

p. 155 'my cold ceiling': Huygens ed. Worp, *Gedichten*, vol. 4 48.

p. 155 'Beroemde, maer': Worp 1976 297.

p. 155 'bewitching little bird': Huygens ed. Worp, *Gedichten*, vol. 3 212.

p. 155 But unquestionably the most intellectual: Mirjam de Baar, 'Schurman, Anna Maria van', in *Digitaal Vrouwenlexicon van Nederland*.

Notes

p. 156 **a finely detailed gravure self-portrait:** Katlijne van der Stighelen and Jeanine de Landtsheer, 'Een suer-soete Maeghd voor Constantijn Huygens', in *De Zeventiende Eeuw* 2009 149–202.

p. 156 **'in everlasting copper' et seq. quote:** quoted in ibid.

p. 156 **'the handless maid':** quoted in ibid.

p. 157 **'mannelijk':** Frans R. E. Blom and Ad Leerintveld, 'Vrouwen-schoon met Mannelicke reden geluckigh verselt', in *De Zeventiende Eeuw* 2009 97–114.

p. 157 **'Nobilissima virginum' et seq. quotes:** see for example KA 45 f98r.

p. 157 **it was usual for women:** van Elk 44; Katlijne van der Stighelen and Jeanine de Landtsheer, 'Een suer-soete Maeghd voor Constantijn Huygens', in *De Zeventiende Eeuw* 2009 149–202.

p. 157 **Susanna, had an intelligence:** Dekker 2000 30.

p. 157 **'one could make':** quoted in van der Vinde 28.

p. 157 **Béatrix de Cusance, the well-connected duchess:** van der Vinde 105; Huysman and Rasch 104.

p. 157 **'I love it more than' et seq. quote:** Rasch 2019 36.

p. 158 **Maria Casembroot, on the other hand:** Keesing 1987 136; van der Vinde 65.

p. 158 **Another young woman friend:** ibid. 72.

p. 158 **In 1682, at the age:** Keesing 1987 176.

p. 159 **'the princess never played better':** quoted in Smit 244.

p. 159 **His attitude of openness:** Katlijne van der Stighelen and Jeanine de Landtsheer, 'Een suer-soete Maeghd voor Constantijn Huygens', in *De Zeventiende Eeuw* 2009 149–202; Rasch 2007 2.

p. 159 **'I am fallen':** Huygens ed. Worp, *Briefwisseling*, vol. 5 186–7.

p. 160 **As a young woman, Margaret Lucas:** Whitaker 96.

p. 160 **'was not a woman that invented':** quoted in ibid. 168.

p. 161 **Cavendish found it advantageous:** van Elk 19; Akkerman and Corporaal.

p. 161 **'this book is ten times':** quoted in Weststeijn 14.

p. 161 **Huygens was able to offer:** Joby 161; Akkerman and Corporaal.

p. 161 **His committed interest:** Nadine Akkerman and Marguérite Corporaal, 'Margaret Cavendish, Constantijn Huygens en de Bataafse tranen', in *De Zeventiende Eeuw* 2009 224–39; Weststeijn 15.

p. 162 **'Holland tears':** Brodsley, Frank and Steeds.

p. 162 **'And that which':** anonymous 1663, quoted for example in Weld 79–80.

p. 162 'the natural reason' et seq. quotes: Huygens ed. Worp, *Briefwisseling*, vol. 5 284.

p. 163 'to myne outward': ibid. 284–6.

p. 163 'Madam, I found': ibid. 286–7.

p. 164 'Thus, Sir, you': ibid. 287.

p. 164 speak in terms of probabilities: Larson 157.

p. 165 Hooke made a comparison: Brodsley, Frank and Steeds.

p. 167 'the Chymicall glasses': Pepys, vol. 3 8–9.

p. 167 Pepys had recorded earlier: ibid., vol. 8 196–7.

p. 167 '100 boys and girls': ibid. 208–9.

p. 167 'The Duchesse hath been': ibid. 242–4.

7 SATURN

p. 169 The identification of four moons: OC15 46.

p. 169 with the assistance of magic-lantern projections: Andriesse 128.

p. 169 which he labelled 'étoile b.': OC15 238.

p. 170 was 'errant': OC15 240.

p. 170 The true satellite: OC15 172–7, 38–46.

p. 170 ADMOVERE: OC1 331–3; HUG 45.

p. 172 'It is 9 days': OC1 340–1.

p. 173 'which we saw both very distinctly': OC22 466.

p. 174 'estimated at 50 thousand écus': OC22 481.

p. 174 He also tracked down: Rudolf Rasch, 'Constantijn en Christiaan Huygens' relatie tot de muziek', in *De Zeventiende Eeuw* 1996 52–63.

p. 175 'some new, light little songs': OC1 347–8.

p. 175 'our convex glasses': OC1 345–6.

p. 175 'Making the lenses': OC1 354–5.

p. 175 'I have new tunes aplenty': OC1 351–2.

p. 176 'in the middle': OC1 345–6.

p. 176 'which is completely new': OC1 357–8.

p. 176 'It is true' et seq. quote: OC1 359.

p. 177 'For the rest of this sojourn': OC1 356–7.

p. 177 'She speaks to me very often': OC1 373–4.

p. 177 'led skilfully by the nose': OC1 374–5.

p. 178 'I was afraid at first': OC1 397–9.

p. 178 'the beautiful things': OC2 142–3.

p. 178 Montmor was a senior aristocrat: OC2 318–20.

p. 179 **Prince Rupert's drops were demonstrated:** Nadine Akkerman and Marguérite Corporaal, 'Margaret Cavendish, Constantijn Huygens en de Bataafse tranen', in *De Zeventiende Eeuw* 2009 224–39.

p. 179 **'I have no doubt that your Friend':** OC2 147–8.

p. 179 **made Huygens feel 'most glorious':** OC2 156–62.

p. 180 **showed looping 'handles':** OC1 481–2.

p. 180 **Or was Saturn itself ellipsoidal:** OC15 198.

p. 180 **an equatorial 'torrid Zone':** OC1 474–6.

p. 180 **Such a ring would appear:** OC15 204.

p. 181 **In March 1656 Huygens communicated:** OC1 464–5, 485–6.

p. 181 **Huygens's working sketches:** HUG 45.

p. 182 **Huygens's proposition was met:** OC15 270.

p. 183 **'That the whole body':** OC15 318.

p. 183 **'the effects that the ring':** OC15 340.

p. 184 **'he grinds in red copper dishes' et seq. quote:** OC1 411–12.

p. 184 **'verrekyckatorum Slypatores' et seq. quotes:** OC1 419–21.

p. 185 **Hevelius had observed Saturn:** van Helden 1974.

p. 185 **'We are told that his sickness':** OC2 32–3.

p. 185 **'not to communicate to anyone':** OC22 520.

p. 186 **'when . . . the Hypothesis of Hugenius':** OC3 415–17.

p. 187 **Huygens knew from direct experience:** OC15 272.

p. 187 **'Perhaps the Dutch sky':** Keesing 1983 56.

p. 187 **other important arguments in his favour:** Centre National de la Recherche Scientifique; Yoder 1988 169.

p. 188 **'translate the seen patterns':** Kemp 2000 43.

p. 188 **For many, however:** van Helden 1974.

p. 189 **'no further towards Science':** quoted in Shapin and Schaffer 37.

p. 189 **Huygens published his full analysis:** Yoder 1988 132.

p. 190 **diagrammatic representation of the planet:** OC15 309.

p. 190 **'It necessarily follows that the same ring':** OC15 298.

p. 191 **This information and other knowledge:** OC15 334–40.

p. 191 **'I received it the day before yesterday' et seq. quote:** OC2 494–7.

p. 191 **'I have begun to read':** OC2 506.

p. 191 **'You set out your hypothesis':** OC2 510–11.

p. 191 **fought off a 'furious headache' et seq. quote:** OC2 523–5.

p. 192 **'I have a favour to ask':** ibid.

p. 192 **Countess Albertine Agnes:** Geert H. Janssen, 'Albertine Agnes van Oranje', in *Digitaal Vrouwenlexicon van Nederland*.

8 TIME AND CHANCE

p. 195 Although a prize for a clock: Bell 35–7.

p. 195 Diodati in turn chose a family friend: Bedini 19.

p. 196 Vincenzo died only seven years later: Edwardes 21.

p. 197 These 'wings' (*alae*) or 'cheeks' (*joues*): Andriesse 152.

p. 198 in March 1655 Huygens himself: OC1 318–20.

p. 199 'In recent days I have invented': OC2 4–5.

p. 199 His design was accurate: Whitestone 2008.

p. 199 a sandglass with hourly gradations: Zumthor 52.

p. 199 a ticking clock in the next room: ibid. 150.

p. 199 the *pes horarius* or 'time foot': Edwardes 181.

p. 199 However, most of the clocks: Whitestone 2008.

p. 200 'so-called science of longitude': OC17 56.

p. 201 he sent copies of *Horologium*: OC2 218–19.

p. 201 'I have been among': OC2 309–10.

p. 202 Hortensius, once a student of Snel and Beeckman: Bots 351.

p. 202 But his interest was renewed: OC2 186–7; Andriesse 159.

p. 202 For a moment it even seemed: OC2 276–7.

p. 202 'so finely adjusted two clocks': OC2 270–4.

p. 203 'I don't know how the sow': OC2 277–8.

p. 203 Coster, for example carried on his trade: ibid.

p. 203 'a most thankless business': OC2 218–19.

p. 203 'more wit than he': OC2 286–7.

p. 204 'robbed of . . . the glory': OC2 467–70.

p. 204 involved in 'trade': Marconnel 12; see for example OC2 266–8 for a request for prices.

p. 204 'I beg you to tell me': OC2 308–9.

p. 204 Huygens's responded promptly: OC2 313–14.

p. 204 Meanwhile, he sought other means: Yoder 1988 73.

p. 204 produce perfectly isochronous oscillation: ibid. 50; Joella G. Yoder, '"Following the Footsteps of Geometry": The Mathematical World of Christiaan Huygens', in *De Zeventiende Eeuw* 1996 83–95.

p. 204 'the happiest of all the discoveries': OC2 522.

p. 206 Having found that one cycloid: Yoder 1988 89.

p. 206 He also described the relation: Andriesse 173; OC17 11; OC17 246n2.

p. 206 The church in Scheveningen: Vermaas 84; van der Weel 34.

p. 207 'Mijn Heer Christiaen' et seq. quote: OC2 125.

p. 207 Huygens described the improvements: OC2 156–62; OC17 32.

p. 208 'As for the construction': OC2 165–7.

p. 208 the value of acceleration due to gravity: Yoder 1988 35, 22, 68.

p. 209 the first person to produce: ibid. 169, 15.

p. 209 'I was the original': OC16 302–11.

p. 209 'He who depends upon himself': Horace 219.

p. 210 Games of chance had a particular status: van Egmond and Mostert 70.

p. 211 University students were said to be: Zumthor 101.

p. 211 Yet the element of randomness: van Egmond and Mostert 192.

p. 211 'thoughtful reasoning is useful' et seq. quote: Klinge 76.

p. 212 If you cleaned up here: Vlieghe 2011 20.

p. 212 Playing dice soon passed: Zumthor 76.

p. 212 'By arranging so many visits': OC1 356–7.

p. 213 In his renowned exploration of play: Huizinga 21–2.

p. 214 'foundations of the calculus': OC1 404.

p. 215 Huygens next turns from the problem: OC14 27.

p. 217 Thanks to van Schooten's enthusiasm: OC14 9.

p. 217 the need to understand matters of risk: Goldgar 222–6.

p. 217 On one occasion, in 1669: OC14 13.

p. 218 'A man of 56 years': OC6 526–31.

p. 218 basset was a card game: OC14 16–17.

p. 219 produced by Cornelis Drebbel: Alpers 13; Tierie 5.

p. 219 Huygens's apparatus displays: OC22 522.

p. 219 A rough diagram: HUG 8 47r.

p. 219 In 1646, for example: Snyder 123–4; Wagenaar.

p. 220 'You would not believe how much': OC4 102–4.

p. 220 In keeping with the melodramatic: HUG 10 76v.

p. 221 'lantern of fear': OC4 266–71.

p. 221 Huygens's drawings show: Rossell.

p. 221 Huygens may also have produced: OC22 197.

p. 221 'so that I can see': OC4 197–8.

p. 222 A few weeks later: Leerintveld, Lemmens and van der Ploeg 151–2.

p. 222 'since I cannot think of' et seq. quote: OC4 111–12.

p. 222 'I am prepared to make': OC4 197–8.

p. 222 After his intense period: Yoder 1988 15.

p. 223 'solemn follies': OC3 65.

9 SCIENTIFIC SOCIETY

p. 225 Christiaan Huygens returned to Paris: OC22 526.

p. 225 His brother Constantijn wrote: OC3 145–6.

p. 225 suits of 'mouse-grey cloth' et seq. quote: OC3 169–70.

p. 225 but still his brother wanted: OC3 172–5.

p. 225 Christiaan took up lodgings: OC22 527.

p. 225 'for there is no way': OC3 226–7.

p. 226 'a surprise and extreme joy': OC2 309–10.

p. 226 'I learn with joy': OC3 212.

p. 226 'saw experiments made with mercury' et seq. quote: OC22
534–45.

p. 227 This was never more so: Brown 82.

p. 227 The troublesome Roberval: ibid. 82–5.

p. 227 'it is still very imperfect': OC2 455–6.

p. 227 On 2 November: OC22 533.

p. 228 'Little bottles in water': OC22 546–7

p. 228 'illustres, including state counsels': OC3 209–11.

p. 228 'We laughed a great deal': OC3 178–80. Constantijn uses a phrase
in Greek taken from Aesop's *Fables*: συόδα τῶν ἀλόγων. I have
translated this as 'synod of illogicality', assuming an error has
rendered συνόδα as συόδα.

p. 228 But it was clearly not able: Bos et al. 59.

p. 229 'although everybody did not agree': OC2 173–6.

p. 229 One of the most important: Brown 86.

p. 229 His letters to Huygens: see for example OC3 253–8.

p. 229 a beautiful daughter, Marianne: OC22 547, 559–60.

p. 230 'heretical' in his views: OC3 431.

p. 230 Chapelain asked him to enquire: OC3 259–60.

p. 231 'hanged and quartered': OC3 246–7.

p. 231 'I do not suppose': OC3 252–4.

p. 231 'accompanied by porpoises': OC22 567.

p. 232 'I dined with that great mathematician': Evelyn, vol. 1 342.

p. 232 In fact, Huygens: RS CMO/1/7 f14.

p. 232 'I . . . don't find': OC3 275–7.

p. 233 In 1658 Wren had proposed: Hooijmaijers 18.

p. 233 'does not seem as distinct': OC22 570.

p. 234 'I learned that the glass tears': OC22 569.

p. 234 'which was verified': OC22 572.

p. 234 'a right pedant': OC22 574.

p. 234 'discoursed for a long time': OC3 276.

p. 234 The conversation inspired Huygens: Shapin and Schaffer 235.

p. 234 On 3 May: Brown 114.

p. 234 'All this by report': OC22 575.

p. 235 'He amused himself, I think': OC3 271–2.

p. 235 'I imagine that you will have been': OC3 268–9.

p. 236 Chapelain believed that Montmor's academy: Brown 94.

p. 236 'It seems our Academy': OC3 272–4.

p. 237 'an assembly of the Choisest': OC7 7–13.

p. 237 He reported back to Chapelain: Bos et al. 59.

p. 237 Huygens had hoped to return: OC3 358–9.

p. 237 but he was obliged: OC3 370–1.

p. 237 He busied himself at home: OC22 582.

p. 237 'goes over in his memory': OC3 304.

p. 237 'I told you that Mademoiselle P.': OC3 408–9.

p. 238 'admirable but I admit': OC3 307–8.

p. 238 Heinrich Oldenburg, as he was once known: Hall 2002 11–47.

p. 239 'to produce any great matter': Oldenburg, vol. 1 227.

p. 239 But his most important task: Hall 2002 56 and passim.

p. 239 A typical letter to Huygens: see for example OC6 519–21.

p. 240 Here, his skill often lay: Hall 2002 200.

p. 241 'Monsieur Grubendol': ibid. 105–7.

p. 241 'I wish always an end': OC5 478.

p. 241 'dangerous desseins and practices': quoted in Hall 2002 109.

p. 241 'writing news to a Virtuoso': Pepys, vol. 8 289–93.

p. 241 After two months, Oldenburg was released: Hall 2002 118–19.

p. 241 sudden death from ague: ibid. 295.

p. 242 'to do some yet new experiments': OC22 71–3.

p. 242 'seventeenth-century "Big Science"': Shapin and Schaffer 38.

p. 243 'The pump does not work' et seq. quote: OC3 370–1.

p. 243 'which Mr. Boyle was not able': quoted in Shapin and Schaffer 236–7.

p. 243 'which died in the same way': OC3 395–6.

p. 244 'soe few, as well as soe judicious': OC4 217–20.

p. 244 Huygens remained the only investigator: Shapin and Schaffer 234–5; Stroup 1981.

p. 244 Huygens himself had initially: OC22 587.

p. 244 Descartes and his followers: Shapin and Schaffer 41.

p. 245 The philosopher Hobbes: ibid. 88.

p. 245 However, another of Huygens's experiments: OC22 587.

p. 245 'anomalous suspension' et seq. quote: Shapin and Schaffer 239.

p. 245 'hydrostatical paradoxes': OC3 328–31.

p. 246 The apparent fact of the new: Shapin and Schaffer 77.

p. 246 But there was an official accolade: RS CMO/1/7 f14.

10 NEW MUSIC

p. 247 Music always sounded: Rasch 2007 125; Dekker 2000 29.

p. 247 At the age of two: Smit 23–4.

p. 248 'It happened that in the presence': Huygens ed. Heesakkers 25.

p. 248 'the most perfect music': quoted in Smit 81.

p. 248 Music served many purposes: Bachrach 180.

p. 249 'to fiddle myself out of bad humour': Huygens ed. Worp,
        Briefwisseling, vol. 6 322–3.

p. 249 'sweeten the bitter displeasure': Rasch 2007 134.

p. 249 A likely portrait of her: van der Vinde 63.

p. 249 They first met in 1642: Smit 211.

p. 250 'a certaine tang': Rasch 2007 808.

p. 250 'a season when I imagine': ibid. 741.

p. 250 'Since all our aural communications': Rasch 2007 2; Rudolf Rasch,
        '"Aensienlicxte der Vrouwen, Aenhoorlixte daer toe": Utricia Ogle
        in de ogen (en oren) van Constantijn Huygens', in *De Zeventiende
        Eeuw* 2009 131–47.

p. 250 'Holland's Orpheus': OC1 530–2.

p. 250 The most important of these: Rasch 2007 122–9.

p. 252 'after-dinner diversions': Huygens ed. Worp, *Briefwisseling*, vol. 4
        447.

p. 252 'new' and 'pleasant to sing': Huygens ed. Worp, *Gedichten*, vol. 4
        152.

p. 252 'created music that nobody': Huygens ed. Heesakkers 30.

p. 252 The *Pathodia* thus may have played: van Helden et al. 331n.

p. 252 His interest in originality: Rasch 2007 133.

p. 253 The Calvinist liturgy: ibid. 125.

p. 253 'not unbecoming for persons': Huygens tr. Kan 23.

p. 253 'born, nurtured and trained': Huygens ed. Zwaan, *Gebruyck* 44.

p. 253 'to rejoin Geneva with Rome': Rasch 2007 394.

p. 253 This did not happen: ibid. 131.

p. 254 but found it a 'miserable instrument': ibid. 1127.

p. 254 'at least so, as the saying goes': ibid. 1266.

p. 254 musical upbringing he had received: ibid. 119.

p. 254 Not only was music: Kuhn 36.

p. 255 Prompted by Beeckman: Vrooman 49.

p. 255 'who sing very well': OC7 112–13.

p. 255 the practical production of music: OC22 665.

p. 255 All that survives of any music: van der Craats.

p. 255 odd snatches of notated melody: see for example HUG 3 93r.

p. 256 one of Christiaan Huygens's first forays: OC20 8–11.

p. 256 The discovery during the seventeenth century: van der Craats.

p. 257 Stevin, for example, favoured: Struik 140n.

p. 257 'sound-houses' et seq. quote: Bacon 485.

p. 257 the spectacular new carillon: OC22 582.

p. 257 After much calculation: OC20 43–58.

p. 257 'I have been busy for a few days': OC3 12.

p. 258 Huygens's hope was not to discover: Cohen 1984 210.

p. 258 Instrument-makers had to introduce: ibid. 215.

p. 258 Huygens did not favour: ibid. 224.

p. 259 his notes are strewn: see for example HUG 27 4r–v.

p. 260 Huygens was aware from his French mentor: Cohen 1984 219–20; Rudolf Rasch, 'Constantijn en Christiaan Huygens' relatie tot de muziek', in *De Zeventiende Eeuw* 1996 52–63.

p. 260 'without being confused by': OC10 171.

p. 260 But in 1669 Huygens returned: van der Craats; Cohen 1984 222.

p. 260 During the eighteenth century: Cohen 1984 217.

p. 261 'that my harpsichord invention': OC6 472–3.

p. 261 'I have at other times': OC20 161.

p. 261 'harmonious upon attentive examination' OC20 161–2.

p. 262 'neither can we presume': Huygens, *Celestial* 88.

p. 262 Constantijn, meanwhile was more concerned: Rudolf Rasch, 'Constantijn en Christiaan Huygens' relatie tot de muziek', in *De Zeventiende Eeuw* 1996 52–63.

p. 262 But Christiaan was not obsessed: ibid.; OC22 407; Cohen 1984 217.

p. 262 'my couplet which gives': Rasch 2007 1239.

p. 262 'I would be led ear-wise': ibid. 1240.

p. 263 a Dutch physicist, Adriaan Fokker: Fokker.

11 THE PARISIAN

p. 266 He went to Montmor's: OC4 474–5.

p. 266 By providing hands-on instruction: Shapin and Schaffer 228.

p. 266 Spurred by what they heard: Brown 119; Taton 24.

p. 266 However, their independent programme: Porter and Teich 63.

p. 266 When fire devastated: Débarbat 146.

p. 266 'In Paris there is nothing new': OC5 69–70.

p. 266 It was the dream of Jean-Baptiste Colbert: George 388.

p. 267 Humourless and irascible: Boll 1–4.

p. 267 His concept of scientific management: ibid. 67–76.

p. 267 Colbert eventually set up: Boll 100.

p. 267 even if they came from abroad: Murat 116.

p. 268 Christiaan was put on an annual pension: Brugmans 57–61.

p. 268 His vision was of an organization: d'Aubert 197–8; Brown 61.

p. 268 Colbert saw Huygens's Dutchness: Centre National de la Recherche Scientifique 43.

p. 268 'I have never done anything': OC4 416–17.

p. 269 'mondanités, jolies femmes': Débarbat 147.

p. 269 Indeed, the most auspicious: Hahn 10; Centre National de la Recherche Scientifique 45.

p. 269 the exposing of '*tromperies*': Bos et al. 60.

p. 269 His new brother-in-law: OC4 464–6.

p. 269 Susanna placed a request: OC5 31–4.

p. 269 He was forced to make: OC4 477–8; OC5 9–11.

p. 269 'a new invention and very convenient': OC5 17–19.

p. 270 'if you haven't entirely forgotten': OC5 24–6.

p. 270 'I beg you to send me': OC5 52–4.

p. 270 Taking delivery – after an argument: OC5 59–60.

p. 271 Sir Robert Moray saw: Bos et al. 70.

p. 271 'the other was so strongly jostled': OC4 295–8.

p. 271 'being suspended and shaken': OC17 160.

p. 272 'It is certain that the clock': OC4 443–5.

p. 272 In November the two clocks: OC18 114.

p. 272 'I hope that for the new voyage': OC4 474–5.

p. 272 'Swaggering, roystering and corrupt': quoted in Ollard 81.

p. 272 'a cunning fellow': Pepys, vol. 2 169.

p. 273 'the cursed beginner': quoted in Ollard 15.

p. 273 Although Holmes did have some interest: ibid. 77.

p. 273　'as if', Huygens complained: OC5 6–7.

p. 273　'However, since you still remain': OC5 8–9.

p. 274　De Witt acknowledged: OC5 23–4.

p. 274　'At least my friends in Holland': OC5 39–43.

p. 274　Moray meanwhile moved skilfully: OC5 92–6.

p. 274　Huygens agreed to Moray's suggestion: OC5 98–102.

p. 274　As it was not possible: OC5 115–17, 126.

p. 275　Huygens had no compunction: OC5 152–4.

p. 275　'It is why I continue': OC5 130–1.

p. 275　At the beginning of December: OC5 156–9.

p. 275　It was another six weeks: OC5 204–6.

p. 275　the *Jersey* and its companion: Yoder 1988 153; OC18 116.

p. 276　'For observing my pendulas': quoted in Ollard 123.

p. 276　'And because they had complete confidence' et seq. quote: OC5 204–5.

p. 276　Was it not a little too good: OC5 164–6.

p. 276　'whether the Captain seems': OC5 224–5.

p. 277　There was every reason to doubt: Ollard 85–97.

p. 277　The war was an immediate impediment: OC5 185–9, 212–15.

p. 277　It was here that William Brouncker: OC5 233–8.

p. 278　After further interviews with Holmes: OC5 268–70, 281–3.

p. 278　Huygens moved swiftly: OC5 263–4.

p. 278　Chapelain, who was negotiating: OC5 267–8.

p. 278　'I will write to thank': OC5 265.

p. 278　'But I will say only': OC5 280.

p. 279　'cannot deny the utility': OC5 276–7.

p. 279　A Huygens pendulum clock: OC6 216–18.

p. 280　'The shaking of the vessel': OC6 501–3.

p. 280　A trial of a seconds-pendulum clock: OC18 636–42; Eric Schliesser and George E. Smith, 'Huygens's 1688 Report to the Directors of the Dutch East India Company', in *De Zeventiende Eeuw* 1996 198–214.

p. 281　Smallpox had swept through: OC5 528–33.

p. 281　'plenty of reasons': OC6 373–5.

p. 281　Huygens was to receive a stipend: Bos et al. 62; Taton 35.

p. 281　He pushed for further information: OC6 16–17.

p. 281　on 8 April Huygens wrote: OC6 21.

p. 282　In the Palace of Versailles hangs a canvas: Verduin; Icke 2009 20.

p. 282　Louis XIV paid only one: Boll 94.

p. 282 'gentlemen, it is not necessary': quoted in Murat 388.

p. 283 'said some greatly pleasing things' et seq. quote: OC6 40–2.

p. 283 Huygens was soon comfortably installed: Boll 94; Andriesse 214.

p. 283 news from home of the plague: OC6 80–2.

p. 283 he attended the anatomical dissection: OC6 103–4.

p. 284 Huygens had always been: Bos et al. 60; Stoffele 114.

p. 284 Perhaps there were Frenchmen: Brugmans 65.

p. 284 Huygens's arrival in Paris: Taton 39.

p. 284 For Huygens personally: Bos et al. 62.

p. 284 Colbert's belief was that all facts: Boll 100; Meyer 230.

p. 285 'jealous of that new glory': Voltaire 1966 ch. 31 39–40.

p. 285 closer in conception: to Bacon's model Boll 98.

p. 285 Colbert envisioned an institution: d'Aubert 198.

p. 285 To eliminate wastage: Hahn 52, 15.

p. 285 He attracted foreign scholars: Boll 127.

p. 285 no further foreign academicians: Marconnel 309.

p. 285 French was the chief: Porter and Teich 66.

p. 286 Since the early membership: Marconnel 314–35.

p. 286 'I wish to be persuaded': OC6 33–4.

p. 287 gathered for its first proper assembly: Taton 5.

p. 287 The members who had met: ibid. 38.

p. 287 natural driver of the agenda: d'Aubert 198.

p. 287 He was in effect: Taton 35; Centre National de la Recherche Scientifique 58.

p. 287 Huygens continued to participate: ibid. 60–4.

p. 288 'to work on natural history' et seq. quotes: OC6 95–6.

p. 288 '1. To identify': OC19 255–7.

p. 291 Early on the morning of 2 July: OC6 56–8.

p. 291 He admired Cassini's measurement: OC6 46–8.

p. 291 'something exceptional': OC6 147–9.

p. 292 he turned for help: OC6 155, 158–9.

p. 292 'I rather admire': OC6 151–2.

p. 292 'composite lens emulating a hyperbolic lens': quoted in Dijksterhuis 2004, *Lenses and Waves* 80.

p. 293 'So then marriage there be': OC6 236–7.

p. 293 'that one cannot see': OC6 205–6.

p. 293 'without doubt the reason can be given': OC6 213–16.

p. 293 'beginning of a treatise': OC6 334–5.

p. 293 The work was read: OC6 356–8.

p. 293 'lest you be pre-empted': OC6 387–9.

p. 294 'I wish I could apply myself': OC6 390–1.

p. 294 The answers were many: Débarbat 153; Centre National de la Recherche Scientifique 46; Leerintveld, Lemmens and van der Ploeg 110–27; Marconnel 276–91.

p. 294 'an agreed universal measure' et seq. quotes: OC5 185–9.

p. 294 And there were yet more waterworks: OC6 211–12, 506–12.

p. 295 the academy failed to live up to: Bos et al. 64.

p. 295 Only sixty-two academicians: Porter and Teich 71.

p. 295 The fly in the ointment was Colbert: George 400.

p. 295 the 'toute liberté' that Huygens once: OC5 373–5.

p. 295 sense of purpose that seemed to spring: OC22 637.

p. 295 The most serious of these: Andriesse 248–51.

p. 295 In 1668 Huygens conducted: OC22 641–2.

p. 296 refers to a 'capitis dolor': OC1 184–5.

p. 296 On this occasion, he was so drained: OC7 35.

p. 296 'Hypochondriac melancholia pure and simple': OC7 22.

p. 297 'yet there was something worse' et seq. quotes: OC7 7–13.

p. 297 The doctor had recommended: OC7 25–6.

p. 297 to express religious doubts: OC7 27–9.

p. 297 Huygens recuperated only slowly: Andriesse 256.

p. 298 France and the Dutch Republic: Israel 784.

p. 298 'where I am employed': OC6 543–4.

12 SCIENCE DURING WARTIME

p. 299 Netscher portrait: Wieseman 99–100.

p. 299 Netscher painted the astronomer: ibid. 102.

p. 300 In the late spring: Dijksterhuis 2018; OC7 58–78.

p. 301 He returned to find: OC22 665.

p. 301 'a certain Doctor Leibnitzius': OC7 46–7.

p. 301 Huygens busied himself: OC7 80–1.

p. 302 'most handsome and magnificent': OC7 78–9.

p. 302 Cassini was appointed as the first director: Murat 194.

p. 302 Cassini was always: Albert van Helden, 'Contrasting Careers in Astronomy: Huygens and Cassini', in *De Zeventiende Eeuw* 1996 96–105.

p. 302 'He . . . does not miss': OC7 348–9.

p. 302 'I observed yesterday in the evening': OC7 114–15.

p. 303 He was doubtful that Cassini: OC7 120–2.

p. 303 In 1669, Huygens had written: Fokko Jan Dijksterhuis, 'Huygens'
Dioptrica', in *De Zeventiende Eeuw* 1996 117–26; Dijksterhuis 2004,
Lenses and Waves 84.

p. 304 Oldenburg sent Huygens a brief description: OC7 124–6.

p. 304 When the Royal Society's: OC7 140–3.

p. 304 'beautiful & ingenious': OC7 134–6.

p. 304 Liaising with Constantijn in The Hague: OC7 151–2.

p. 305 Encouraged by Oldenburg: OC7 140–3; Bos et al. 75.

p. 305 'who hath done so much': OC7 207–8.

p. 305 'ingenious' and 'very likely': OC7 165–7, 185–7.

p. 305 This time Newton responded: Hall 1992 124.

p. 305 Newton was doubtless exasperated: ibid. 126.

p. 306 'the most saturated colours': OC7 242–4.

p. 306 'I can assure you that' et seq. quotes: OC7 265–7.

p. 306 After more than a year: Hall 2002 170.

p. 306 'seeing that he holds': OC7 302–3.

p. 306 However, as a Cartesian: Bos et al. 74. For example, Huygens
wrote to Leibniz about the cause of colours on 29 May 1694.

p. 307 Financial collapse: Israel 796–9.

p. 307 'everything that is happening' et seq. quote: OC7 181–4.

p. 307 'nothing which could do me ill': OC7 216.

p. 307 Even at this time: OC7 252–4.

p. 307 Huygens feared his homeland: OC7 199–200.

p. 307 The Orangists: Israel 793–4.

p. 307 But Christiaan could at least: OC7 137–40.

p. 307 'If the State is saved': OC7 193–5.

p. 308 Their bodies were dragged: Israel 803.

p. 308 'The story of Monsieur the Pensionary' et seq. quote: OC7
218–20.

p. 308 De Witt's republican regime: Israel 893.

p. 309 'Good figures.': OC6 304.

p. 309 It was the father, too: Snyder 173.

p. 309 Largely self-taught: Cook 290.

p. 309 To construct a microscope: Snyder 108.

p. 310 The best lens that survives: Ruestow 14.

p. 310 Scepticism about bead lenses: ibid. 10–14.

p. 310 'The motion of most of these': van Leeuwenhoek, vol. 1 138–66.

p. 310 'before six beats': ibid., vol. 2 281–93.

p. 311 added 'van' to his name: Ruestow 160.

p. 311 'some small vanity': OC8 295–6.

p. 311 Leeuwenhoek rejected the concept: Ruestow 223n.

p. 311 'the seed of plants': OC9 354–6.

p. 311 A little later, Johannes Swammerdam: Ruestow 27.

p. 311 His work included an attempt: van Berkel, van Helden and Palm 63.

p. 311 Patterned shells and minerals: Goldgar 86.

p. 312 Dutch paintings seemed to require: Alpers 25.

p. 312 They had revealed: Bunge 61.

p. 312 Christiaan first became aware: OC7 313–16; Fournier.

p. 312 'seemed to convert everything' et seq. quote: OC7 399–400.

p. 313 Even using these aids: Ruestow 153–4.

p. 313 'little animals which appeared' et seq. quotes: OC8 22–7.

p. 313 'trifling observations': OC8 21.

p. 313 Christiaan Huygens probably learned: Fournier; Ruestow 24–6.

p. 314 Christiaan may have improved: Andriesse 306.

p. 314 Constantijn even had one: Keesing 1983 100.

p. 314 Huygens reminded his French colleagues: OC8 96–7.

p. 314 'none of the honour': OC8 112.

p. 314 'I doubt not that the French': OC8 224–5.

p. 315 'Cursed and banned from': This infamous decree survives in many versions translated first into Portuguese and then Dutch and other languages from its original Hebrew.

p. 316 '[A] good-looking young man' et seq. quote: quoted in Nadler 155.

p. 316 'the noblest and most lovable': Russell 552.

p. 316 'abominable heresies': quoted in Nadler 120.

p. 317 The Huygens brothers regarded Spinoza: ibid. 221.

p. 317 This unlikely companionship: Nadler 221; Klukhuhn 59.

p. 318 'the Jew of Voorburg' and 'the Israelite': OC6 155, 168.

p. 318 'Grinding and polishing lenses': Nadler 184.

p. 318 'a spherical surface' OC5 535–9.

p. 319 'Thus expert action manifests': Baltas 55.

p. 320 'We have become the thing': Sennett 174.

p. 320 'invariably accompanied by a feeling' et seq. quotes: Baltas 56.

p. 320 In a series of letters: OC6 2–3, 24–6, 36–9. These letters, included in the *Oeuvres Complètes* under the presumption that they may have been written to Huygens, are now believed to have been addressed to Johannes Hudde.

p. 321 **And, while Huygens was always eager:** see for example OC6 158–9.

p. 321 **actually desperate:** see for example Spinoza to Oldenburg, OC5 535–9.

13 FEUDS AND TRIALS

p. 323 **'bold discoverer' et seq. quote:** OC22 668.

p. 323 **'Start it going':** Huygens ed. Worp, *Gedichten*, vol. 8 39–40.

p. 323 **'one of the masterpieces':** Yoder 1988 5.

p. 323 **second only to Newton's:** The remark is from the historian of mathematics Florian Cajori, quoted in Merton 171.

p. 324 **'the recognised leader':** Andriesse xiii.

p. 324 **'complete for a long time':** OC3 118–20.

p. 324 **But while Chapelain:** OC3 114–15.

p. 324 **Above the proof Huygens wrote:** Andriesse xii.

p. 325 **'Centrifugal Force':** OC18 360–5.

p. 325 **'universal perpetual measure' et seq. quote:** OC18 348–9.

p. 325 **Chapelain checked over the text:** OC18 34.

p. 325 **'the private areas' et seq. quote:** OC18 77, 81.

p. 325 **He distributed copies:** OC18 321.

p. 325 **'full of very subtile':** OC7 325–32; OC22 682.

p. 325 **Newton sought to draw:** Andriesse 268.

p. 326 **John Wallis felt:** Bos et al. 76.

p. 326 **'I find, since':** quoted in Mordechai Feingold, 'Huygens and the Royal Society', in *De Zeventiende Eeuw* 1996 22–36.

p. 326 **On 12 June 1673:** OC7 303–5.

p. 326 **Huygens wrote back:** OC7 314.

p. 326 **Pressed further, Huygens admitted:** Hall 2002 193–4.

p. 326 **Hooke and Huygens were of very:** Inwood 6–7; Biagioli.

p. 327 **'very fine and of great consequence' et seq. quote:** OC7 382–3.

p. 328 **The peace between Huygens and Hooke:** Jardine 2003 199.

p. 328 **At the same time, he communicated:** OC7 399–400.

p. 328 **He rated the new mechanism:** OC22 693.

p. 328 **'If your boy is beautiful':** OC7 430–1.

p. 329 **'Hooke has also requested a privilege':** OC7 454–6.

p. 329 **That summer, Hooke raced:** Inwood 201.

p. 329 **'R. Hooke invenit':** This inscription may be viewed at the Science Museum, London.

p. 329 Huygens, meanwhile, set about: OC7 425–6.

p. 329 He entrusted the safe carriage: OC7 467–8.

p. 330 'M. Hooke's proceedings': ibid.

p. 330 'As for M. Hooke': OC7 469–70.

p. 330 This Huygens duly did: OC7 529–30.

p. 330 Brouncker was already won over: OC7 533–4.

p. 330 He launched a furious attack: OC7 516–17.

p. 330 He also compiled his own: OC7 518–26.

p. 331 'I claim nothing of the glory': OC7 498.

p. 331 'I could not imagine why': OC7 405–6.

p. 332 Colbert awarded him the privilege: OC7 419–20.

p. 332 In England, the unauthorized dissemination: Hall 2002 197.

p. 332 Hooke was not awarded: Jardine 2003 202.

p. 332 Interest in the balance-spring: Jardine 1999 163.

p. 332 Both Hooke and Huygens emerged: ibid. 175.

p. 332 If it is impossible: Jardine 2008 268.

p. 332 Hooke continued to blame: Hall 2002 183.

p. 333 'hath a great Zeale': Oldenburg, vol. 6 435–6.

p. 333 On 24 November 1680: Centre National de la Recherche Scientifique 149–50.

p. 333 'Echo of the water' et seq. quotes: OC19 374.

p. 334 Nevertheless, he reasoned correctly: OC10 507–1.

p. 335 The Danish geometer Rasmus Bartholin: Dijksterhuis 1961 460.

p. 335 'white and transparent': OC8 18–19.

p. 336 Rømer's discovery that light travels: OC8 30–1.

p. 336 'light extends circularly': OC13 742.

p. 336 Having established the angles: Dijksterhuis 2004, 'Once Snell Breaks Down'; Dijksterhuis 2004, *Lenses and Waves* 176, 159.

p. 336 'EUREKA. Cause marvellous refraction': OC19 427.

p. 337 Besides, he was about to: Dijksterhuis 2004, *Lenses and Waves* 204–7.

p. 337 'My son, the Parisian' et seq. quotes: OC8 71–3.

p. 338 The country lacked a Paris: van Berkel, van Helden and Palm 44.

p. 338 'being reduced by the floodings': OC8 71–3.

p. 338 His difficult son: ibid.; Israel 827.

p. 338 'some little stews': OC8 86–8.

p. 338 'any exceptional Alteration': OC8 93–5.

p. 339 Back in Paris: Dijksterhuis 2004, *Lenses and Waves* 207.

p. 339 'Experimentum Crucis': OC10 409–15.

p. 339 When Huygens came to describe: Andriesse 292.

p. 339 'like cannonballs or arrows' et seq. quote: OC19 461–3.

p. 339 the Huygens principle: Dijksterhuis 2004, *Lenses and Waves* 174.

p. 340 'particular waves': ibid. 166.

p. 341 In the end, it was his: OC19 396.

p. 342 The development of his theory: Dijksterhuis 2004, *Lenses and Waves* 213.

p. 342 Huygens was not to publish: Centre National de la Recherche Scientifique 14, 159.

p. 342 'Wat lightgh": Huygens ed. Worp, *Gedichten*, vol. 8 255.

p. 343 Comets were omens: Schama 1987 147–8.

p. 343 'That each knows so well' et seq. quotes: Huygens ed. Worp, *Gedichten*, vol. 8 259–62.

p. 344 Pierre Bayle, for example: Colie.

p. 344 Constantijn Huygens, too, tended to: Jorink 109.

p. 344 Huygens saw a corona appear : Bots 409, 428.

p. 345 'For some time': OC8 312–13.

p. 345 'What does it mean': OC22 94–5.

p. 345 He had believed: OC22 714.

p. 345 'smoke or vapour': OC19 283–310.

p. 346 opposition to mechanistic interpretations: Porter and Teich 76.

p. 346 'sculptor has not done': OC8 211–13.

p. 346 'new discoveries among the stars' et seq. quotes: OC8 196–9.

p. 347 'I hope, my Brother': OC8 278–9.

p. 347 But Colbert still had projects: d'Aubert 198; Centre National de la Recherche Scientifique 61.

p. 348 a mechanical brass planetarium: OC21 111–32, 184.

p. 348 'our bourgeois philosopher': OC8 158–60.

p. 348 'the great man of the age': OC8 295–6.

p. 348 One goal was to create: Zuidervaart 2011 44.

p. 349 'a small boy sent to me': OC9 51–2.

p. 350 Huygens felt himself to be competing: OC21 191–9; Andriesse 323.

p. 350 'simpler', but 'less perfect' et seq. quote: OC9 94.

p. 350 In private, though, Huygens: OC9 99–100.

p. 351 'usually the best is that': OC21 263.

p. 351 'To polish by hand': OC21 267–87.

p. 352 'most precisely according to': OC8 376–8.

p. 352 But no encouraging word: OC22 726.

p. 353 'I do not know what change' et seq. quote: OC8 456–8.

p. 353 Constantijn interceded repeatedly: OC8 457n, 483–4, 550–1.

p. 353 'the good opinion France has had': OC8 552–3.

p. 353 'to Monsr. Huijgens' et seq. quotes: OC8 457n.

p. 353 To make the rebuff quite plain: OC22 727.

p. 353 'intrigues of certain people': OC8 553–4.

p. 354 'It was as if': OC9 4–5.

p. 354 'Louis has great need': OC9 112–13.

p. 354 Huygens was not alone: Hahn 20.

p. 354 After the deaths of useful intermediaries: OC9 117–20.

p. 355 His father proposed: OC22 727.

p. 355 Instead, Constantijn saw to it: Andriesse 337.

p. 355 Huygens often felt the project: OC9 20–2.

p. 355 He was no more enthusiastic: OC9 30–1.

p. 355 'piety, despair or foolishness': OC4 271–3.

p. 356 'a plan for the cloister': OC22 548n191.

p. 356 There were other favourites: Keesing 1983 63; OC22 600.

p. 356 a romance with Haasje Hooft: OC7 126–7.

p. 356 During his long, final residency: Proust and Proust 245–6; OC22 387n; Leerintveld, Lemmens and van der Ploeg 183.

p. 356 'One more September': Huygens ed. van Strien 48.

p. 356 'Cease murderous years': Huygens ed. Worp, *Gedichten*, vol. 8 178.

p. 356 He left behind him: Rasch 2007 213; Joby 84.

p. 357 'He died in 1687': OC9 456–7.

p. 357 'the matters that arise': OC9 130–3.

p. 357 'done near the end': OC22 921, 754.

p. 357 The house on the Plein: OC9 243–5

p. 358 The father's other estates: ibid.

p. 358 The house at Hofwijck: Andriesse 348.

p. 358 'I imagine that by doing': OC9 252.

p. 358 'I have not gone round': OC9 295–6.

14 LIGHT AND GRAVITY

p. 359 Prince William's fleet: OC9 303; Huygens ed. Dekker 49–53.

p. 360 Fortunately, the 50,000 printed copies: Dekker 2013 103.

p. 360 In the frantic days: Huygens ed. Dekker 53.

p. 360 'a great and glorious enterprise': Huygens, *Journaal*, vol. 1 1.

p. 360 The armada – four times the size: Israel 851.

p. 360 'Protestant wind' et seq. quotes: Huygens ed. Dekker 56–7.

p. 361 The Dutch army, 21,000 strong: Israel 849.

p. 361 'the high mountains and deep valleys' et seq. quote: Huygens ed. Dekker 57–8.

p. 362 'polite' but 'extremely frightened' et seq. quote: ibid.

p. 362 'to warm myself': ibid. 65.

p. 362 'many good paintings by Titian' et seq. quote: ibid. 68.

p. 363 'Now your arrival in London' et seq. quotes: OC9 304–5.

p. 363 'When shall we work together': OC9 308.

p. 363 'I would wish to be in Oxford': OC9 304–5.

p. 364 'many bonfires were lit': Huygens ed. Dekker 69.

p. 364 He visited the shop: Huygens ed. Dekker 75–8.

p. 364 The Dutch were horrified: Dekker 2013 103.

p. 364 Huygens was officially: ibid. 53, 47.

p. 364 In February 1689 he was approached: ibid. 55.

p. 365 'The Lords and Commons': Huygens ed. Dekker 79.

p. 365 All that spring Huygens suffered: ibid. 84.

p. 365 He cannot have been: ibid. 79.

p. 365 'It is the air here . . .': ibid. 85.

p. 365 'I am losing the greatest friend': Japikse, part 2, vol. 3 89.

p. 366 another page, Arnold van Keppel: Noordam 114.

p. 366 William's sexuality had never been: Schama 1987 601.

p. 366 Nor was it a great scandal: Noordam 121.

p. 366 'I am always till my last sob yours': Japikse, part 1, vol. 1 38.

p. 366 Besides, other factors: Noordam 115.

p. 367 'The King had encouraged him': Huygens ed. Dekker 89.

p. 367 'very Melancholy and Discontented': anonymous 1689.

p. 367 'It is only England that will profit': OC9 308–9.

p. 367 'so little fondness for studies': OC9 310–12.

p. 368 'not a single soul' et seq. quote: OC9 312–13.

p. 368 He travelled to Greenwich: Huygens ed. Dekker 95.

p. 368 Newton had sent a personal copy: Hall 1992 216.

p. 368 Furthermore, the ground had been: Bos et al. 78.

p. 369 'Vortices destroyed by Newton': OC21 437.

p. 369 Huygens felt unable to embrace: Martins.

p. 370 Despite the two men's differing philosophies: Hall 1992 232.

p. 370 Though he had long been: ibid.; Bos et al. 78; Huygens ed. Dekker 96.

p. 371 '24 Wednesday.': OC9 333–4.

p. 371 'only too certain': OC9 370–1.

p. 372 'infinite inclusion of animals and plants': OC9 361–2.

p. 372 Constantijn visited lens-grinders: Huygens ed. Dekker 106–8.

p. 372 On one occasion, he was amused: OC9 379–81.

p. 372 Christiaan sometimes thought it would: OC9 416.

p. 372 'a great cooling': OC9 353–4.

p. 372 'disconcerted to see': Huygens, *Journaal*, vol. 1 286.

p. 372 'which had very bad paintings' et seq. quote: Huygens ed. Dekker 121.

p. 372 'The King was hit by a cannonball': Huygens, *Journaal*, vol. 1 295.

p. 373 'Nobody spoke with any assurance' et seq. quote: ibid. 296–7.

p. 373 'our English soldiers committed' et seq. quote: ibid. 298.

p. 373 'The house are fairly good': ibid. 301.

p. 373 William's Irish campaign: ibid. 317, 321.

p. 373 Christiaan Huygens's many-times postponed: Dijksterhuis 2004, *Lenses and Waves* 222, 3.

p. 374 'One may ask why' et seq. quotes: OC19 453–5.

p. 375 'light expands successively': OC19 473.

p. 375 'I do not believe we know': OC7 298–301.

p. 375 Huygens also abandoned: OC20 479; Martins.

p. 375 'was found having tried to smuggle': OC9 370–1.

p. 375 'You will see that I have': OC9 357–9.

p. 375 Leibniz, Pierre Bayle and Johannes Hudde: OC9 366–7.

p. 376 Another recipient was the French Protestant: OC9 428–33.

p. 376 'perfectly beautiful and worthy': OC10 605–8.

p. 377 Huygens was opposed to all: Barth.

p. 377 'I have said nothing': OC9 470–3.

p. 377 Locke had made contact: Andriesse 353.

p. 377 'with much pleasure': OC9 391–3.

p. 377 a lineage of theories: Duhem, vol. 10 178.

p. 378 Leonardo da Vinci also proposed: Dijksterhuis 1961 459.

p. 378 early experiments on light diffraction: McConnell 50; Ronchi 447.

p. 378 Young even experienced abuse: Gribbin 2002 406.

p. 378 Confirmation that Huygens's theory: Kuhn 216.

p. 378 Huygens himself did not even live: Dijksterhuis 2004, *Lenses and Waves* 224; Hall 1992 351.

p. 380 Mathematicians such as Simon Stevin: Kemp 2000 29.

p. 380 an 'ethical imperative': ibid. 31.

p. 380 **Significantly, it was the elder:** Huygens ed. Worp, *Briefwisseling*, vol. 2 118.

p. 381 **'a figure will assist':** OC1 47–6.

p. 381 **even deleting the word 'space':** Westfall 1971 150.

p. 381 **light falling on a 'toothed surface':** OC19 407–15.

p. 381 **Huygens's visual mentality:** Yoder 2013 14; Yoder 1988 145.

p. 382 **geometry would often remain:** Yoder 1988 63.

p. 382 **This most mathematically expressible:** scientific visualization is extensively explored in Miller.

p. 382 **It is no surprise to learn:** Andriesse xv; C. D. Andriesse, 'The Melancholic Genius', in *De Zeventiende Eeuw* 1996 3–13.

p. 383 **Italian astronomers muttered:** Keesing 1983 56.

p. 383 **'Een ligt schijnende in duystere plaatsen':** Berkvens-Stevelinck, Israel and Posthumus Meyjes 143–6.

p. 383 **An antinomian tract:** Knuttel 71.

p. 384 **The English portraitist Joshua Reynolds:** Reynolds.

p. 384 **'This contact with reflections':** Venturi 24.

p. 385 **Monet wrote to Pissarro:** van Tilborgh 36.

p. 385 **'The sky above and the water below':** ibid. 27–8.

p. 385 **'a country at anchor' et seq. quote:** de Goncourt and de Goncourt 380–1.

p. 385 **its 'daily deluge':** Marvell 112–16.

p. 386 **In the 1970s:** Klukhuhn 196; de Kroon and de Kroon.

p. 387 **In the riverine landscapes:** Fuchs 141.

p. 387 **In August 1633 he bought:** Smit 171.

15 OTHER WORLDS

p. 389 **'I recall with pleasure':** OC21 779.

p. 390 **'a company of idle unreasonable stuff':** Huygens, *Celestial* 102.

p. 390 **They were merely the latest names:** Dick 3.

p. 391 **'those whose Ignorance or Zeal':** Huygens, *Celestial* 5.

p. 391 **'It is hardly possible':** OC21 680–1.

p. 391 **'We advance nothing here':** OC21 686–9.

p. 392 **'mountains and plains of the Moon':** OC21 680–1.

p. 392 **'we should sink them below':** Huygens, *Celestial* 21.

p. 393 **'[W]e may have':** ibid. 57.

p. 393 **'the main and most diverting' et seq. quote:** ibid. 37.

p. 393 **'rational Creatures possess'd':** ibid. 57.

p. 393   "tis a very ridiculous opinion': ibid. 77.

p. 393   'yet that they should have': ibid. 100.

p. 394   'first attempt to mount': Ball 254.

p. 394   but he expressed far greater scepticism: Davidson 177.

p. 395   Huygens, who did more than anyone: Shapiro 43.

p. 395   'some will say': Huygens, *Celestial* 63.

p. 396   'Mr. des Cartes': OC10 399–406.

p. 396   'constantly confuses me': OC10 398–9.

p. 396   It is a characterization: see for example Bell 5, or Daintith and Gjertsen.

p. 397   To be a Cartesian in the Dutch Republic: Israel 585.

p. 397   a strong proponent of Descartes's ideas: Murat 126.

p. 397   Yet it is clear already that: Hahn 52.

p. 398   He may also have been: Jorink and van Miert 121–36.

p. 398   As early as 1652: Marconnel 195; OC16 9.

p. 398   It was not blind allegiance: Yoder 1988 34–5.

p. 399   It would be more accurate: Hooykaas 12.

p. 399   'false taste' for Descartes: Hall 1992 255.

p. 399   Whereas Descartes had no hesitation: Vrooman 115.

p. 399   The successes of Dutchmen: Bunge 61.

p. 399   'very jealous of the fame': OC10 399–406.

p. 400   'do much injury': OC10 238–40.

p. 400   He for one did not: Yoder 1988 177–8.

p. 400   He built upon the most robust: ibid. 42.

p. 400   Heir to Galileo and Bacon: Dijksterhuis 2004, *Lenses and Waves* 9.

p. 400   'a mere problem solver': ibid. 263.

p. 401   Christiaan Huygens is sitting out: Keesing 1983 142.

p. 401   'greatly changed': OC9 336–7.

p. 401   'not ill-disposed': Huygens, *Journaal*, vol. 2 229.

p. 401   'They key to my heart': Huygens ed. Zwaan, *Hofwijck*, line 1387.

p. 401   In September 1689: OC9 346–7.

p. 401   'nothing to aspire to here': OC9 336–7.

p. 401   writing repeatedly to Constantijn: OC9 333–6, 336–7, 344.

p. 402   'With a second letter': OC9 336–7n1.

p. 402   Constantijn told his brother: OC9 345–6.

p. 402   In September 1689: OC9 336–7.

p. 403   The *Brandenburg* reached the Cape: OC18 642–52.

p. 403   including from 'some ignorants': OC10 268–70.

p. 403   'that from S. Jago': OC10 389–90.

p. 403 But the readings on the return: OC18 648.

p. 403 The *Brandenburg*'s captain bridled: OC10 396–8.

p. 403 'misunderstanding' and 'miscalculation' et seq. quote: OC10 423–4.

p. 404 In London, too, astronomers struggled: OC9 414–15.

p. 404 'For since Mr. Cassini claims': OC10 487–8.

p. 404 Leibniz tried to induct Huygens: OC9 448–52.

p. 404 '[W]hat I like most': OC10 225–30.

p. 405 'What you tell me': OC10 238–40.

p. 405 another curve, known as the tractrix: Joella G. Yoder, '"Following in the Footsteps of Geometry": The Mathematical World of Christiaan Huygens', in *De Zeventiende Eeuw* 1996 83–95.

p. 405 but he was still troubled: OC9 482–7.

p. 405 Huygens found the *Principia*: OC10 257–9.

p. 406 'Monsieur Newton should be most happy': OC10 241–2.

p. 406 However, Fatio lacked Oldenburg's tact: Hall 1992 234.

p. 407 'You would have been able': OC10 139–43.

p. 408 divergent opinions coexisted: see for example OC10 609–15.

p. 408 The template of civil exchange: Thomas 136.

p. 408 a baffled 'Why?': OC10 316–21 note h.

p. 408 'I shall try to remember': OC10 316–21.

p. 408 what he himself called his 'tristitia': OC10 719n1.

p. 409 'I think however': OC10 662–4.

p. 409 The letter showed no dimming: OC10 708–10.

p. 409 Christiaan added a personal codicil: OC10 719n1.

p. 409 'that brother Christiaan' et seq. quotes: Huygens, *Journaal*, vol. 2 472–3.

p. 410 'was getting the most deplorable ideas': ibid. 476.

p. 410 'My wife writes': ibid. 477.

p. 410 This was the same diagnosis: OC10 719n1.

p. 410 'Real madness seemed to underlie' et seq. quotes: Huygens, *Journaal*, vol. 2 485–6.

p. 410 'if I wanted to see him': ibid. 487.

p. 411 Constantijn did not insist: OC10 719n1.

p. 411 'he began to curse': Huygens, *Journaal*, vol. 2 493.

p. 411 'spoke with him for a long time' et seq. quote: ibid. 504.

p. 411 'I can well forgive him': OC7 27–9.

p. 412 'De Gloria' observes: OC21 517–21.

p. 412 'De Morte' begins: OC21 522–3.

p. 412 'Who would not wish' et seq. quotes: ibid.

p. 413 Huygens was hardly alone: Keesing 1983 157.

p. 413 This vision must have been: OC21 667.

p. 413 'it far surpasses man' et seq. quotes: OC21 341–3.

p. 414 '[I]t is fatal' et seq. quote: OC10 719–22.

p. 414 The Lordship of Zeelhem: Keesing 1993.

p. 414 In his will, Christiaan instructed: OC22 775–8; HUG 44 16–26.

p. 414 'because he had no': ibid.; Keesing 1983 162.

p. 415 'her oldest daughter': ibid.

AFTERWORD

p. 418 From here, he might have gone on: Dijksterhuis 1961 378.

p. 418 Albert Einstein acknowledged his debt: quoted in French 267. On
10 August 1954 Einstein wrote to Michele Angelo Besso: 'The
[special] theory [of relativity] cannot dispense with the concept of
an inertial system, a concept that cannot be supported from the
viewpoint of the theory of knowledge. (The inconsistency of this
concept was illuminated very clearly by [Ernst] Mach, but it had
already been recognized with less clarity by Huygens and
Leibniz.)'

p. 422 'the raising in The Hague' et seq. quote: quoted in René van
Rijckevorsel; 'Het Huygens-monument', *Algemeen Handelsblad*, 14
January 1908.

p. 423 'Sublime mind!' et seq. quote: Boullée 107.

p. 424 'as if by magic' et seq. quote: ibid.

p. 424 'who destroyed the Cartesian system' et seq. quote: Voltaire 2007
47.

p. 424 'passed into mythical realms' et seq. quotes: Fara 255.

p. 425 'Nature and Nature's Laws': Pope 67.

p. 425 Rijsbrack and Roubiliac both: Fara 56.

p. 425 Voltaire produced a popular gloss: ibid. 135.

p. 425 'In the whole' et seq. quote: anonymous, 'On the Heads and
Intellectual Qualities of Sir Isaac Newton and Lord Bacon',
Phrenological Journal, vol. 18, no. 29 new series (1845).

p. 426 Newton himself did what he could: Inwood 5.

p. 426 although Wootton – or his indexer: Wootton.

p. 426 A search in the library catalogue: searches made on 21 January
2020.

p. 427 'The parallel between Newton': Westfall 1993 60.

p. 427 'took Britain by storm': ibid. 193.

p. 428 'the place where they grow tulips': van Herk and Kleibrink 85 ref. 4.

p. 429 'the fact that Holland has': Sagan 164n.

p. 429 the 'gateway drug': Thomas Levenson, tweet, 27 June 2017.

p. 430 For this reason, Titan is now: NASA Release 19-052 (2019).

p. 430 Data sent back confirmed: NASA, 'NASA's Cassini Data Show Saturn's Rings Relatively New' (17 January 2019).

p. 430 Here, it found geysers: J. H. Waite et al., 'Cassini Finds Molecular Hydrogen in the Enceladus Plume: Evidence for Hydrothermal Processes', *Science*, vol. 356 (2017) 155–9.

p. 431 'we were always': Huygens, *Celestial* 2.

Bibliography

WORKS BY CHRISTIAAN HUYGENS

Huygens, Christiaan, *The Celestial Worlds Discover'd* (London: Timothy Childe, 1698)

Huygens, Christiaan, ed. Hollandse Maatschappij der Wetenschappen, *Oeuvres Complètes de Christiaan Huygens*, 22 vols. (The Hague: Martinus Nijhoff, 1888–1950)

Huygens, Christiaan, ed. Rudolf Rasch, *Le cycle harmonique (Rotterdam 1691) Novus cyclus harmonicus (Leiden 1724) with Dutch and English translations* (Utrecht: Diapason, 1986)

Huygens, Christiaan, tr. Richard J. Blackwell, *The Pendulum Clock or Geometrical Demonstrations Concerning the Motion of Pendula as Applied to Clocks* (Ames: Iowa State University Press, 1986)

WORKS BY CONSTANTIJN HUYGENS SR

Huygens, Constantijn, ed. J. A. Worp, *Gedichten*, 9 vols. (Groningen: J. B. Wolters, 1892–99)

Huygens, Constantijn, ed. J. A. Worp, *De Briefwisseling van Constantijn Huygens*, 6 vols. ('s-Gravenhage: Martinus Nijhoff, 1911–17)

Huygens, Constantijn, tr. A. H. Kan, *De Jeugd van Constantijn Huygens door hemzelf beschreven*, 2nd edn (Rotterdam: Ad. Donker, 1971)

Huygens, Constantijn, ed. F. L. Zwaan, *Gebruyck of Ongebruyck van 't Orgel* (Amsterdam: North-Holland, 1974)

Huygens, Constantijn, ed. F. L. Zwaan, *Hofwijck* (Jerusalem: Chev, 1977)

Huygens, Constantijn, ed. L. Strengholt, *Zee-straet* (Zutphen: W. J. Thieme & cie, 1981)

Huygens, Constantijn, ed. C. L. Heesakkers, *Mijn Jeugd*, 2nd edn. (Amsterdam: Querido's, 1994)

Bibliography

Huygens, Constantijn, tr. Peter Davidson and Adriaan van der Weel,
  A Selection of the Poems of Sir Constantijn Huygens (1596–1687)
  (Amsterdam: Amsterdam University Press, 1996)
Huygens, Constantijn, ed. Tod van Strien, *Korenbloemen: gedichten van
  Constantijn Huygens* (Amsterdam: Querido's, 1996)
Huygens, Constantijn, ed. Harrie M. Hermkens, *Trijntje Cornelis: een volkse
  komedie van de Gouden Eeuw* (Amsterdam: Prometheus, 1997)
Huygens, Constantijn, tr. Frans R. E. Blom, *Mijn leven verteld aan mijn
  kinderen in twee boeken* (Amsterdam: Prometheus/Bert Bakker, 2003)
Huygens, Constantijn, ed. Ad Leerintveld, *De Zeestraat van 's-Gravenhage
  naar Scheveningen* (Den Haag: Valerius Pers, 2004)

Works by Constantijn Jr and Lodewijk Huygens

Huygens, Constantijn, *Journaal van Constantijn Huygens den zoon*, 3 vols.
  (Utrecht: Kemink & Zoon, 1876–88)
Huygens, Constantijn, ed. Rudolf Dekker, *The Diary of Constantijn Huygens
  Jr, Secretary to Stadholder-King William of Orange* (Amsterdam:
  Panchaud, 2015)
Huygens, Lodewijck, ed. A. G. H. Bachrach and R. G. Collmer, *The
  English Journal 1651–1652* (Leiden: Brill, 1982)

Web resources

Many of the above resources, and others, have been digitized and are
available on the excellent website of the Digitale Bibliotheek voor de
Nederlandse Letteren, www.dbnl.org.
Other major online resources consulted:
http://resources.huygens.knaw.nl (Royal Dutch Academy of Sciences)
www.let.leidenuniv.nl/Dutch/Huygens (Constantijn Huygens's verse)
http://adcs.home.xs4all.nl (Dutch history of science)
http://ckcc.huygens.knaw.nl/epistolarium/ (Dutch Republic correspondents)
http://emlo.bodleian.ox.ac.uk (Early Modern Letters Online)
Biographical resources:
http://resources.huygens.knaw.nl/retroboeken/nnbw
http://www.biografischportaal.nl
http://resources.huygens.knaw.nl/vrouwenlexicon
http://rkd.nl
www.oxforddnb.com

Bibliography

GENERAL

Académie royale des sciences, *Mémoires de l'Académie Royale des Sciences, 1666–1699*, vol. 10 (Paris: Compagnie des Libraires, 1730)

Akkerman, Nadine and Marguérite Corporaal, 'Mad Science Beyond Flattery: The Correspondence of Margaret Cavendish and Constantijn Huygens', *Early Modern Literary Studies*, vol. 14, no. 2 (2004) 2–21

Alpers, Svetlana, *The Art of Describing* (London: John Murray, 1983)

Anderson, Christy, Anne Dunlop and Pamela H. Smith, eds., *The Matter of Art: Materials, Practices, Cultural Logics, c. 1250–1750* (Manchester: Manchester University Press, 2015)

Andriesse, C. D., *Huygens: The Man Behind the Principle*, tr. Sally Miedema (Cambridge: Cambridge University Press, 2005)

anonymous, *The Ballad of Gresham College* (London: 1663)

anonymous, *A Sad and Lamentable Account of the Strange and Unhappy Misfortune of Mr. John Temple* (London: Bartholomew, 1689)

Bachrach, A. G. H., *Sir Constantine Huygens and Britain: 1596–1687*, vol. 1 (Leiden: Leiden University Press, 1962)

Bacon, Francis, ed. Brian Vickers, *The Major Works* (Oxford: Oxford University Press, 1996)

Ball, Philip, *Curiosity: How Science Became Interested in Everything* (London: Bodley Head, 2012)

Baltas, Aristides, *Peeling Potatoes or Grinding Lenses: Spinoza and Young Wittgenstein Converse on Immanence and its Logic* (Pittsburgh: University of Pittsburgh Press, 2012)

Barth, Michael, 'Huygens at Work: Annotations in his Rediscovered Personal Copy of Hooke's Micrographia', *Annals of Science*, vol. 52 (1995) 601–13

Bedini, Silvio A., *The Pulse of Time: Galileo Galilei, the Determination of Longitude, and the Pendulum Clock* (Florence: Leo S. Olschki, 1991)

Beeckman, Isaac, ed. C de Waard, *Journal tenu par Isaac Beeckman de 1604 à 1634*, vol. 3 (The Hague: Martinus Nijhoff, 1945)

Bell, A. E., *Christiaan Huygens and the Development of Science in the Seventeenth Century* (London: Edward Arnold & Co., 1950)

Berg, Maxine and Elizabeth Eger, *Luxury in the Eighteenth Century: Debates, Desires and Delectable Goods* (Basingstoke: Palgrave Macmillan, 2008)

Bergvelt, Ellinoor, Michiel Jonker and Agnes Wiechmann, *Schatten in Delft: burgers verzamelen, 1600–1750* (Delft: Stedelijk Museum Het Prinsenhof/Zwolle: Waanders, 2002)

Bibliography

Bergvelt, Ellinoor and Renée Kistemaker, eds., *De wereld binnen handbereik: Nederlandse kunst- en rariteitenverzamelingen, 1585–1735* (Amsterdam: Amsterdam Historical Museum, 1992)

Berkvens-Stevelinck, C., J. Israel and G. H. M. Posthumus Meyjes, *The Emergence of Tolerance in the Dutch Republic* (Leiden: Brill, 1997)

Biagioli, Mario, 'Etiquette, Interdependence, and Sociability in Seventeenth-Century Science', *Critical Inquiry*, vol. 22 (1996) 193–238

Boerhaave Museum, *Christiaan Huygens: een quaestie van tijd* (Leiden: Boerhaave Museum, 1979)

Boll, Jacob, *The Information Master: Jean-Baptiste Colbert's Secret State Intelligence System* (Ann Arbor: University of Michigan Press, 2009)

Bos, H. J. M., M. J. S. Rudwick, H. A. M. Snelders and R. P. W. Visser, eds., *Studies on Christiaan Huygens: Invited Papers from the Symposium on the Life and Work of Christiaan Huygens, Amsterdam, 22–25 August 1979* (Lisse: Swets & Zeitlinger, 1980)

Bots, Hans, ed., *Constantijn Huygens: zijn plaats in geleerd Europa* (Amsterdam: Amsterdam University Press, 1973)

Boullée, Etienne-Louis, ed. Helen Rosenau, *Architecture, essai sur l'art* (London: Tiranti, 1953)

Boyle, Robert, *New Experiments, Physico-Mechanicall, touching the spring of the Air and its effects* (Oxford: Thomas Robinson, 1660)

Brodsley, Laurel, Charles Frank and John W. Steeds, 'Prince Rupert's Drops', *Notes and Records of the Royal Society of London*, vol. 41 (1986) 1–26

Broekman, Inge, *De rol van de schilderkunst in het leven van Constantijn Huygens (1596–1687)* (Hilversum: Verloren, 2005)

Broekman, Inge, *Constantijn Huygens, de kunst en het hof* (Amsterdam: Amsterdam University Press, 2010)

Brown, Harcourt, *Scientific Organizations in Seventeenth Century France (1620–1680)* (Baltimore: Williams and Wilkins, 1934)

Brüderle-Krug, Nicole, *An Engraved Römer for Huygens* (Amsterdam: Rijksmuseum, 2018)

Brugmans, Henri L., *Le séjour de Christian Huygens à Paris* (Paris: E. Droz, 1935)

Buckley, Wilfred, *European Glass* (London: Ernest Benn, 1926)

Bunge, Wiep van, *From Stevin to Spinoza: An Essay on Philosophy in the Seventeenth-Century Dutch Republic* (Leiden: Brill, 2001)

Burnett, D. Graham, *Descartes and the Hyperbolic Quest: Lens Making*

Machines and Their Significance in the Seventeenth Century
(Philadelphia: American Philosophical Society, 2005)

Burton, Robert, ed. Holbrook Jackson, *The Anatomy of Melancholy* (New
York: New York Review of Books, 2001)

Cats, Jacob, ed. J. van Vloten, *Alle de werken van Jacob Cats*, 4 vols. (Zwolle:
De Erven J. J. De Tijl, 1862)

Centre National de la Recherche Scientifique, *Huygens et la France: Table
ronde du CNRS, Paris, 27–29 Mars 1979* (Paris: Librairie
Philosophique J. Vrin, 1982)

Chareix, Fabien, *La philosophie naturelle de Christiaan Huygens* (Paris: J. Vrin,
2006)

Cohen, H. F., *Quantifying Music: The Science of Music at the First Stage of the
Scientific Revolution, 1580–1650* (Dordrecht: D. Reidel, 1984)

Cohen, I. Bernard, *Revolution in Science* (Cambridge, MA: Harvard
University Press, 1985)

Colie, Rosalie L., 'Constantijn Huygens and the Rationalist Revolution',
Tijdschrift voor Nederlandse Taal-en Letterkunde, vol. 73 (1955) 193–209

Cook, Harold. J., *Matters of Exchange: Commerce, Medicine, and Science in the
Dutch Golden Age* (New Haven: Yale University Press, 2007)

Crenshaw, Paul, *Rembrandt's Bankruptcy: The Artist, His Patrons, and the Art
Market in Seventeenth-Century Netherlands* (Cambridge: Cambridge
University Press, 2006)

Crenshaw, Paul, 'The Catalyst for Rembrandt's *Satire on Art Criticism*',
Journal of Historians of Netherlandish Art, vol. 5.2 (2013) DOI:
10.5092/jhna.2013.5.2.9

d'Aubert, François, *Colbert: La vertu usurpée* (Paris: Perrin, 2010)

D'Elia, Alfonsina, *Christiaan Huygens: Una biografia intellettuale* (Milan:
Franco Angeli, 1985)

Daintith, John and Derek Gjertsen, eds., *A Dictionary of Scientists* (Oxford:
Oxford University Press, 1999)

Daley, Koos, *The Triple Fool: A Critical Evaluation of Constantijn Huygens'
Translations of John Donne* (Nieuwkoop: De Graaf, 1990)

Davids, Karel, *The Rise and Decline of Dutch Technological Leadership:
Technology, Economy and Culture in the Netherlands, 1350–1800*, 2 vols.
(Leiden: Brill, 2008)

Davidson, Keay, *Carl Sagan: A Life* (New York: Wiley, 1999)

de Goncourt, Edmond and Jules de Goncourt, *Journal des Goncourt:
Mémoires de la vie littéraire*, vol. 1: 1851–1861 (Paris: Bibliothèque-
Charpentier, 1891)

Bibliography

de Kroon, Maarten and Pieter-Rim de Kroon, *Hollands Licht* (Hilversum: AVRO, 2004) DVD

de Kruyter, Claas Wybe, *Constantijn Huygens' Ooghentroost: een interpretieve studie* (Meppel: Boom, 1972)

de Vries, Jan, *Barges and Capitalism: Passenger Transportation in the Dutch Economy 1632–1839* (Amsterdam: Amsterdam University Press, 2006)

de Vries, Jan, *The Industrious Revolution: Consumer Behavior and the Household Economy, 1650 to the Present* (Cambridge: Cambridge University Press, 2008)

Débarbat, Suzanne, 'Huygens, l'Académie des Sciences et l'Observatoire de Paris' in *Proceedings of the International Conference 'Titan: From Discovery to Encounter'* (Noordwijk: ESA Publications Division, 2004) 145–56

Dekker, Rudolf, *Childhood, Memory and Autobiography in Holland: from the Golden Age to Romanticism* (Basingstoke: Macmillan, 2000)

Dekker, Rudolf M., *Humour in Dutch Culture of the Golden Age* (Basingstoke: Macmillan, 2001)

Dekker, Rudolf, *Family, Culture and Society in the Diary of Constantijn Huygens Jr, Secretary to Stadholder-King William of Orange* (Leiden: Brill, 2013)

Descartes, René, ed. Charles Adam and Paul Tannery, *Oeuvres de Descartes*, 12 vols. (Paris: J. Vrin, 1964–71)

Dick, Steven J., *Plurality of Worlds: The Origins of the Extraterrestrial Life Debate from Democritus to Kant* (Cambridge: Cambridge University Press, 1982)

Dijksterhuis, E. J., *Simon Stevin* (The Hague: Martinus Nijhoff, 1943)

Dijksterhuis, E. J., tr. C. Dikshoorn, *The Mechanization of the World Picture* (Oxford: Clarendon Press, 1961)

Dijksterhuis, Fokko Jan, *Lenses and Waves: Christiaan Huygens and the Mathematical Science of Optics in the Seventeenth Century* (Dordrecht: Kluwer, 2004)

Dijksterhuis, Fokko Jan, 'Once Snell Breaks Down: From Geometrical to Physical Optics in the Seventeenth Century', *Annals of Science*, vol. 61 (2004) 165–85

Dijksterhuis, Fokko Jan, 'Hudde, de wiskunde en de stad', *Studium*, vol. 11 (2018) 62–77

Doorman, G., *Patents for Inventions in the Netherlands During the 16th, 17th and 18th Centuries* (The Hague: Martinus Nijhoff, 1942)

Duhem, Pierre Maurice Marie, *Le système du monde: histoire des doctrines cosmologiques de Platon à Copernic* (Paris: Hermann, 1900)

Dupré, Sven and Christoph Lüthy, eds., *Silent Messengers: The Circulation of Material Objects of Knowledge in the Early Modern Low Countries* (Berlin: Lit, 2011)

Edwardes, Ernest L., *The Story of the Pendulum Clock* (Altrincham: John Sherratt and Sons, 1977)

Emmer, P. C., *De Nederlandse slavenhandel 1500–1850* (Amsterdam: Uitgeverij De Arbeidspers, 2000); tr. Chris Emery, as *The Dutch Slave Trade, 1500–1850* (New York: Berghahn Books, 2006)

Evelyn, John, ed. William Bray, *The Diary of John Evelyn*, 2 vols. (London: M. W. Dunne, 1901)

Eyffinger, A., ed., *Huygens herdacht* (The Hague: Koninklijke Bibliotheek, 1987)

Fara, Patricia, *Newton: The Making of a Genius* (New York: Columbia University Press, 2002)

Farmelo, Graham, ed., *It Must be Beautiful: Great Equations in Modern Science* (London: Granta, 2002)

Feynman, Richard, *The Character of Physical Law* (London: Penguin, 1992)

Fokker, Adriaan D., 'Christiaan Huygens' oktaafverdeling in 31 gelijke diëzen', *Caecilia en de muziek*, vol. 98 (1941) 149–52

Fournier, Marian, 'Huygens' Microscopical Researches', *Janus*, vol. 68 (1981) 199–209

Frankenhuis-van Scheijen, C., 'Zuilichem, de heerlijkheid, het kasteel en driemaal Constantijn Huygens', *Tussen De Voorn en Loevestein*, vol. 14 (May 1978) 9–21

Franklin, James, *The Science of Conjecture: Evidence and Probability before Pascal* (Baltimore: Johns Hopkins University Press, 2001)

French, A. P., *Einstein: A Centenary Volume* (London: Heinemann, 1979)

Fuchs, R. H., *Dutch Painting* (London: Thames & Hudson, 1978)

Gaukroger, Stephen, John Schuster and John Sutton, eds., *Descartes' Natural Philosophy* (London: Routledge, 2000)

George, Albert J., 'The Genesis of the Académie des Sciences', *Annals of Science*, vol. 3 (1938) 372–401

Gerson, H., tr. Yda D. Ovink, *Seven Letters by Rembrandt*, transc. Isabella H. van Eeghen (The Hague: L. J. C. Boucher, 1961)

Godfrey, Eleanor S., *The Development of English Glassmaking 1560–1640* (Oxford: Clarendon Press, 1975)

Goldgar, Anne, *Tulipmania: Money, Honor, and Knowledge in the Dutch Golden Age* (Chicago: University of Chicago Press, 2007)

Gribbin, John, *In Search of Schrödinger's Cat: Quantum Physics and Reality* (New York: Bantam, 1984)

Bibliography

Gribbin, John, *Science: A History* (London: Allen Lane, 2002)

Grootendorst, Albert W., Jan Aarts, Miente Bakker and Reinie Erné, eds., *Jan de Witt's Elementa Curvarum Linearum Liber Secundus* (London: Springer, 2010)

Guerlac, Henry, *Newton on the Continent* (Ithaca, NY: Cornell University Press, 1981)

Guthke, Karl S., *The Last Frontier: Imagining Other Worlds from the Copernican Revolution to Modern Science Fiction* (Ithaca, NY: Cornell University Press, 1990)

Haeseker, Barend, *Constantijn Huygens 'Vileine hippocraten'* (Rotterdam: Erasmus, 2010)

Hahn, Roger, *The Anatomy of the Scientific Institution: The Paris Academy of Sciences, 1666–1803* (Berkeley: University of California Press, 1971)

Hall, A. Rupert, *From Galileo to Newton* (New York: Dover, 1981)

Hall, A. Rupert, *Isaac Newton: Adventurer in Thought* (Cambridge: Cambridge University Press, 1992)

Hall, Marie Boas, *Henry Oldenburg: Shaping the Royal Society* (Oxford: Oxford University Press, 2002)

Harris, L. E., *The Two Netherlanders: Humphrey Bradley and Cornelis Drebbel* (Cambridge: W. Heffer, 1961)

Heijbroek, J. F., ed., *Met Huygens op reis: tekeningen en dagboeknotities van Constantijn Huygens jr. (1628–1697), secretaris van stadhouder-koning Willem III* (Zutphen: Terra, 1982)

Helmers, Helmer J. and Geert H. Janssen, *The Cambridge Companion to the Dutch Golden Age* (Cambridge: Cambridge University Press, 2018)

Hermans, Theo, ed., *A Literary History of the Low Countries* (Rochester, NY: Camden, 2009)

Higgitt, Rebekah, *Recreating Newton: Newtonian Biography and the Making of Nineteenth-Century History of Science* (London: Pickering & Chatto, 2007)

Hofman, H. A., *Constantijn Huygens (1596–1687)* (Utrecht: HES, 1983)

Holmes, Richard, *The Age of Wonder: How the Romanic Generation Discovered the Beauty and Terror of Science* (London: Harper, 2008)

Hooft, P. C., *Nederlandsche historien* (Amsterdam: Wetstein & Scepérus, 1703)

Hoogstraeten, Samuel van, *Inleyding tot de hooge schoole der schilderkonst: anders der zichtbaere werelt* (Rotterdam: F. van Hoogstraeten, 1678)

Hooijmaijers, Hans, et al., *Christiaan Huygens: Facets of a Genius* (Noordwijk: ESA Publications, 2004)

Hooykaas, R., *Experientia ac ratione: Huygens tussen Descartes en Newton*
(Leiden: Rijksmuseum voor de Geschiedenis van de
Natuurwetenschappen en van de Geneeskunde, 1979)

Horace, tr. Christopher Smart, *The Works of Horace, Translated Literally into*
English Prose, vol. 2 (London: J. Newberry, 1756)

Huisken, Jacobine, *Jacob van Campen: het klassieke ideaal in de Gouden Eeuw*
(Amsterdam: Architectura & Natura Pers, 1995)

Huizinga, Johan, *Homo Ludens: A Study of the Play-Element in Culture*
(London: Routledge, 1998)

Hunt, John Dixon, *The Dutch Garden in the 17th Century: 12th Colloquium*
on the History of Landscape Architecture (Washington DC: Dumbarton
Oaks Research Library and Collection, 1990)

Hunter, Michael, *The Royal Society and its Fellows 1660–1700*, 2nd edn
(London: British Society for the History of Science, 1994)

Huysman, Ineke and Rudolf Rasch, *Béatrix en Constantijn: De briefwisseling*
tussen Béatrix de Cusance en Constantijn Huygens, 1652–1662
(Amsterdam: ING/Boom, 2009)

Icke, Vincent, *Christiaan Huygens in de onvoltooid verleden toekomende tijd*
(Groningen: Historische Uitgeverij, 2005)

Icke, Vincent, *De ruimte van Christiaan Huygens* (Groningen: Historische
Uitgeverij, 2009)

Inwood, Stephen, *The Man Who Knew Too Much: The Strange and Inventive*
Life of Robert Hooke, 1635–1703 (London: Macmillan, 2002)

Israel, Jonathan I., *The Dutch Republic: Its Rise, Greatness and Fall, 1477–*
1806 (Oxford: Clarendon Press, 1998)

Jaeger, F. M., *Cornelis Drebbel en zijne tijdgenooten* (Groningen: P.
Noordhoff, 1922)

Japikse, Nicolaas, ed., *Correspondentie van Willem III en van Hans Willem*
Bentinck, eersten Graaf van Portland, 2 vols. (The Hague: Martinus
Nijhoff, 1927–37)

Jardine, Lisa, *Ingenious Pursuits: Building the Scientific Revolution* (London:
Little, Brown, 1999)

Jardine, Lisa, *On A Grander Scale: The Outstanding Career of Sir Christopher*
Wren (London: HarperCollins, 2002)

Jardine, Lisa, *The Curious Life of Robert Hooke: The Man Who Measured*
London (London: HarperCollins, 2003)

Jardine, Lisa, *The Awful End of Prince William the Silent: The First*
Assassination of a Head of State with a Handgun (London:
HarperPerennial, 2005)

Bibliography

Jardine, Lisa, *Going Dutch: How England Plundered Holland's Glory* (London: HarperCollins, 2008)

Jardine, Lisa, *Temptation in the Archives: Essays in Golden Age Dutch Culture* (London: UCL Press, 2015)

Joby, Christopher, *The Multilingualism of Constantijn Huygens (1596–1687)* (Amsterdam: Amsterdam University Press, 2014)

Jorink, Eric, *Reading the Book of Nature in the Dutch Golden Age, 1575–1715* (Leiden: Brill, 2010)

Jorink, Eric and Dirk van Miert, eds., *Isaac Vossius (1618–1689) between Science and Scholarship* (Leiden: Brill, 2012)

Kamphuis, G., 'Enkele aantekeningen bij de biografie van Constantijn Huygens door Jacob Smit', *Tijdschrift voor Nederlandse Taal- en Letterkunde*, vol. 97 (1981) 276–96

Keesing, Elisabeth, *Constantijn en Christiaan* (Amsterdam: Querido's, 1983)

Keesing, Elisabeth, *Het volk met lange rokken* (Amsterdam: Querido's, 1987)

Keesing, Elisabeth, 'Wanneer was wie heer van Zeelhem?', *De Zeventiende Eeuw*, vol. 9 (1993) 63–4

Kemp, Martin, *The Science of Art: Optical Themes in Western Art from Brunelleschi to Seurat* (London: Yale University Press, 1992)

Kemp, Martin, *Visualizations: The Nature Book of Art and Science* (Oxford: Oxford University Press, 2000)

Klinge, Margret, *David Teniers de Jonge* (Antwerp: Snoeck-Ducaju & Zoon, 1991)

Klukhuhn, André, *Licht: De Nederlandse Republiek als bakermat van de Verlichting* (Amsterdam: De Bezige Bij, 2016)

Knuttel, W. P. C., *Verboden boeken in de Republiek der Vereenigde Nederlanden* (The Hague: Martinus Nijhoff, 1914)

Kuhn, Thomas S., *The Essential Tension: Selected Studies in Scientific Tradition and Change* (Chicago: University of Chicago Press, 1977)

Larson, Katherine R., *Early Modern Women in Conversation* (Basingstoke: Macmillan, 2011)

Lau, Katherine Inouye and Kim Plofker, 'The Cycloid Pendulum Clock of Christiaan Huygens', in Amy Shell-Gellasch, ed., *Hands On History: A Resource for Teaching Mathematics*, MAA Notes, no. 72 (2007) 145–52

Leerintveld, Ad, *Constantijn Huygens: De collectie in de Koninklijke Bibliotheek* (Amersfoort: Bekking & Blitz, 2013)

Leerintveld, Ad, Charlotte Lemmens and Peter van der Ploeg, eds., *Constantijn en Christiaan Huygens: een gouden erfenis* (Eindhoven: Lecturis, 2013)

Bibliography

Leeuw, Karl Maria Michael de, *Cryptology and Statecraft in the Dutch Republic* (Amsterdam: Amsterdam University Press, 2000)

Mak, Geert, tr. Liz Waters, *A Portrait and a Family: Rembrandt, Jan Six and the Rise and Fall of the Kings of the Republic* (Amsterdam: Atlas Contact, 2017; uncorrected bound proof)

Marconnel, M. H., 'Christiaan Huygens: A Foreign Inventor in the Court of Louis XIV, His Role as a Forerunner of Mechanical Engineering' (unpublished thesis, Open University, 1999)

Margócsy, Dániel, *Commercial Visions: Science, Trade, and Visual Culture in the Dutch Golden Age* (Chicago: University of Chicago Press, 2014)

Martins, Roberto de A., 'Huygens's Reaction to Newton's Gravitational Theory', in Judith V. Field and Frank A. J. L. James, eds., *Renaissance and Revolution: Humanists, Scholars, Craftsmen and Natural Philosophers in Early Modern Europe* (Cambridge: Cambridge University Press, 1993) 203–13

Marvell, Andrew, ed. Elizabeth Story Donno, *Andrew Marvell: The Complete Poems* (Harmondsworth: Penguin, 1972)

McConnell, Anita, *A Survey of the Networks Bringing a Knowledge of Optical Glass-Working to the London Trade, 1500–1800* (Cambridge: Whipple Museum of the History of Science, 2016)

Meijer Drees, Marijke, *Andere landen, andere mensen: de beeldvorming van Holland versus Spanje en Engeland omstreeks 1650* (The Hague: Sdu Uitgevers, 1997)

Merton, R. K., *Science, Technology & Society in Seventeenth-Century England* (New York: Harper & Row, 1970)

Meyer, Jean, *Colbert* (Paris: Hachette, 1981)

Miller, Arthur I., *Imagery in Scientific Thought: Creating 20th Century Physics* (Cambridge, MA: MIT Press, 1984)

Murat, Inès, *Colbert* (Paris: Fayard, 1980); tr. Robert Francis Cook and Jeannie van Asselt, *Colbert* (Charlottesville: University Press of Virginia, 1984)

Nadler, Steven, *Spinoza: A Life* (Cambridge: Cambridge University Press, 1999)

Neri, Antonio, tr. Christopher Merret, *The Art of Glass* (London: Octavian Pulleyn, 1662)

Noordam, D. J., *Riskante relaties: vijf eeuwen homoseksualiteit in Nederland, 1233–1733* (Hilversum: Verloren, 1995)

Nooteboom, Cees, *In de bergen van Nederland* (Amsterdam: De Bezige Bij, 2010)

Bibliography

Oldenburg, Henry, ed. A. R. Hall and M. B. Hall, *The Correspondence of Henry Oldenburg*, 13 vols. (London: Taylor & Francis, 1965–86)

Ollard, Richard, *Man of War: Sir Robert Holmes and the Restoration Navy* (London: Hodder and Stoughton, 1969)

Page, Willie F., *The Dutch Triangle: The Netherlands and the Atlantic Slave Trade, 1621–1664* (New York: Garland, 1997)

Pal, Carol, *Republic of Women: Rethinking the Republic of Letters in the Seventeenth Century* (Cambridge: Cambridge University Press, 2012)

Parker, Geoffrey, *Global Crisis: War, Climate Change and Catastrophe in the Seventeenth Century* (New Haven: Yale University Press, 2012)

Pepys, Samuel, *The Diary of Samuel Pepys*, ed. Robert Latham and William Matthews, 11 vols. (London: Bell & Hyman, 1970)

Ploeg, Willem, *Constantijn Huygens en de natuurwetenschappen* (Rotterdam: Nijgh & van Ditmar, 1934)

Pope, Alexander, ed. Pat Rogers, *Alexander Pope* (Oxford: Oxford University Press, 1994)

Porter, Roy and Mikuláš Teich, *The Scientific Revolution in National Context* (Cambridge: Cambridge University Press, 1992)

Poundstone, William, *Carl Sagan: A Life in the Cosmos* (New York: Henry Holt, 2000)

Prak, Maarten, *The Dutch Republic in the Seventeenth Century* (Cambridge: Cambridge University Press, 2005)

Proust, Jacques and Marianne Proust, eds., *Le puissant royaume de Japon: la description de François Caron (1636)* (Paris: Chandeigne, 2003)

Rasch, Rudolf, ed., *Driehonderd brieven over muziek van, aan en rond Constantijn Huygens* (Hilversum: Verloren, 2007)

Rasch, Rudolf, ed., *Duizend brieven over muziek van, aan en rond Constantijn Huygens* (Utrecht: Houten, 2019)

Raymond, Joad and Noah Moxham, *News Networks in Early Modern Europe* (Leiden: Brill, 2016)

Reynolds, Joshua, Sir, Edmond Malone and Thomas Gray, *The Works of Sir Joshua Reynolds, Knt*, vol. 1 (London: T. Cadell and W. Davies, 1797)

Roberts, Benjamin, *Sex and Drugs before Rock'n'Roll: Seventeenth-Century Dutch Golden Age Youth Culture* (Amsterdam: Amsterdam University Press, 2017)

Roberts, Lissa, Simon Schaffer and Peter Dear, eds., *The Mindful Hand: Inquiry and Invention from the Late Renaissance to Early Industrialisation* (Amsterdam: Koninklijke Nederlandse Akademie van Wetenschappen, 2007)

Bibliography

Ronchi, Vasco, ed., *Scritti di Ottica* (Milan: Edizioni Il Polifilo, 1968)

Rossell, Deac, 'The Magic Lantern and Moving Images Before 1800', *Barockberichte*, vol. 40/41 (2005) 686–93

Roth, Leon, *The Correspondence of Descartes and Constantijn Huygens, 1635–1647* (Oxford: Clarendon Press, 1926)

Ruestow, Edward G., *The Microscope in the Dutch Republic: The Shaping of Discovery* (Cambridge: Cambridge University Press, 1996)

Russell, Bertrand, *History of Western Philosophy* (London: George Allen & Unwin, 1979)

Sagan, Carl, *Cosmos* (New York: Random House, 1980)

Sass, Else Kai, *Comments on Rembrandt's Passion Paintings and Constantijn Huygens's Iconography* (Copenhagen: Munksgaard, 1971)

Schama, Simon, *Rembrandt's Eyes* (London: Allen Lane, 1999)

Schama, Simon, *The Embarrassment of Riches: An Interpretation of Dutch Culture in the Golden Age* (London: Collins, 1987)

Schinkel, A D., *Bijdrage tot de kennis van het karakter van Constantijn Huygens* (Rotterdam: W. Messchert, 1842)

Schoot, Joas van der, 'Interpreting the Kosmotheoros (1698). A Historiographical Essay on Theology and Philosophy in the Work of Christiaan Huygens', *De Zeventiende Eeuw*, vol. 30 (2014) 20–39

Schwartz, Gary, *Rembrandt: His Life, His Paintings* (London: Viking, 1985)

Schwartz, Gary, *Rembrandt's Universe: His Art, His Life, His World* (London: Thames & Hudson, 2006)

Scott, J. F., *The Scientific Work of René Descartes* (London: Taylor & Francis, 1976)

Sennett, Richard, *The Craftsman* (London: Allen Lane, 2008)

Shapin, Steven and Simon Schaffer, *Leviathan and the Air-Pump: Hobbes, Boyle and the Experimental Life* (Princeton: Princeton University Press, 2011)

Shapiro, Barbara J., *Probability and Certainty in Seventeenth-Century England* (Princeton, NJ: Princeton University Press, 1983)

Sluijter, Eric Jan, 'Ownership of Paintings in the Dutch Golden Age', in Ronni Baer, ed., *Class Distinctions: Dutch Painting in the Age of Rembrandt and Vermeer* (Boston: Museum of Fine Arts, 2015), 89–111

Smit, Jacob, *De grootmeester van woord- en snarenspel: het leven van Constantijn Huygens* (The Hague: Martinus Nijhoff, 1980)

Smith, Pamela H., *The Body of the Artisan: Art and Experience in the Scientific Revolution* (Chicago: University of Chicago Press, 2004)

Bibliography

Smith, Pamela H. Amy R. W. Meyers and Harold J. Cook, *Ways of Making and Knowing: The Material Culture of Empirical Knowledge* (Ann Arbor: University of Michigan Press, 2011)

Smith, Pamela H., and Benjamin Schmidt, *Making Knowledge in Early Modern Europe: Practices, Objects, and Texts, 1400–1800* (Chicago: University of Chicago Press, 2007)

Snyder, Laura J., *Eye of the Beholder: Johannes Vermeer, Antoni van Leeuwenhoek, and the Reinvention of Seeing* (New York: W. W. Norton, 2015)

Sorbière, Samuel de, *A Voyage to England* (London: J. Woodward, 1709)

Sparberg, E. B., 'Misinterpretation of Theories of Light', *American Journal of Physics*, vol. 34 (1966) 377–89

Sterne, Laurence, *The Life and Opinions of Tristram Shandy, Gentleman* (London: R. and J. Dodsley, 1760)

Stoffele, Bram, 'Christiaan Huygens: A Family Affair' (unpublished Master's thesis, Utrecht University, 2006)

Stroup, Alice, 'Christiaan Huygens & the Development of the Air Pump', *Janus*, vol. 68 (1981) 129–58

Stroup, Alice, 'Christian Huygens et l'Académie royale des sciences', *La Vie des Sciences*, vol. 13, (1996) 333–41

Struik, Dirk J., *The Land of Stevin and Huygens* (Dordrecht: D. Reidel, 1981)

Swan, Claudia, *Art, Science, and Witchcraft in Early Modern Holland: Jacques de Gheyn II (1565–1629)* (Cambridge: Cambridge University Press, 2005)

Szydlo, Zbigniew, *Water Which Does Not Wet Hands: The Alchemy of Michael Sendivogius* (Warsaw: Polish Academy of Sciences, 1994)

Taton, René, *Les origines de l'Académie Royale des Sciences* (Paris: Université de Paris, 1965)

Temple, William, *Observations upon the United Provinces of the Netherlands* (London: Sa. Gellibrand, 1673)

Teylers Museum, *Going Dutch: The Invention of the Pendulum Clock*, Proceedings of a Symposium at Teylers Museum, Haarlem, 3 December 2011 (Zaanse Schans: Museum van het Nederlandse Uurwerk, 2013)

Thomas, Keith, *In Pursuit of Civility: Manners and Civilization in Early Modern England* (New Haven: Yale University Press, 2018)

Tierie, Gerrit, *Cornelis Drebbel (1572–1633)* (Amsterdam: H. J. Paris, 1932)

Tummers, Anna, *The Eye of the Connoisseur: Authenticating Painting by*

Rembrandt and his Contemporaries (Amsterdam: Amsterdam University Press, 2011)

van Berkel, Klaas, *Isaac Beeckman on Matter and Motion: Mechanical Philosophy in the Making* (Baltimore: Johns Hopkins University Press, 2013)

van Berkel, Klaas, Albert van Helden and Lodewijk Palm, *A History of Science in the Netherlands* (Leiden: Brill, 1999)

van de Craats, J., 'Christiaan Huygens en de muziek', *De Zeventiende Eeuw*, vol. 7 (1991) 7–16

van der Vinde, Lea, *Vrouwen rondom Huygens* (Voorburg: Huygensmuseum Hofwijck, 2010)

van der Weel, Heleen, *De Oude Kerk op Scheveningen* (The Hague: Kruseman, 1982)

van Dorsten, J. A., 'Huygens en de Engelse "Metaphysical Poets"', *Tijdschrift voor Nederlandse Taal- en Letterkunde*, vol. 76 (1958) 111–25

van Egmond, W. S. and M. Mostert, eds., *Spelen in de middeleeuwen: over schaken, dammen, dobbelen en kaarten* (Hilversum: Verloren, 2001)

van Elk, Martine, *Early Modern Women's Writing: Domesticity, Privacy and the Public Sphere in England and the Dutch Republic* (Cham, Switzerland: Macmillan, 2017)

van Gelder, H. E., *Ikonografie van Constantijn Huygens en de zijnen* (The Hague: Martinus Nijhoff, 1957)

van Gelder, H. E., 'De jonge Huygensen en Rembrandt', *Oud Holland*, vol. 73 (1958) 238–41

van Helden, Albert, '"Annulo Cingitur": The Solution of the Problem of Saturn', *Journal of the History of Astronomy*, vol. 5 (1974) 155–73

van Helden, Albert, Sven Dupré, Rob van Gent and Huib Zuidervaart, eds., *The Origins of the Telescope* (Amsterdam: KNAW Press, 2010)

van Helden, Anne C., and Rob H. van Gent, 'The Lens Production by Christiaan and Constantijn Huygens', *Annals of Science*, vol. 56 (1999) 69–79

van Herk, Gijsbert and Hermann Kleibrink, *De Leidse sterrewacht: 4 eeuwen wacht bij dag en bij nacht* (Zwolle: Waanders/De Kler, 1983)

van Leeuwenhoek, Antonie, ed. Lodewijk C. Palm et al., *Alle de brieven van Antoni van Leeuwenhoek* (Amsterdam: Swets & Zeitlinger, 1939–99)

van Rijckevorsel, René, 'Het Huygens-monument', *Jaarverslag* (Haarlem: Koninklijke Hollandsche Maatschappij der Wetenschappen, 2016) 37–9

van Strien, Ton, 'Huygens als vertaler van John Donne', *De Nieuwe Taalgids*, vol. 85 (1992) 247–52 van Strien, Ton and Kees van der Leer, *Hofwijck: Het gedicht en de buitenplaats van Constantijn Huygens* (Zutphen: Walburg Press, 2002)

van Tilborgh, Louis, *Monet in Holland* (Zwolle: Waanders, 1986)

van Tricht, H. W., *De Briefwisseling van Pieter Corneliszoon Hooft*, 3 vols. (Culemborg: Willink/Noorduijn, 1976)

van Zanten, Laurens, *Spiegel der gedenckweerdighste wonderen en geschiedenissen onses tijds* (Amsterdam: Johannes van den Bergh, 1661)

Venturi, Lionello, *Les archives de l'impressionisme*, vol. 1 (Paris: Durand-Ruel, 1939)

Verduin, C. J., 'A Portrait of Christiaan Huygens Together with Giovanni Domenico Cassini', in *Proceedings of the International Conference 'Titan: From Discovery to Encounter'* (Noordwijk: ESA Publications Division, 2004) 157–70

Vermaas, J. C., *Geschiedenis van Scheveningen*, vol. 2 (The Hague: Couvreur, 1976)

Vlieghe, Hans, 'Constantijn Huygens en de Vlaamse schilderkunst van zijn tijd', *De Zeventiende Eeuw*, vol. 3 (1987) 191–210

Vlieghe, Hans, *David Teniers the Younger (1610–1690): A Biography* (Turnhout: Brepols, 2011)

Vogelaar, Christiaan, *Jan van Goyen* (Leiden: Stedelijk Museum De Lakenhal, 1996)

Vollgraff, J. A., 'Biografie Christiaan Huygens' (unedited typescript, n.d., Leiden University Library, BPL 2781)

Voltaire, *Le Siècle de Louis XIV* (Paris: Garnier-Flammarion, 1966)

Voltaire, ed. John Leigh, tr. Prudence L. Steiner, *Philosophical Letters, or Letters Regarding the English Nation* (Indianapolis: Hackett, 2007)

Vrooman, Jack Rochford, *René Descartes: A Biography* (New York: Putnam's, 1970)

Wagenaar, W. A., 'The True Inventor of the Magic Lantern: Kircher, Walgenstein or Huygens?', *Janus*, vol. 66 (1979) 203–17

Weld, C. R., *A History of the Royal Society*, vol. 1 (London: Parker, 1848)

Westfall, Richard, *Force in Newton's Physics: The Science of Dynamics in the Seventeenth Century* (London: Macdonald, 1971)

Westfall, Richard, *The Life of Isaac Newton* (Cambridge: Cambridge University Press, 1993)

Westfall, Richard S., *Never at Rest: A Biography of Isaac Newton* (Cambridge: Cambridge University Press, 2010)

Weststeijn, Thijs, *Margaret Cavendish in de Nederlanden: filosofie en schilderkunst in de Gouden Eeuw* (Amsterdam: Amsterdam University Press, 2008)

Whitaker, Katie, *Mad Madge: Margaret Cavendish, Duchess of Newcastle, Royalist, Writer and Romantic* (London: Chatto & Windus, 2003)

White, Christopher, *Rembrandt* (London: Thames & Hudson, 1984)

Whitestone, Sebastian, 'The Identification and Attribution of Christiaan Huygens' First Pendulum Clock', *Antiquarian Horology*, vol. 31 (2008) 201–22

Whitestone, Sebastian, 'Christiaan Huygens' Lost and Forgotten Pamphlet of His Pendulum Invention', *Annals of Science*, vol. 69 (2012) 91–104

Whitestone, Sebastian, 'Galileo, Huygens and the Invention of the Pendulum Clock', *Antiquarian Horology*, vol. 38 (2017) 365–84

Wieseman, Marjorie E., *Caspar Netscher* (Doornspijk: Davaco, 2002)

Willach, Rolf, *The Long Route to the Invention of the Telescope* (Philadelphia: American Philosophical Society, 2008)

Willemyns, Roland, *Dutch: Biography of a Language* (Oxford: Oxford University Press, 2013)

Wootton, David, *The Invention of Science: A New History of the Scientific Revolution* (London: Allen Lane, 2015)

Worp, J. A., 'Constantijn Huygens' journaal van zijn reis naar Venetië in 1620', *Bijdragen en Mededeelingen van uit Historisch Genootschap*, vol. 15 (1894) 62–153

Worp, J. A., 'De jeugd van Christiaan Huygens, volgens een handschrift van zijn vader', *Oud Holland*, 13 (1913) 209–35

Worp, J. A., *Een onwaerdeerlycke vrouw: brieven en verzen van en aan Maria Tesselschade* (Utrecht: HES, 1976)

Yoder, Joella G., *Unrolling Time: Christiaan Huygens and the Mathematization of Nature* (Cambridge: Cambridge University Press, 1988)

Yoder, Joella, ed., *Catalogue of the Manuscripts of Christiaan Huygens Including a Concordance with His Oeuvres Complètes* (Leiden: Brill, 2013)

De Zeventiende Eeuw, vol. 3, no. 2 (1987) (Special issue: Constantijn Huygens)

De Zeventiende Eeuw, vol. 12, no. 1 (1996) (Special issue: Christiaan Huygens)

De Zeventiende Eeuw, vol. 25, no. 2 (2009) (Special issue: Vrouwen rondom Huygens)

Zuidervaart, Huib J., 'The "Invisible Technician" Made Visible: Telescope Making in the Seventeenth and Early Eighteenth-century Dutch Republic', in Alison D. Morrison-Low, et al. eds., *From Earth-Bound to Satellite: Telescopes, Skills and Networks* (Leiden: Brill, 2011) 41–102

Zuidervaart, Huib J., '"Scientia" in Middelburg rond 1600: De
        wisselwerking tussen academici, kooplieden en handwerkers in het
        vroegmoderne Middelburg: essentieel voor de creatie van 's werelds
        eerste verrekijker?', *Archief Mededelingen van het Koninklijk Zeeuwsch
        Genootschap den Wetenschappen* (2017) 43–100

Zuidervaart, H. J. and L. Nellissen, *De echt uitvinder van de verrekijker*
        (Middelburg: Koninklijk Zeeuwsch Genootschap der Wetenschappen,
        2008)

Zumthor, Paul, tr. Simon Watson Taylor, *Daily Life in Rembrandt's Holland*
        (London: Weidenfeld and Nicolson, 1962)

Index

Page numbers in **bold** refer to illustrations.

Index

Index

Index